Cristian Tolsa

The Orphic Astrologer Critodemus

Untersuchungen zur antiken Literatur und Geschichte

Herausgegeben von
Marcus Deufert, Heinz-Günther Nesselrath
und Peter Scholz

Band 155

Cristian Tolsa

The Orphic Astrologer Critodemus

Fragments with Annotated Translation and Commentary

DE GRUYTER

ISBN 978-3-11-132876-8
e-ISBN (PDF) 978-3-11-132914-7
ISSN 1862-1112

Library of Congress Control Number: 2023940599

Bibliographic information published by the Deutsche Nationalbibliothek
The Deutsche Nationalbibliothek lists this publication in the Deutsche Nationalbibliografie; detailed bibliographic data are available on the internet at http://dnb.dnb.de.

Typesetting: Integra Software Services Pvt. Ltd.
Printing and binding: CPI books GmbH, Leck

www.degruyter.com

To Mar, Clàudia, and Ariadna

The research for this book has been supported by the following fellowships and grants:

Humboldt Postdoctoral Fellowship (Alexander von Humboldt-Stiftung, Universität Osnabrück, 2018–2019).

Juan de la Cierva-incorporación Fellow (Universitat de Barcelona, 2019–2021).

Research project *Social aspects in medieval Arabic astrological sources: their legacy and discontinuity in the Greek tradition*. Grant PID2021-126415NB-I00 funded by MCIN/AEI/10.13039/501100011033 and by "ERDF A way of making Europe".

Abbreviations

CCAG I	Alexander Olivieri et al., *Catalogus Codicum Astrologorum Graecorum. Bd. I. Codices Florentini*, Brussels 1898.
CCAG VII	Franz Boll, *Catalogus Codicum Astrologorum Graecorum. Bd. VII. Codices Germanici*, Brussels 1908.
CCAG VIII 1	Franz Cumont, *Catalogus Codicum Astrologorum Graecorum. Bd. VIII. Codices Parisini Teil 1*, Brussels 1929.
CCAG VIII 3	Pierre Boudreaux, *Catalogus Codicum Astrologorum Graecorum. Bd. VIII. Codices Parisini Teil 3*, Brussels 1912.
CCAG VIII 4	Pierre Boudreaux and Franz Cumont, *Catalogus Codicum Astrologorum Graecorum. Bd. VIII. Codices Parisini Teil 4*, Brussels 1921.
FGrHist	Felix Jacoby, *Die Fragmente der griechischen Historiker. Teil 3, Geschichte von Staedten und Voelkern. – C, Autoren ueber einzelne Laender – Band 1. Aegypten – Geten [Nos. 608a–708]*, Leiden 1958.
GH	Otto Neugebauer and Henry Barlett van Hoesen, *Greek Horoscopes*, Philadelphia 1987.
HAMA	O. Neugebauer, *History of Ancient Mathematical Astronomy*, Berlin/Heidelberg 1975.
OF	Alberto Bernabé, *Poetae epici graeci. Testimonia et fragmenta. Pars II fasc. 2: Orphicorum et Orphicis similium testimonia et fragmenta*, Leipzig 2005.
RE III 1	Georg Wissowa (ed.), *Realencyclopädie der classischen Altertumswissenschaft. Dritter Band, Fünfter Halbband, Barbarus–Campanus*, Stuttgart 1897.
RE XI 2	Georg Wissowa (begun by), Wilhelm Kroll (ed.), *Realencyclopädie der classischen Altertumswissenschaft. Elfter Band, Zweiundzwanzigster Halbband, Komogrammateus–Kynegoi*, Stuttgart 1922.

https://doi.org/10.1515/9783111329147-202

Preface

This book collects the 20 surviving fragments from Critodemus, arranged thematically in 8 chapters. They are presented in the original Greek (except for one fragment in Latin, and one mostly numerical fragment) with a simplified *apparatus criticus* noting the most relevant variants, followed by an annotated translation and commentary essays. Most fragments are quotations from astrological authors such as Vettius Valens, Firmicus Maternus, or Hephaestio, whose manuscripts have been adequately collated in modern editions, hence the decision not to make the book sensu stricto an edition. I have, however, reassessed the variants and consulted the manuscripts in case of doubt.

Greek astrology will appear to many an unusual subject for an academic book, justified only when directly associated with more serious subject matter such as politics or poetry. However, there should hardly be a need today to justify the study of ancient astrology per se. The tradition of astral prediction has a long history in ancient Mesopotamia, and it became a popular phenomenon in the central Graeco-Roman region in the late Hellenistic and Roman imperial eras. It thus deserves close exploration, at the same level, for example, as ancient religion, regardless of our views on the modern practice. It is true that the texts of ancient astrology are frequently neither elegant nor pleasant to read, but we should study them seriously and carefully, in the same way that we examine documentary papyri to discover the most diverse matters related to ancient history. In any case, the varied and complex ways in which astrology was approached by the ancient practitioners demonstrate that the subject was important to them, and this should be the sufficient reason for their contemporary study if we wish to obtain a clearer and more complete picture of the Graeco-Roman era. In a deeper sense, the ancient concern with astrology reveals a certain worldview that emerged in the late Hellenistic period, related to the way in which new forms of knowledge inherited from Mesopotamia and Egypt were adapted into Greek culture.[1]

But how is a mostly unknown author relevant to the history of astrology as a whole? Indeed, Critodemus is a shadowy figure sometimes invoked in the ancient sources as one of the earliest astrological authorities. His main witnesses are the better-known and well-preserved astrological manuals of Vettius Valens (second century AD) and Firmicus Maternus (fourth century. AD), and more importantly for chronological reasons, Pliny the Elder's *Natural History*. In particular, the treat-

1 This is reminiscent of the ancient astrologers' introductory defenses of astrology, such as Ptolemy *Tetrabiblos* I 1-3, or the first book of Firmicus Maternus' *Mathesis*. See also a defense of the modern study of the history of astrology, focused on its importance for the history of astronomy (the author's main interest), in Neugebauer 1951.

https://doi.org/10.1515/9783111329147-203

ment of Critodemus by Valens informed the latter's treatise in many ways: therefore, understanding Critodemus and understanding Valens often go hand in hand.

Critodemus by himself is crucial to understanding the early history of Hellenistic astrology, since he belongs to its earliest period, namely the late second century or early first century BC. Critodemus is also the first Greek astrological author attested to have written under his own name, eschewing the widespread tradition of adopting a pen name of a divine figure, mostly taken from Egyptian lore, such as Hermes Trismegistus, Nechepsos, or Petosiris. Nevertheless, he seems to have resorted to the Orphic tradition for the characterization of part of his own astrological work. This should not be surprising, given that the name of Orpheus, a mediator between gods and humans like Hermes, was attached to some astrological texts in the late Hellenistic period. As we shall see, Critodemus may have adapted doctrines from these texts to personal astrology, combining them with other elements of the Orphic tradition that served his presentation of personal horoscopic astrology to a wider Greek-speaking audience.

Another interesting aspect of the study is that the surviving fragments allow us to reconstruct the transformations undergone by Critodemus' manual, thus illustrating the particular nature of the transmission of astrological texts in antiquity. Critodemus' work, like its most obvious antecedent, the poem of Nechepsos and Petosiris, was probably originally written in iambic trimeters. It was, however, paraphrased soon after its composition – at least in the parts closest in form to a list – into a highly homogeneous text. Furthermore, the compilation seems to have been enlarged with some additions in the form of horoscope examples and tables.

Finally, the different textual situations of the fragments make Critodemus an invaluable case study on the diversity of reception practices in later Graeco-Roman astrology. The fourth-century antiquarian Hephaestio offers the simplest example, since his primary aim was to preserve ancient doctrines, but the treatment of Vettius Valens is intriguing: Valens, though he praised the astrological theories of Critodemus, did not aim to salvage or collect the doctrines that could be useful to him but instead presented them in an intricate and specifically modified way.

This book thus explores the far-reaching thread represented by Critodemus, by connecting, contextualizing, and analyzing a diverse set of Hellenistic astrological practices, from the general presentational strategies to the technical aspects of astrology. It will therefore be of interest to a broad spectrum of researchers of the ancient Greek world, from classicists and historians of ancient Greek astronomy and astrology to historians of the Hellenistic world and scholars interested in the history of ancient Greek books or religion.

The book is organized into two main parts: the first contains the fragments that concern us for the chronological, geographical, and more broadly cultural setting of Critodemus and his work (Chapters 1–3), while the second presents the more technical astrological fragments (Chapters 4–8). In Appendix II, the reader will find a presentation of the basic astrological doctrines and procedures that will hopely be useful for the comprehension of the second part.

Chapter 1 takes a close look at the mentions of Critodemus in Pliny's *Natural History*, which were mediated by M. Terentius Varro, setting a *terminus ante quem* for our author in the mid-first century BC. Varro's attribution to Critodemus of a statement on the time span of Babylonian observations is probably spurious, but its analysis sheds light on the authors and natural philosophical theories of the third and second centuries BC that formed the background for the irruption of astrology in the Greek-speaking milieu, as well as on their unusual treatment in the hands of Varro.

In Chapter 2, I gather together the fragments from astrological authors, providing their particular opinions on Critodemus' work as a whole. The account of Vettius Valens is by far the most detailed, showing a surprisingly ambiguous stance which tells us something both about himself and about the nature of Critodemus' work. Valens quotes its title (*Vision*) and first lines, which are comparable to the openings of other astrological texts not just for their metaphorical character but also for the possibly autobiographical information that they convey (in this case, about Critodemus' travels in Egypt).

Chapter 3 focuses on the astrological oaths that are alluded to by Valens in his criticisms. They are clearly inspired by Orphic texts. Both Valens (despite his negative judgment) and Firmicus seem to offer their own versions of Critodemus' astrological oath. Common features can be tentatively singled out and attributed to Critodemus.

As for the technical fragments, Chapter 4 contains the fragments on the "distributions", the partition of human life in intervals of 10 years and 9 months, which were later incorporated in different forms by Valens, Firmicus, and Hephaestio into their own astrological treatises.

In Chapter 5, there is a long fragment on the so-called Egyptian "terms", a subdivision of the zodiacal signs in unequal sections assigned to the planets which was widely used in Greek astrology, deriving from Babylonian practice. Critodemus was apparently original in describing the effects of each one of these sections.

Chapter 6 presents three fragments dealing with theories on the manner and time of death. These were topics of the highest importance in Hellenistic astrological practice.

Finally, Chapters 7 and 8 deal with tables that were appended at some point to Critodemus' work. Specifically, Chapter 7 contains a tabular fragment establishing a theory of critical years and bodily affections, likely based on original material in Critodemus' book, as well as a fragment on the course of illness from Dorotheus' manual possibly deriving from Critodemus. Chapter 8 deals with two large tables for calculating the length of life, one of which was used for the insertion of a different system of "terms".

Beyond the binarism testimonia/fragments

The nature of the texts related to Critodemus is so diverse that the usual distinction between testimonia and fragments in classical editions would make little sense here. To list just a few examples, F 1 is the mention of Critodemus in an index (from Pliny's *Natural History*); F 6 gathers astrological oaths in Valens and Firmicus clearly inspired by that (or those) of Critodemus, but which are obviously their authors' own creation; F 14 and 16 are tables appended to Critodemus' work at a later stage; and F 20 is Valens' own very particular use of F 16. So it becomes clear that, even if the real Critodemus is always in the background, he rarely appears unmediated, in what classical editions term "a fragment". In fact, even his most technical fragments suffered the aforementioned transformation into a stylized and standardized astrological summary. F 3, the beginning of his work, is probably the closest we have to Critodemus' own words, but even there it is clear that the original iambic trimeter has been altered as a result of paraphrasing and prosification.

Therefore, rather than distinguishing between testimonia and fragments, I offer before each fragment a description of its textual status.

Note on the use of existing editions

I will generally use the most recent editions of the ancient works for quotations and references – notably Pingree 1986 for Vettius Valens and Hübner 1998 for Ptolemy's *Tetrabiblos* – with the following citing conventions: Roman numerals for the book, followed by Hindu-Arabic numerals separated by a dot for the chapter and section, for example "Ptolemy I 1.2". It is important to keep in mind that sections are sometimes numbered continuously throughout one book (as in Mayhoff's Pliny or Long's Diogenes Laertius), but they more often begin anew for each chapter (as in Hübner's Ptolemy, Pingree's Valens, or Krohn's Vitruvius).

Contents

Introduction

State of the art

For a general overview, I refer the reader to the excellent exposition by Stephan Heilen of the history of this small area of classical scholarship.[1] Turning specifically to Critodemus, our author has received little attention, in part presumably because of the difficulties of situating him chronologically and separating his material from that of Valens. Neugebauer and van Hoesen raised doubts about the validity of the *terminus ante quem* represented by Pliny (before his own writing in the AD 70s), alleging that the date of many horoscopes in the work of Vettius Valens would situate Critodemus as late as the end of the first or the beginning of the second century AD.[2] Pingree later attempted to solve the issue by arguing that most of these horoscopes have no clear connection with this author.[3] He clearly was on the right track, but in the case of the horoscopes used together with the tables, it is likely that they were in fact taken from the work of Critodemus (see discussion of F 9 and 19). Their date would certainly, as Neugebauer and van Hoesen explained, be too late for Pliny (one of the horoscopes is from AD 68), but, like the tables, these horoscopes were probably not original to Critodemus' work. These problems in dating our author obviously arise from the complexities and the peculiar fluidity in the transmission of ancient astrological texts (more on this below).

Pingree conjectured that Critodemus was from the period after 30 BC, when Egypt came to be under Roman rule, based on Critodemus' familiarity with Egyptian astrology, and because he apparently used the term θηριομαχεῖν ("to fight with beasts", related to gladiatorial contests), but, again, by the point in the text at which the term appears, the material is probably no longer from Critodemus (see discussion below, F 13).

The only monograph on Critodemus is a degree thesis by Heinz Peter, which for the first time collected the fragments (and translated them into German), providing a summary commentary; I owe to Peter the identification of the horoscopes of birth and death in F 9, those of Emperor Nero, and a redating of the horoscope of F 20 to AD 2.[4]

1 Heilen 2015 I, 3–9. Cf. also the updated catalogue of ancient horoscopes in the same volume, 204–333.
2 *GH*, 185–186.
3 Pingree 1978, 426.
4 See Peter 2001, 122–123, for my fragment F 9, where he also rightly observes, *pace* Pingree, that the table mentioned by Valens must be a multiplication table; see ibid., 149 for the horoscope in F 20. If we accept that Critodemus was known to Varro (Chapter 1), none of these horoscopes can be original to Critodemus' work.

https://doi.org/10.1515/9783111329147-001

Astral prediction in Mesopotamia

We can say that, in a relatively restricted sense of the concept, astrology takes off in Hellenized Egypt toward the mid- or late second century BC. However, divination by means of astronomical observation has a much older history in the ancient agrarian societies of the Mediterranean region and the Near East. Astronomical omens were probably already in use in the earliest period of Mesopotamian history, in the third millennium BC, along with other techniques such as the inspection of animal entrails, as methods of divination to predict the outcome of critical sociopolitical or natural events for the country. The series *Enūma Anu Enlil* ("When Anu and Enlil"), which contains thousands of predictions for the king and the state based on astronomical and meteorological phenomena, began its history in the early part of the second millennium BC. The clay tablets containing this text, formed of conditional clauses with the structure "if A, then B", were copied, completed, and used until the end of cuneiform culture in late Hellenistic times. The clauses may have been based on a system of deduced correlations between observations and subsequent events: to give a hypothetical example, if there had been frequent observations of a good harvest following a lunar eclipse close to such-and-such star, a clause of the following type may have been established: "If a lunar eclipse is observed close to such-and-such star, then there will be a good harvest".[5]

This divinatory practice was no doubt a powerful driving force behind the development of astronomical methods and theories: for astral omens to be created and used, the celestial bodies had to be mapped to some extent in time and space. Astrological omens and their associated astronomical theories were also naturally intertwined with the development of calendars. At the root of such calendar-making practices was the discovery that much of the material basis for such predictions – namely the motions of the celestial bodies – possessed some kind of regularity, in some cases more obviously than in others. This was a crucial difference with other kinds of divination such as extispicy (liver observation) and paved the way for the gradual development of the specialized science of mathematical astronomy. Toward the end of the second millennium BC, the first astronomical compendium, MUL.APIN ("The plough"), was composed, which, in addition to astral omens, contained essential data about the position and motion of the fixed stars, the Sun, and the Moon.

5 Lehoux 2002 argues that the overlap between Mesopotamian judicial texts and omens, both at the structural level (list of cases set out as conditional clauses) and in the wording (the judicial council of the gods involved in divination) could indicate that omens were interpreted as records of past decisions of the gods.

It is thought that, at least during the late Babylonian period, only an extremely small class of scribes dealt with these matters, called "scribes of *Enūma Anu Enlil*" in administrative documents of the temple of Marduk in Arsacid Babylon (third century BC).[6] Part of this tradition reached Egypt at the time of, and probably thanks to, the Persian domination (sixth to fifth centuries BC) of both countries, but one can hardly say that it caught on easily.[7] We have to wait until the late Hellenistic period to witness an effective transmission of astral divination and astronomical theory and techniques, which, not by coincidence, takes place in a context of greater Hellenization in Seleucid Babylon. Babylonian culture and institutions, unlike those of other Hellenistic kingdoms such as Ptolemaic Egypt, had remained relatively unchanged for one and a half centuries after Alexander's conquest.[8] Our evidence is inevitably biased because Aramaic and Greek papyri and parchments have perished, but, perhaps significantly, only very few Greek inscriptions from Hellenistic Babylon survive.[9] Of course, there are traceable adaptations to Greek rule, and maybe the end of the third century can be interpreted as a turning point because of the substantial increase of Greek names in the record, despite the possibility that they were simply prestige names used by the local elite.[10] A major breakthrough is signaled by the introduction of a Greek colony in Babylonia by Antiochus IV (175–164 BC),[11] which precedes by only a short period the first datable specimens of horoscopic astrology and techniques in Hellenistic Egypt, such as Hypsicles' table of rising times.

Before that, in the late Babylonian period, throughout the second half of the first millennium BC, traditional astral prediction concerning the future of the state and the king was supplemented with a new kind of prediction on the outcome of

6 Rochberg 1998, 5.

7 The best witness is the Vienna papyrus D 6286, which contains a treatise on eclipses as well as lunar omens, published in Parker 1959, probably deriving from a sixth- or fifth-century source. Cf. *HAMA*, 568. Another omen text that originated in this time could be the *brontologion* (predictions by thunder) attributed to Hermes Trismegistus in *CCAG* VII, 226–230. It is divided in months, and the predictions depend on whether thunders occur by day or night and on their approximate position in the sky (e.g., Eastern or Western horizon). It was reworked in the Roman imperial era, since it shows the Roman names of the months. A fundamental piece of evidence is that it refers to Egypt as a subject country (ἡ Αἴγυπτος τοὺς ἄρχοντας ἑαυτῆς χειρώσεται) and to the Persian king. Furthermore, it does not mention the zodiacal signs, just like the eclipse treatise in the Vienna papyrus. Cf. also *CCAG* VII, 167–171 on earthquakes, also attributed to Hermes Trismegistus. It is worth noting that the influence of Babylonian astronomy and its period relations was felt in the Greek world by that time, too (cf. the 59-year lunisolar period of Oenopides of Chios and the Metonic cycle).

8 Sherwin-White 1987.

9 Kosmin 2015, 174.

10 Beaulieu 2018a, 264–265; Boiy 2006, 289.

11 Van der Speck 2009.

the lives of particular individuals, based on the appearance of the sky at their time of birth. Our only information about this comes from the approximately 30 surviving horoscopes, also called proto-horoscopes to distinguish them from the later Greek tradition. It is believed that the scribes responsible for the new type of prognostications belonged to the same class, since one of these horoscopes is cast for Anu-bēlšunu, whose name appears at the colophon of many astronomical tablets and who was father to a scribe of *Enūma Anu Enlil* (he was surely one himself too) who copied an astrological text with depictions of constellations.[12] But these scholars did not only use such techniques for themselves and their families: two horoscopes show the names of the clients (the "natives", in the astrologers' terminology), which are Greek in form. There is one Aristokrates (no. 10), for whom bravery and high rank are predicted, and one Nikanor (no. 12), who could possibly be identified with an official in the Babylonian Esagila, of the same name and period.[13]

Two crucial innovations in astronomical methods occurred in this period. One was the zodiac, a new system of reference formed by 12 equal sections of the Sun's yearly path (the ecliptic), replacing the older, more cumbersome systems of the normal stars. The older method was advantageous for direct observation, since it tracked the passage of the celestial bodies through a group of well-known stars and constellations, but the new system was much better for calculation. This leads to the second innovation: the development of sophisticated mathematical schemes for the prediction of planetary, as well as solar and lunar, positions, mainly through step and zigzag functions. The older, simpler, and yet fairly accurate method of using diaries of astronomical observations and well-known periods to predict future positions – in which the celestial bodies returned to the same position and repeated their whole motion pattern – was not abandoned, but it is possible that the new horoscopic astrology predominantly used the mathematical schemes.[14]

Since these practices were restricted to temple scholars, we can perhaps connect their developments in the fifth century BC to changes in Babylonian archival and religious institutions following the revolts of 484 BC against Persian rule. If past historians seem to have exaggerated the magnitude of the Persian revenge, it remains true that there is a significant shift in the status of the archives, some of which were suddenly closed after that year, while the surviving ones were often owned by local clients of the Persian governors;[15] in Sippar, one of the main rebel

12 Rochberg 1998, 6 and 80.

13 Rochberg 1998, 4, 84, and 88.

14 Steele 2008, 64, argues this case from the remarkable coincidence that the abovementioned Anu-bēlšunu authored a text containing data calculated with mathematical schemes for the years 252–241 BC, and so perhaps with the aim of casting his own horoscope, for the year 249 BC.

15 Waerzeggers 2018.

strongholds, it has been shown that state officials suddenly withdrew the incomes of the city temples without explanation.[16] Also, while it is unclear to what extent Uruk supported the revolts, major changes in the religious organization of the city in that period, such as the dismantling of the main city temple, that of Eanna, and the foundation of the new temples of Irigal and Reš, have been related to the political atmosphere.[17] It would not be implausible, then, to interpret the emergence of the new personal astrology and its associated methods – no longer focused on the prediction of the future of the Babylonian country – as a scholarly response or adaptation to closer monitoring under Persian rule.

Astrology in late Hellenistic Egypt

The period shortly after 150 BC can be taken as a rough starting point for Hellenistic horoscopic astrology in Egypt. The mathematician Hypsicles, active around that time, is the author of a short treatise on the rising times of the zodiacal signs. The work marks the first time that this system of Babylonian origin is used outside of Mesopotamia.[18] At the same time, the knowledge it contains was fundamental for the calculation of the ascendant point, one of the most prominent features of Hellenistic astrology (hence "horoscopic", since the ascendant point was called ὡρόσκοπος). For this reason, it has been suggested that the writings attributed to Nechepsos and Petosiris, which were already known to Varro (mid-first century BC, cf. Chapter 1), and which are the main antecedent of Critodemus' work, were produced around that date or shortly thereafter in the second half of the second century BC.[19] This probably took place in Alexandria, since we know that the treatise only discussed the rising times for this latitude.[20]

We have reasons to believe that texts on astral divination written in Demotic Egyptian were numerous in the late Hellenistic period, but our perspective is strongly biased toward Greek material because of the Greek literary tradition, the late ancient compilations of Greek astrological material, and the early European interest (and scholarly capacity) in the excavation of Greek material from Oxy-

16 Jursa 2018.
17 Beaulieu 2018b, 192.
18 On Babylonian rising time schemes, cf. Steele 2017.
19 Heilen also thinks that the fact that Nechepsos and Petosiris F 6 Riess predict wars between Greek states could imply that the oldest version predates the defeat of Corinth in 146 BC: Heilen 2011, 24. However, the fragment contains nontechnical mundane astrology loosely attributed by Hephaestio to the "ancient Egyptians", which could also refer to Hermes Trismegistus.
20 Heilen 2011, 25: cf. Val. III 13.6.

rhynchus. We do, however, have good examples of Demotic horoscopes,[21] Demotic mundane astrology,[22] and Demotic astronomical texts.[23] But nevertheless, even considering its wider attestation, the Greek archaeological evidence is not representative of the chronology of Hellenistic astrology, since most of the material comes from the second and third centuries AD, which is partly attributable to the greater survival of papyri from those centuries. For example, it is reasonable to suppose that astrological practice peaked in the first century AD, given the numerous references in Roman literature, as well as the relatively large number of astrological authors and treatises attested for the period, even though the archaeological record for that period is modest.[24]

The specifically Egyptian component of early Hellenistic astrology (second to first centuries BC) is evident in the pseudepigraphic names of astrological authors, the best known of them is Hermes Trismegistus. The Egyptian god of wisdom Thoth had been interpreted by the Greeks as an Egyptian Hermes as early as Plato (who calls him Theuth in *Phaedrus* and *Philebus*). While "great" was a common epithet of Egyptian gods, it is possible that Thoth was first appended a second "great" for his status as lord of Hermopolis. One of the subsequent variants, "thrice great", begins to appear in several independent textual witnesses from the second century BC on.[25] This Hermes famously appeared later, in the second or third century AD, as the author of a series of philosophical dialogues fundamentally influenced by Platonism but exhibiting authentic Egyptian elements; in the last comprehensive study of the corpus, Bull defends the thesis that the treatises were composed in Greek by Hellenizing members of the Egyptian priestly class.[26] The same is plausibly the case of the Ptolemaic astrological texts. Since the Egyptian elements are quite specific, it is not far-fetched to suppose that the Egyptian priestly class, which is most likely to have preserved the knowledge of Egyptian literature and myths, was behind them. This becomes especially clear when we consider the tradition of Nechepsos

21 To date, they have not been edited in a single work. Cf. Ross 2019, 509–526. For a recent analysis of a set of Demotic horoscopes, see Escolano-Poveda 2022.

22 Cf., e.g., the text with omens depending on the zodiacal signs (Gemini or Sagittarius) of the planets in Hughes 1951.

23 Cf. the first century BC text on lunar eclipses in Negebauer et al. 1981.

24 See the excellent edition with translation and commentary of the Oxyrhynchus astronomical papyri as well as the papyrus horoscopes (Jones 1999). For the chronological distribution of the Oxyrhynchus material, cf. ibid., 6–7.

25 See Bull 2018, 33–35. The epithet is attested in Egyptian and Greek in ostraka from the Saqqara archive of Hor (168 BC), transliterated from Egyptian into Greek letters in Hermopolis sometime in the second century BC, and in the text known as the *Breathing Permit of Hor*, of around the same time.

26 Bull 2018, 12.

and Petosiris. As Ryholt has demonstrated, "Nechepsos" derives from the common denomination "Necho the wise" (*Necho Psōs*) of Pharaoh Necho II, of the Saite Dynasty (26th), who reigned from 610 BC: an association existed between Necho and astrology because his predecessor Psammetichus I died at the time of a lunar eclipse. Petosiris, on the other hand, is probably meant to be the sage, trickster, and magician Petese known from many pieces of Demotic fiction: in so far as the story can be reconstructed, Petese/Petosiris found a hidden astrological papyrus which he presented to King Nechepsos, who later composed a work of his own inspiration.[27] The depth of detail in the narrative clearly suggests that the composers of this material had intimate knowledge of the Egyptian tradition.[28]

Now, why specifically Egypt? It is well known that around the same period, in the mid-second century BC, the astronomer Hipparchus, working on the island of Rhodes, had access to Babylonian astronomical data. Therefore, Babylonian astral texts could also be found outside this country, but Egypt was the obvious place for the practice of astrology. Firstly, there was a previous Egyptian tradition of adapting Babylonian astral omens, from the time of Persian domination in the sixth/fifth century. Other reasons that contribute to explaining the implantation of this tradition are the strong presence of star lore in Egyptian cultural imagery and the high status of magic and divination in the Egyptian literary tradition. We might consider, for example, the *Setne stories* I (from perhaps the third century BC), which feature the search for a book of magic authored by Thoth.[29] The Egyptian literary conventions allowed for the use of divine names for this kind of wisdom, a kind of knowledge that was difficult to present in an empirical-demonstrative exposition. It is not an easy task to find elsewhere in the Greek world other prestigious cultural

27 The first part of the story is gathered from the introductory lines of a recently discovered demotic astrological manual: Ryholt 2011. One of the few attestations of the name of the king is in the Vienna papyrus D 6286 (original of the sixth/fifth century BC) as a source of information for eclipse omina. See also Quack 2002 on Petosiris/Petese, and Heilen 2011.

28 Petese is best known for the Demotic fiction named "The Dream of Nectanebo or the Prophecy of Petese", in which King Nectanebo sees in a dream the god Onuris (Anhur) complaining to Isis that his temple has not been completed: see Ryholt 2002. The same Nectanebo is the main character in book I of the *Alexander Romance*, in which he plays the role of an astrologer prophesizing to Alexander the Great's mother-to-be Olympias, while also engendering the Macedonian king. The Arabic descriptions and Renaissance pictures of Claudius Ptolemy showing him crowned as King of Egypt are probably not only due to his name but also to this Egyptian *topos* of astrologer-kings.

29 Cf. Rutherford 1997 for the influence of this piece of Demotic fiction and others as models for the *Aethiopica* of Heliodorus. Cf. the famous passages on the books of Hermes used in Egyptian temples in Clem. *Strom.* VI 4.37 and on the thousands of books allegedly authored by Thoth in Iambl. *De myst.* 8.1.

milieux that could effectively transmit divinatory texts: the second-best candidate is probably the Orphic tradition.

Several aspects of Hellenistic astrology, apart from the mode of presentation, are unmistakably native Egyptian contributions. One of them is the zodiacal subdivision of the decans, groups of stars close to and along the zodiac marking 10-day periods with divine names and listed in different kinds of ancient Egyptian sources from the First Intermediate Period (ca. 2200 BC) on, such as diagonal star tables and star lists on astronomical tomb ceilings.[30]

It is possible that the astrological board was another Egyptian innovation. Only a handful of these objects has survived, in which the main zodiacal divisions (signs, terms, and decans) are depicted as concentric segmented rings carrying the names and/or images of their elements, and upon which gems were placed to symbolize the positions of the celestial bodies and other meaningful points such as the ascendant, the midheaven, or the Lot of Fortune. The best-preserved board, found at Grand (Vosgues), dates from the second century AD and, remarkably, shows the same specific iconography in the center – the Sun and the Moon – as the fictional board of the astrologer-king Nectanebo in the *Alexander Romance*.[31]

The astrological board can be related both functionally and historically to the Egyptian milieu. On the one hand, a fundamental practice in Hellenistic Egyptian astrology which has not been attested in Babylon is the systematic observation of the so-called aspects between two planets on the astrological chart; that is, the combination of the effects of two celestial bodies qualified according to their relative position. If the planets are in square aspect (i.e., at 90°) or in opposition (180°), the negative character of one of the planets has a negative influence on the effects of

30 Cf. Symons 2007 for the diagonal tables of the decans in Egyptian tomb ceilings and their relationship with the civic calendar of 365 days, where she challenges the long-accepted view that they were used as clocks to track time at night (*HAMA*, 561); and Symons 2014 more generally on the ancient Egyptian decans in many kinds of sources. The most detailed account of the decans in Greek astrological texts is that of Hephaestio I 1, in which he lists all the decans by name and attributes several effects and characteristics to the people born with the ascendant in each of them.

31 In the *Alexander Romance* (I 12), the astrologer-king comically encourages Alexander's mother-to-be Olympias, with the board in hand, to delay her delivery so that the ruler's ascendant may be more favorable. The description of the astrological board is in I 4. Cf. Nawotka 2017, 25–26 and 52–53. Nawotka situates the origin of the narrative of book I in the Ptolemaic age because of the popularity of the stories about Nectanebo by that time, but also argues that, because the description of the board matches the appearance of the Grand board, the passage containing the description of this object could have been inserted in the same later period (second century AD). On the Grand board and its Egyptian origin, see Goyon 1993. Plutarch, *Rom.* 12.3 calls astrology "the method of the board" (τῆς περὶ τὸν πίνακα μεθόδου). Cf. also the letter of a Roman soldier in Mons Claudianus asking for the astrological board and the little stars to be sent to him (πέμψας μοι τὸ πινάκιον καὶ τὰ ἀστέρια, O. Claud. 4.803). See also Chapter 6 on the word πίναξ in relation with Critodemus F 11.

Astrological board from Grand (second century AD). © Institute for the Study of the Ancient World/ Guido Petruccioli, photographer. Musée départemental d'art ancien et contemporain, Épinal (Vosges, France).

the other; if they are in triangular aspect (60°), the combination is interpreted as positive.[32] A global, diagrammatic vision of the chart like the one provided by the board was undoubtedly most useful to the astrologer for this kind of judgment, as well as for the more complex practice of comparing two charts.

Senet board of Amenophis III. Brooklyn Museum, Charles Edwin Wilbour Fund.

On the other hand, popular Egyptian board games like *senet* may have partially inspired the creation of the astrological board in late Hellenistic Egypt. The *senet* board consisted of three rows of ten squares each, along which each of two players had to move their tokens after throwing a set of flat casting sticks that served as a

32 On the absence of the *aspects* in Mesopotamian astrology, cf. Koch-Westenholz 1995, 167.

randomizer, until one player had moved all their tokens off the board. Some aspects of the game were related to death and the civic calendar, with which astrology is concerned at the most basic level. In the *Book of the Dead*, chapter 17, playing the game of *senet*, is associated with the departure of the dead from the necropolis.[33] One myth connects the last five squares, often depicted with special signs, with the last 5 days of the Egyptian civic calendar, the so-called epagomenal days.[34] More generally, the association between *senet* and the calendar may also derive from the fact that all the months had 30 days, which is the total number of squares on the board.[35]

The chronology of Critodemus: late second/early first century BC

From the analysis of the fragments in Chapter 1 (F 1–2), we can plausibly assume that the work of Critodemus was known by Marcus Terentius Varro (dead 27 BC), who was the main source on astrological matters for Pliny the Elder's *Natural History*, where our author is first mentioned. We have no other reliable information with which to date Critodemus, but we may confidently identify some fragments (cf. Chapters 7 and 8) as later insertions. Other fragments (cf. Chapters 4–6) also show that the text was at some point heavily modified and standardized.

Horoscopes, that is, birth charts featuring the positions of celestial bodies at the person's birth (and sometimes at other points of life), are often useful for dating astrological texts, since there is often just one possible date in antiquity matching the given planetary positions within the span of a few centuries.[36] Horoscopes thus provide at least a *terminus post quem* for the text, but modern interpreters tend as a norm to take them as more precise chronological indicators, since astrologers often used horoscopes cast for individuals living in their own lifetime for their manuals, coming from their own archive. In our case, we have two double horoscopes, dated

---

33 This passage might be the ultimate origin of the story of the descent into the underworld of King Rhampsinitus (Rhameses III), where he plays dice with Demeter, in Herodotus II 122: cf. Crist et al. 2016, 138–140.

34 Plutarch *Is. Osir.* 12 (355D–E).

35 Cf. Crist et al. 2016, 139. The denomination "house of . . . (god's name)" for these last squares in POxy 470 (third century AD) shows intriguing parallels with the technical features and language of Hellenistic astrology, since the planets and the luminaries were attached to the items in the zodiacal subdivisions (the individual signs, decans, etc.), and the governing celestial body in question was called the "house-ruler" (οἰκοδεσπότης).

36 This is easily calculated with astronomical tables or specialized software. The planetary positions in the charts discussed in this book have been computed with the free software Skychart/ Cartes du Ciel.

AD 37/68 (date of birth and death) (Chapter 4, F 9) and AD 2/36 (birth and death) (Chapter 8, F 20). The situation is quite similar for both sets: they both appear in the work of Vettius Valens, where they are given as examples of doctrines using tables from Critodemus, but evidently according to Valens' own methods (see respective comments *ad locum*). Valens possibly took these horoscopes from Critodemus' work, where they had been used to illustrate a particular theory, possibly on transits (the passage of a celestial body through the same position as it had at birth).

Therefore, we are left with the *terminus ante quem* of Varro, roughly situating Critodemus at the beginning of the first century BC or at the end of the second, assuming that he was active after the didactic astrological poem of Nechepsos and Petosiris. There are two main arguments to support the latter hypothesis: (1) Critodemus' preface seems to be interdependent with the preface of the poem of Nechepsos/Petosiris (cf. F 3, Chapter 2); and (2) Nechepsos and Petosiris are often the names attributed by ancient authors to the first Hellenistic comprehensive compilation(s) of horoscopic astrology, which is the topic Critodemus addresses. Furthermore, Critodemus seems to allude in the same fragment (F 3) to his own trip to Egypt, which may be reliable information given the early date in the history of Hellenistic astrology and the necessarily scant distribution of astrological theory elsewhere in the Greek world, which may have made travel necessary. Therefore, it makes sense to consider Critodemus as subsequent to Nechepsos and Petosiris.[37]

Situating Critodemus geographically: Rhodes?

The manuscript evidence does not provide any indication as to the place of Critodemus' origin or activity.[38] However, the name Κριτόδημος/Κριτόδαμος (the Doric variant) has relatively few attestations in the inscriptional record, which helps us situate our astrologer geographically, considering the narrow chronological window between the mid-second century and the mid-first century BC. In the fol-

37 Already *RE* XI 2, s.v. "Kritodemos 4" (Boll) establishes a relationship between Critodemus and Nechepsos-Petosiris because of the similar presentation of astrology as a vision in the two works (cf. Chapter 2).

38 See the apparent (but only apparent) indication of origin attached to a personal name that could be that of Critodemus in Dorotheus V 41, the Arabic version of our F 15 (Chapter 7). Hephaestio says that the distributions were devised by "some of the ancient Egyptians" (τινες τῶν ἀρχαίων Αἰγυπτίων, Heph. II 29.1), probably not referring to Nechepsos and Petosiris, since everywhere else he simply writes "the ancient Egyptians" or "the Egyptian sages", He is probably referring to intermediary sources (Chapter 4).

lowing table, I group the entries by region according to the *Lexicon of Greek Personal Names* (retrieved 15/02/2023, total of 59 individuals):

Quantity	Region	Settlements	Datable instances (by century, all BC)
17	Attica	Athens	Fifth–fourth (9), third (3), second (2), first (1)
9	Rhodes	Rhodes, Kamiros, Lindos, Ialysos	Fourth/third (1), third (4), second (1), first (2)
8	Euboia	Eretria	Fourth/third (7), second (1)
3	Delos	Delos	Third (3)
3	Thessaly	Gonnoi, Krannon	Third (2), second (1)
3	Lokris	Physkeis, Myania	Second (3)

It should not be surprising that the name is not attested in Rhodes before the Hellenistic period, since most inscriptions come from this period of cultural and political relevance. By contrast, the relatively small presence of the name in Attica in our author's time should be regarded as relevant, since there are abundant Attic inscriptions for that era. The same is true for Euboia. It is also significant that no Egyptian papyri or inscriptions bear the name, nor any inscription from Asia Minor.[39] This leaves Rhodes as a strong candidate.

In F 3, probably the fragment closest in form to the original text, Critodemus describes his own arrival to a "harbor free of perils" (ἀκίνδυνος λιμήν) and a "most secure resting place" (μονὴ ἀσφαλεστάτη) after "having set sail" (πέλαγος προσόρμησας) and "traversed deserted land" (ἔρημον ὀδεύσας). While literary tropes were surely at play (cf. commentary ad loc., Chapter 2), Critodemus could be alluding to a stay in Egypt to familiarize himself with astrological learning by attending lectures and studying books such as those of Nechepsos and Petosiris. Of course, one cannot rule out the possibility that Critodemus' trip was entirely fictional, but intellectuals frequently traveled between Alexandria and other Greek cities in the Mediterranean at that time.

As for the "place of safe arrival", the role of Athens in policing the Aegean was taken over by the city of Rhodes in the Hellenistic period, with the aim of securing commercial traffic through its port. The Rhodian fleet was specifically designed for fighting piracy, as revealed in the ancient sources.[40] It is thus at least plausible that

39 In the literary sources, only four authors mention the name: a doctor from Kos (Arrian, *Alex. Anab.* VI 11.1), a victor in Olympia from Arcadia (Paus. VI 8.5), and a patronymic in Demosthenes (*In Theocr.* 35, *In Neaeram* 25) and Lysias (*Aristoph.* 15).

40 See, e.g., Strab. XIV 2.5: ἀφ' ἧς ἐθαλαττοκράτησε πολὺν χρόνον καὶ τὰ ληστήρια καθεῖλε ("as a result it [Rhodes] controlled the seas for a very long time and destroyed piracy"), Ael. Aristid. *Or.* 25.4. See Gabrielsen 2004, 228–229.

Critodemus was alluding to the city of Rhodes in his characterization of the harbor and the place of his arrival.

Rhodes shared in late Hellenistic times the traditional roles of Athens and Alexandria as cultural centers. The island was attractive for its proximity to Alexandria and to important Ionian cities such as Pergamum, as well as for its wealth, brought in by commerce. Specifically, astronomy occupied a relatively prominent place in the intellectual activity of the island. The great astronomer Hipparchus established himself there toward the middle of the second century BC, as attested by astronomical observations used by Claudius Ptolemy; the Stoic philosopher Posidonius, interested in astronomical matters, established his school in Rhodes, famously attracting prominent Roman intellectuals and politicians to his lectures. The astronomical writer Geminos, likely active in the first half of the first century BC, was probably from Rhodes.[41] In the first century AD, it is on this island that the future emperor Tiberius met the astrologer Thrasyllus. It is thus entirely plausible that the intellectual climate of the island nurtured the astrologer Critodemus.

Orphic astrology

One presentational device (the oath, Chapter 3) and possibly one important doctrine (the *chronocrators*, Chapter 4) reveal the influence of the Orphic tradition in Critodemus. This, as mentioned above, is to be interpreted in relation to the traditional pseudepigraphical framing of early Hellenistic astrology. Hermes Trismegistus was among the first names to which astrological works were attached in late Hellenistic Egypt. Other pseudepigraphic figures were Nechepsos and Petosiris, or Asclepius.[42] We have already discussed the use of divine names as a means of incorporating into the Greek literary and scientific tradition a new divinatory corpus

---

41 Evans and Berggren 2007, 16, conclude that the evidence is suggestive of Rhodes, although not conclusive. Geminos frequently uses Rhodes as an example in his astronomical explanations: while other astronomers from other places (like Ptolemy) do this on occasion, Geminos does so consistently, and he also seems to expect his readers to be familiar with Rhodian geography. He wrote an abridgement or commentary on Posidonius' *Meteorology* (23), and since the most likely dating makes him a younger contemporary of Posidonius, some say he may have been his pupil, but there are strong reasons to reject this claim (26). Geminos (*Introduction* 14.12) informs us that armillary spheres and celestial globes were frequently built specifically for the latitude of this island, perhaps because of its central position in latitude within the Greek-speaking world.

42 Cf. Valens IX 3.5. Asclepius was probably related to the philosophical tradition of Hermes Trismegistus, in which the name of Asclepius often appeared as the receiver of the learnings: see discussion in Critodemus F 5.

of knowledge without an evident empirical basis. Orpheus, the mythical Thracian singer, was another authorial voice that was available for this strategy.

Since as early as the fifth century BC, we hear about the role of Orpheus as a founder of rituals, liturgic songs, theogonic poems, and special philosophical and religious tenets about the immortality of the soul, sometimes in relation with Dionysus.[43] The fourth-century-BC Derveni papyrus testifies to the early development of a tradition of philosophical and allegorical commentary on the Orphic theogonies from within the Orphic milieu.[44]

We may trace the Orphic influence in Critodemus to two kinds of materials. One kind of Orphic poem was concerned with the description of parts of rituals, which were presented allegorically in cosmological speculations. In this category, we have titles such as *Enthronement*, *Sacrifice*, *Purification*, or *Tightening*, and some lines from a poem called *Oaths* which probably influenced Critodemus' oaths of silence (see Chapter 3). The title of Critodemus' work, *Vision*, might also be interpreted in the light of this tradition.

Over time, many writings on natural and magical questions containing less theological material came to be attached to the name of Orpheus, undoubtedly in order to secure transmission through the prestige of the name. This is not to say that a connection to Orpheus did not exist: Orpheus, with the enchanting character of his music, was conceptualized as a most effective divine mediator and could therefore be considered an appropriate receiver of divine wisdom. For this reason, he was the purported author of poems on the magical properties of stones, on medical herbs and magical medicine, or divination. It is in this last category that we can place the astrological fragments attributed to Orpheus that have come down to us: a poem on good and bad days in the lunar month for agricultural tasks (*Ephemerides*); on finding the proper timing for agricultural endeavors (*Georgika*, also attributed to the fourth-century astrologer Maximus) or endeavors in a more general sense, not circumscribed to a particular type (*Initiatives*, also attributed to the Egyptian god Amon); on fugitive slaves, a popular topic in Hellenistic astrology (again also attributed to Maximus); on the meaning of earthquakes depending on

---

43 Cf., e.g., Plato *Prot.* 316d (τελετάς τε καὶ χρησμῳδίας, "rituals and religious songs"). Plato (*Rep.* 364e–365a) informs us that the so-called *orpheotelestai* were private purveyors of purifications and sacraments who employed books by Orpheus to perform the rites, and also that these purifications and sacraments had been adopted by some of the great cities; cf. West 1983, 10. See Bernabé 2008 on the role attributed to Orpheus in the institution of the Eleusinian mysteries. Cf. Plat. *Crat.* 400c for Orphic views on the soul; and in the same work, 402b, a quotation of an Orphic theogony. See the numerous articles collected in Bernabé and Casadesús 2009. For a study of the Orphic literary poems, see West 1983, and see Morand 2001 on the late collection of *Orphic hymns*. See Karanika 2010 on the role of Orpheus as ritual advisor in Apollonius' *Argonautika*.
44 See, e.g., Betegh 2004.

the sign of the Sun (also attributed to Hermes Trismegistus); on the implications of ingresses of the planets into a new sign, or on the meteorological and agricultural events marking each year in a recurring cycle of 12 years (*Dodekaeterides*).[45] The dating of this corpus is complex because we cannot tie the examples to a historical author, and because their textual tradition is much more fluid than that of literary texts (we have many different versions of some of the texts), but it seems likely that "scientific" texts began to be attributed to Orpheus in the late Hellenistic period.[46] Specifically, the *Dodekaeterides* were probably known to Varro and could have influenced Critodemus in the creation of his theory of *chronocrators*, the time lords governing the different periods of life (see Chapter 4).

The transformations of Greek and Graeco-Roman astrological writing

The ancient textual genre of the astrological manual shows specific characteristics that situate it quite apart from most of the ancient Greek and Roman texts with which scholars are usually concerned. The astrological genre was, for good reasons, generally appreciated not for the quality of its literary style but for the usefulness of its content. This naturally informed a transmission whose greatest concern was not necessarily faithfulness to the original texts. The fact that part of this tradition was, as we have seen, attached to divine names in order to secure the transmission, paradoxically encouraged appropriation and modifications including the change of the author's name: as we have seen, some texts were attributed either to Hermes or Orpheus, or to Orpheus and Amon. Changes in the content were facilitated by the simple style of these texts: the lists of astrological effects were easy to enlarge, and more detailed explanations of the doctrines, sometimes in the form of horoscope examples, were probably welcome. Vettius Valens' request not to modify his treatise, inserted into his own astrological oaths (Chapter 3), clearly indicates that astrologers were aware of such practices.

A decisive factor distinguishing astrological from other kinds of scientific texts whose transmission was less fluid is that astrological texts were mostly used and transmitted internally: that is, by astrologers. This was, for example, not the case

45 The Orphic scientific fragments have been collected and translated into Spanish in Martín Hernández 2015. For the Orphic *Lithika*, see 18–73 (hexameters) and 78–101 (the prose paraphrasis); 157–164 for medicine and herbal fragments; 165–171 for fragments on oracles and divination; 119–126 for the *Ephemerides*; 127–131 for the *Georgika*; 129–154 for the *Initiatives*; 131 for fugitives; 132–136 for earthquakes; 137–149 for ingresses; 108–118 for the *Dodekaeterides*.
46 Martín Hernández 2006, 113; West 1983, 33.

with an otherwise comparable genre such as ancient medicine: even if part of its readership was formed by doctors and other medical professionals, medical texts were frequently written with a more general audience in mind, which brings them closer to the literary genres.[47] Such distance between authors, readers, and copyists and its embodiment in higher stylistic features helped maintain a stable transmission. On the other hand, astrological readers, even when they were not as experienced as the author, were not on such a different level that they could not contribute themselves with relevant changes.

This explains Valens' preoccupation with this issue and his express petition not to alter his manual. Another compensating mechanism that limited to some degree the modifications of astrological texts was the verse form. The treatises of Dorotheus of Sidon and Manilius (first century AD), were written in hexameters. The work of Nechepsos and Petosiris was originally a poem in iambic trimeter, and Critodemus seems to have used this metrical form too. The metrical pattern no doubt prevented the indiscriminate insertion and suppression of material, although there were certainly other intentions behind this practice. An ancient geographical author adduces that the iambic verse form provides clarity and functions as an aide-mémoire:[48] the latter would be especially appropriate for the memorization of the long lists typical of astrological texts. However, the surviving evidence shows that metrical astrological texts were often paraphrased in prose form.[49] The poem of Nechepsos and Petosiris is an apt example: even if metrical traces are evident (especially in F 1 Riess), no ancient author quotes a fully metrical line. The obvious conclusion is that the paraphrase(s) completely substituted the original, metrical form. This was probably the case with Critodemus, too, as we will see (especially Chapter 2, but also Chapters 4–6).

---

47 For rhetorical display in the Hippocratic corpus, see Jouanna 2012; see Johnson 2012 for a study of the diverse readership of Galen.

48 Cf. Ps.-Scymnus 33–35 (author of a geographical poem from roughly the same era as Critodemus, also in iambic trimeter).

49 Sometimes works in verse form were enlarged while maintaining the metrical structure: this is the case of the poem of Manetho (probably another pseudepigraphic name for the Egyptian priest and historian), in hexameters, which took form over several periods with parts written by different authors: cf. Pingree 1978, 435–436, and the new edition with translation and commentary Lightfoot 2021. Astrological treatises in prose were also paraphrased: we have the so-called Proclus paraphrase of Ptolemy's *Tetrabiblos*, and as many as four epitomes of Hephaestio.

First part: **Fragments related to contextualization**

Chapter 1
Critodemus and Berossus in Pliny, via Varro (F 1–2)

A couple of mentions of Critodemus in Pliny's *Natural History* provide us with several insights. First and foremost, they set a *terminus ante quem* for Critodemus' activity by Pliny's lifetime, the 70s AD. In fact, I will argue that Pliny probably only knew Critodemus through his source Varro, active around the mid-first century BC, which would situate the work of the astrologer toward or before the beginning of the first century BC. This is consistent with the texts in Chapter 2, which class Critodemus among the first astrological authorities, probably after the Egyptian pseudepigraphs of Nechepsos and Petosiris (second half of the second century BC).

Concerning the content of Pliny's testimony, I present what I believe are solid grounds to refute the authenticity of F 2B. My arguments are based on the playful character of Varro's attribution of similar lore to second-century-BC authors and mythical figures.

F 1. A Greek authority for Pliny book II: Pliny I 2c

Type of fragment: List of Greek authorities for book II of Pliny's *Natural History*, compiled by himself, mentioning Critodemus alongside Thrasyllus, the astrologer friend of emperor Tiberius (first century AD). Critodemus is not, however, mentioned in the text of book II, as is the case of many authorities, who are consigned in the indexes. The section to which Critodemus' work was relevant can nevertheless be determined with some confidence.

Note on the text: There are no significant variants in the manuscripts and, considering the nature of the fragment, I do not provide the original Latin.

https://doi.org/10.1515/9783111329147-002

Foreign authors

Hipparchus, Timaeus,[1] Sosigenes,[2] Petosiris, Nechepsos,[3] Pythagoreans, Posidonius, Anaximander, Epigenes,[4] Eudoxus, Democritus, **Critodemus**, Thrasyllus, Serapion the geometer,[5] Euclides, Coeranus the philosopher,[6] Dicaearchus,[7] Archimedes, Onesicritus,[8] Eratosthenes, Pytheas,[9] Herodotus, Aristotle, Ctesias,[10] Artemidorus of Ephesus,[11] Isidorus of Charax,[12] Theopompus.[13]

Context in Pliny

The indexes of the *Natural History* should be taken seriously, since they are the result of Pliny's self-declared plan to write down, for each of the 36 books of his work, a list of topics followed by a list of sources. This material forms an integral

1 Pliny cites him as an authority along with Cydenas and Sosigenes in *NH* II 37 for the maximum elongation of Venus and Mercury (distance with the Sun): cf. *HAMA*, 804. He appears as an astrologer again in *NH* XVI 82. Possibly the same as the astrologer of the same name cited in Critodemus F 3B.

2 Astronomer who, according to Pliny, assisted Julius Caesar in creating the Julian calendar: *NH* XVIII 57.210–212.

3 These are the first Graeco-Egyptian astrological authorities, frequently cited in later sources. King Nechepsos is the purported author of an astrological didactic poem written in iambic trimeters ca. mid-second century BC, responding to Petosiris, another fictitious name, who had reportedly introduced him to astrology. For the history of the names and edition with commentary of the metrical fragments, cf. Heilen 2011.

4 Philosopher with a strong interest in astronomy and meteorology said in Seneca's *Natural Questions* to have studied with the "Chaldeans" (VII 4.1: *apud Chaldaeos studuisse se*), probably just meaning "astrologers", as is common in Roman authors: cf. Seneca *NQ* II 32.7, where the term seems to be used in this sense. Seneca reports Epigenes' views on comets. Since Epigenes is said to have criticized the Chaldeans' opinions on comets, and since the poem of Nechepsos and Petosiris contains a section on comets, it is possible that Epigenes was familiar with the latter: cf. Keyser 1994.

5 The Latin term is *gnomonicus*, which involves the study of angles related to spherical geometry, i.e., astronomy and geography. He is perhaps the astronomer friend of Cicero (*Ep. ad Att.* II 4.1), possibly identifiable with a Serapio mentioned by Theon of Alexandria in his commentary on Ptolemy's *Handy Tables*: cf. *HAMA*, 584.

6 Stoic of Nero's time: cf. Tac. *Ann.* XIV 59.

7 Geographer friend of Theophrastus.

8 Historian who accompanied Alexander the Great.

9 Explorer and geographer, fourth century BC.

10 Asclepiad doctor of the fifth century BC, personal physician of Artaxerxes II; also historian-geographer, author of *Persica* and *Indica*.

11 Geographer of around 100 BC.

12 Geographer of the first century BC/AD, author of the *Parthian Stations*, a commercial itinerary between Antioch and India.

13 Historian/geographer of the fourth century BC, author of *Hellenica*.

part of book I. However, not all of the authors listed in the indexes are cited in the text, nor should we believe that Pliny consulted works by all the authors in the lists. As he acknowledges in his explanation, he mostly draws from a selection of 100 authors (I pr.17: *ex exquisitis auctoribus centum inclusimus XXXVI voluminibus*), which unfortunately is not disclosed, although we may presume that they were, for the most part, Roman authors.[14]

This is important to keep in mind when assessing F 1, which does not correspond to any mention of Critodemus' name in the text of the book. Nevertheless, we can make a reasonable guess about the place in book II in which Pliny may have introduced material from Critodemus, because the indexes are arranged roughly in the order of the topics addressed in the corresponding book.

From a cursory look at the contents of book II, we can see that the first chapters contain expositions on astronomical topics, whereas the last ones deal with geographical mirabilia. Correspondingly, we find astronomers at the beginning of the Greek list (Hipparchus, Sosigenes, and Eudoxus) and geographers/historians toward the end (Herodotus, Artemidorus of Ephesus, etc.). In the following table, I list the authors closest to Critodemus in the index, along with the chapter in which they are cited (in bold script) or may have been used, even if indirectly:

Authors as in the index	Chapter	Topic of the chapter(s)
Hipparch	8	Nature of the stars and planetary motion
Timaeus	**8**	
Sosigenes	**8**	
Nechepsos and Petosiris	8, **23**	Nature of the stars / Dimensions of the world
Pythagoreans	**22**	Harmony of the stars
Posidonius	**23**	
Anaximander	**8**	
Epigenes	24	Comets
Eudoxus	8, **48**	Nature of the stars / Nature of the winds
Democritus	**7**	On god
Critodemus	8	
Thrasyllus	8	
Serapio gnomonicus	72	Gnomonics
Euclid	72	

14 It is significant that in the indexes of each book he separates Latin and Greek sources in two subsequent lists – naming the Greeks "foreign" (*externis*) – and perhaps that in the preface he mocks the titles of Greek works, contrasting them with the more down-to-earth Roman titles.

On closer inspection, we can easily perceive that the ordering of the authors in the index is rather haphazard: from the set Anaximander-Timaeus-Sosigenes, all of whom are mentioned in Chapter 8, dealing with the nature of the stars and the motion of the planets, only Timaeus and Sosigenes appear together in the index. On the other hand, Democritus, who is already mentioned in Chapter 7, is listed much further down, just after Eudoxus, who would fit well in Chapter 8.

In the long Chapter 8, Pliny introduces not only basic astronomical theory on the motion of the planets but also the so-called natural qualities of the planets, which is basic astrological lore covered at the beginning of most astrological manuals. In the case of Thrasyllus, although the astrological manual has not survived, we are fortunate to have what appears to be an exhaustive summary of it, which indicates that he addressed this topic.[15] Pliny could have known Thrasyllus' work firsthand, even if he wrote in Greek. He mentions Thrasyllus in XXXII 19 in an iatromathematical context (astrological medicine), specifically regarding the antagonism of the crab and the serpent, and he is indexed two more times as a source, likely in relation to similar topics.[16] Thrasyllus had been an important political and scholarly figure in the Pliny's near Roman past, close enough to function as direct source, writing in a variety of genres that made him attractive to an encyclopedist like Pliny. By contrast, direct knowledge of Critodemus' work cannot be assumed a priori. In any case, the content of an astrological treatise such as Critodemus would also fit most appropriately in Chapter 8.

F 2. On human inventions: Pliny I 7c / VII 56.192–193

Type of fragment: This is divided into two parts. F 2A is the index of Greek authorities for Pliny's *Natural History* VII, citing Critodemus in the last place. F 2B is the corresponding passage in book VII mentioning Critodemus toward the end of the book. The doctrine attributed to the astrologer assigns a specific number of years to the antiquity of Babylonian celestial observations, but this is problematic because Pliny's source, Varro, can be shown to have parodied this sort of astronomical speculation by inventing data and attributing it to both real and mythical authors.

---

15 *CCAG* VIII 3, 100, 20–21: καὶ περὶ τῶν φύσεων τῶν πλανωμένων διαλαμβάνει.

16 Plin. XXXII 19, 55 "Thrasyllus informs us that there is nothing so antagonistic to serpents as crabs; that swine, when stung by a serpent, cure themselves by eating them; that when the Sun is in Cancer, snakes suffer tortures" (*Thrasyllus auctor est nihil aeque adversari serpentibus quam cancros; sues percussas hoc pabulo sibi mederi; cum sol sit in cancro, serpentes torqueri*). Thrasyllus is also mentioned in the indexes of book IX, on the natural history of fish; and of book XXXI, on remedies from water.

A. Pliny I 7c

Note on the text: As in F 1, the manuscript tradition is fairly uniform, and the nature of the text allows us to go directly to the translation.

Foreign authors
Herodotus, Aristeas,[17] Baeton,[18] Isigonus,[19] Crates,[20] Agatharchides,[21] Calliphanes,[22] Aristotle, Nymphodorus,[23] Apollonides, Phylarchus,[24] Damon,[25] Megasthenes,[26] Ctesias,[27] Tauron,[28] Eudoxus, Clitarchus,[29] Duris, Artemidorus, Hippocrates the doctor, Asclepiades the doctor, Hesiod, Anacreon, Theopompus, Hellanicus, Damastes,[30] Ephorus,[31] Epigenes, Berossus, Petosiris, Nechepsos, Alexander Polyhistor, Xenophon, Callimachus, Democritus, Diyllus the historian,[32] Strato who wrote against the *Discoveries* of Ephorus, Heraclides Ponticus, Asclepiades who wrote the *Sub-*

17 Aristeas of Proconnesus, according to Herodotus (IV 13–16), visited the Issedones, who told him about the Hyperboreans.
18 Apparently an official in charge of measuring distances in Alexander's marches: cf. *NH* VI 22.69.
19 A writer of paradoxa, first century BC or AD.
20 Crates of Mallus, Stoic philosopher and grammarian, related to the Pergamene library, second century BC.
21 Cnidian historian and geographer, second century BC.
22 Otherwise unknown, cited in *NH* VII 2.15 providing unusual geographical-ethnical information, after Crates and Agatharchides and before Aristotle as in the index.
23 Cited after Aristotle in *NH* VII 2.16; Syracusan author of the Hellenistic period also cited by Athenaeus.
24 Historian, contemporary of Aratus, criticized by Polybius.
25 Otherwise unknown, providing geographical information in *NH* VII 2.
26 Also cited in *NH* VII 2, third-century-BC author of an *Indika.*
27 Fifth-century-BC Cnidian historian and physician (an Asclepiad) at the service of Artaxerxes II, also used in *NH* VII 2.
28 Otherwise unknown, cited in *NH* VII 2 before Eudoxus.
29 One of the historians of Alexander the Great.
30 Historian, contemporary of Herodotus, from Sigeum.
31 Author of a universal history, fourth century BC.
32 Author of a universal history, fourth century BC, apparently the continuation of that of Ephorus.

jects of tragedy,[33] Philostephanus,[34] Hegesias,[35] Archemachus,[36] Thucydides, Mnesigiton,[37] Xenagoras,[38] Metrodorus of Scepsis,[39] Anticlides,[40] **Critodemus.**

B. Pliny VII 56.192–193

192 *Litteras semper arbitror Assyriis*[a] *fuisse, sed alii apud Aegyptios a Mercurio, ut Gellius, alii apud Syros repertas volunt, utrique*[b] *in Graeciam attulisse e Phoenice Cadmum sedecim numero, quibus Troiano bello Palameden adiecisse quattuor hac figura ΖΥΦΧ*[c]*, totidem post eum Simoniden melicum ΨΞΩΘ, quarum omnium vis in nostris recognoscitur. Aristoteles decem et octo priscas fuisse et duas ab Epicharmo additas XZ quam a Palamede mavult.* **193** *Anticlides in Aegypto invenisse quendam nomine Menen tradit, XV annorum ante Phoronea, antiquissimum Graeciae regem, idque monumentis adprobare conatur. e diverso Epigenes apud Babylonios DCCXX annorum observationes siderum coctilibus laterculis inscriptas docet, gravis auctor in primis; qui minimum, Berosus et Critodemus, CCCCXC. ex quo apparet aeternus litterarum usus. in Latium eas attulerunt Pelasgi.*

a *Assyriis*: *Assyrias* $\mathbf{F^2a^2d}$
b *utrique*: *uirique* $\mathbf{DR^1F^2a^1}$
c ΖΥΦΧ: this and the other sequences of 4 Greek letters differ widely in the manuscripts

192 I consider that there have always been letters among the Assyrians, but some, like Gellius,[41] say they were invented in Egypt by Mercury, and others in Syria;[42]

---

33 Mythographer of the fourth century BC.
34 Historian, pupil of Callimachus, wrote a work on islands whose section on Cyprus was widely used.
35 Of Magnesia, historian and rhetorician (of the Asiatic style) known to Strabo, Varro, Agatharchides, Dionysius of Halicarnassus, and Cicero.
36 Historian who wrote about his native island, Euboea.
37 Grammarian also mentioned in Plutarch *QG* 19.
38 Historian from Heraclea Pontica quoted by Dionysius of Halicarnassus.
39 Philosopher who turned to politics, friend of Mithridates VI of Pontus, second to first century BC, known mainly from Strabo and Plutarch.
40 Historian who wrote *On returns* (by the Greeks from different expeditions), *Deliaca* (on Delos), and on Alexander the Great.
41 This is obviously not Aulus Gellius, the second-century-AD author of *Attic Nights*, but Gnaeus Gellius, a second-century-BC historian and chronographer of Rome.
42 For a detailed study of the complex distinction between Syria and Assyria, see Andrade 2014; Nesselrath 2016, 4–5.

both groups, in turn, say that they were introduced in Greece from Phoenicia by Cadmus,[43] being sixteen in number, and that to them Palamedes added four with this shape, ΖΥΦΧ,[44] at the time of the Trojan war, after which Simonides the melic poet added ΨΞΩΘ, completing the set that we see in our time.[45] Aristotle rather assumes that the original number was eighteen and that Epicharmus, not Palamedes, added two, ΧΖ.[46] **193** Anticlides transmits that a certain Menes in Egypt[47] invented them, 15,000 years[48] before Phoroneus, a Greek king of great antiquity,[49] and he tries to prove this from the records.[50] From another side, Epigenes, a most serious author, informs us that in Babylon there have been celestial observations for a period of 720,000 years, inscribed on baked bricks; Berossus and Critodemus, who give the lowest number, have 490,000,[51] from which it would seem that there has always been use of letters. They were introduced in Latium by the Pelasgi.[52]

---

43 Herodotus V 58 already says that Cadmus transferred the letters to the West.
44 These Greek letters and those that follow differ widely in the manuscripts (cf. Mayhoff's edition).
45 The theory is first found in Theophrastus fr. 735: 16 original letters and 2 sets of 4 coming thereafter, introduced by Palamedes (the Trojan hero, credited in Pl. *Phaedr.* 261b with the invention of rhetoric) and Simonides the lyric poet, respectively.
46 This opinion is not found in the extant corpus, but a scholion on Dionysius Thrax (p. 185.6–7 Hilgard) transmits that some (unnamed) considered Epicharmus, not Palamedes, the inventor of the group ΖΘΧΦ.
47 Founder of the first dynasty of Egypt and unifier of Lower and Upper Egypt according to the classical tradition.
48 Pliny as a norm only assumes *milia* when giving high numbers in letters: cf. e.g. *ut centum quindecim milibus passuum Arabicus sinus distet ab Aegyptio mari, Caspius vero CCCLXXV a Pontico* (II 173). This is clearly implied here because otherwise we would expect *annos* instead of *annorum*.
49 This probably does not mean that Phoroneus invented the letters for the Greeks, but situates Menes (and the invention of writing in Egypt) chronologically in relation with Greek king lists. Phoroneus was the son of Inachus (a god), and therefore the first human king of the Argives (cf. Pausanias II 15), in the same way that Menes was the first human king of Egypt (Manetho fr. 7, Herodotus II 99).
50 Here, *monumenta* probably means just written records (i.e., chronicles in this case), as elsewhere in Pliny's work, e.g., XIII 4.20: *in M. Ciceronis monumentis invenitur unguenta.* Cf. also XIII 21.69 *postea publica monumenta plumbeis voluminibus, mox et privata linteis confici coepta aut ceris.*
51 Cf. above for the implied *milia*. In any case, it would be unthinkable that Pliny intended the numbers here to be 720 and 490 years, respectively, since he knew well that Egyptian civilization was many thousands of years old (XXXV 15), and nevertheless he gives precedence to Mesopotamia in the invention of writing.
52 The indigenous population living in Greece before the Greeks arrived, according to the Greek tradition. Pliny perhaps alludes to the Arcadian cultural hero Evander, who, Hyginus says (fable 277), introduced the alphabet to Italy, since the Pelasgi were associated with that region (cf., e.g., Herodotus I 146; Pelasgus was also the father of King Lycaon in Arcadia: Strabo V 2.4).

Context in Pliny

Pliny's book VII is divided into 59 chapters devoted to anthropological matters. In chapter 56, Pliny shapes his discussion in a nonsystematic order, following either similarities between inventions or the sequence of inventions by the same mythical figure. Thus, he begins with buying and selling, which he associates with Liber, and continues with the diadem which is attributed to the same deity; the topic of commerce leads to corn (Ceres) and laws (also Ceres); laws prompt letters (where Critodemus is cited), and the Babylonian script on bricks leads to brick kilns (Euryalus and Hyperbius) and mortar.

The context of Pliny's discussion on the origin of writing can to some extent be traced back to earlier writers. Thus, Theophrastus (fr. 735 Fortenbaugh) attributes the invention of letters to the Egyptians, a new development to the Phoenicians, and the transmission to the Greeks to Cadmus. Theophrastus attached the subsequent invention of the remaining eight letters to the Trojan hero Palamedes and to the poet Simonides in two sets of four letters each.[53] The story, then, seems to have been developed by Latin-speaking authors in order to explain the history of Latin letters. For example, we find a version in the grammarian Hyginus (fable 277) who, like Pliny, appends the part on the arrival of the alphabet in Italy. Among the authors cited by Pliny, judging from the wording it seems that only Anticlides attempted to provide chronological arguments.[54]

The other authors (Epigenes, Berossus, and Critodemus) appear in the text as directly related to the antiquity of celestial observations, and therefore they are probably not meant to have discussed the history of writing. Indeed, in chapter 48 of the same book, Pliny again uses the pair Epigenes and Berossus for astronomical and astrological data. This is probably why Berossus and Epigenes appear much higher than Critodemus in the list of authorities (F 2A):

Chapter	Topic	Authors cited (among others)
37	Men excelling in the arts	Berossus
48	Human lifespan	Berossus, Epigenes, Nechepsos, and Petosiris
56	Inventions	Berossus, Epigenes, and Critodemus

53 See Fortenbaugh 2014, 177–186 for a commentary on this fragment attributed to a work titled *Robe*, which might in fact not be by Theophrastus. The presence of Simonides and Epicharmus as inventors of Greek letters has been interpreted as the result of early Athenian scholarship accounting for the arrival in Athens of written works by these authors: cf. Willi 2013.

54 Here we have an independent witness, namely a scholion on Dionysius Thrax (p. 183, 1 Hilgard), which informs us that Anticlides attributed the invention of writing to the Egyptians.

The cluster-like form of the attestations, as well as their relative obscurity, could suggest that Pliny used the same source for Berossos and Epigenes, and perhaps for Critodemus, too.

Berossus

As we shall see, other Greek and Roman authors apart from the ones in F 2B similarly attach incredible antiquity to Babylonian records of astronomical observations. In other sources, the numbers are always different, so the fact that Berossus and Critodemus would have given the same number (490,000) is important. Berossos, author of a lost *Babyloniaca* – a history and chronology of Babylon in Greek language – was a priest of Bel in the Esagila temple of Babylon in the mid-third century BC.[55] According to the Roman architect Vitruvius, he founded a sort of astrological school on the island of Cos (IX 6.2):

> *Eorum autem inventiones reliquerunt inque <ea re> sollertia acuminibusque fuerunt magnis, qui ab ipsa natione Chaldaeorum profluxerunt. primusque Berosus in insula et civitate Coo consedit ibique aperuit disciplinam, post ei studens Antipater iterumque Achinapolus, qui etiam non e nascentia sed ex conceptione genethlialogiae rationes explicatas reliquit.*
>
> The talent, the ingenuity, and reputation of those who come from the country of the Chaldeans, is manifest from the discoveries they have left us in writing. Berossus was the first of them to settle in the island and state of Cos, establishing a school. Afterwards came Antipater and Achinapolus, which latter not only gave rules for predicting a man's fate by a knowledge of the time of his birth, but even by that of the moment wherein he was conceived.[56]

The story seems to be a foundational myth explaining the cultural transmission of astrology into the Graeco-Roman world from Mesopotamia. The ideal school setting it depicts is quite exceptional in the field of ancient astrology, in which we mostly find isolated figures. Some scholars have argued convincingly that it is implausible that a priest of Bel in Babylon would have had sufficient incentive to emigrate to the West.[57] One may add that, even allowing for this possibility, the tiny island of Cos does not seem particularly attractive for such a move. This would be different to the case of the Babylonian astronomer Sudines, whom Strabo (XVI 1.6) situates

---

55 He appears as *šatammu*, high priest and administrator, in a document dated to 253 BC. Cf. Bach 2013; Geller 2014, 221.

56 Trans. Gwilt.

57 Geller 2014, and already *RE* III 1, s.v. "Berossos 4" (Schwartz). Accepting the possibility of migration to the West, cf. Burstein 1978, 5–6; Verbrugghe and Wickersham 1996, 14–15; and De Breucker 2013, 19.

at the court of king Attalus I of Pergamum in approximately the same period (ca. 250 BC).

The figure of Berossus could have been employed to recount the beginning of astrological practice in the Hellenistic world in a more historical fashion than that of the Egyptian pseudepigraphs. Attributing the origin of the astrological discipline to a Babylonian author, even if there is no hint that the historical Berossus had any astrological training, was probably deemed necessary in some historical contexts, given the discrepancy between the incontestably Babylonian origin of most astrological doctrines and the conspicuous absence of Babylonian names in the Graeco-Roman astrological sources.[58]

Pliny's statements in his first mention of Berossus are enlightening and relevant to the question (VII 37.123):

> *Variarum artium scientia innumerabiles enituere, quos tamen attingi par sit florem hominum libantibus: astrologia Berosus, cui ob divinas praedictiones Athenienses publice in gymnasio statuam inaurata lingua statuere; grammatica Apollodorus, cui Amphictyones Graeciae honorem habuere; Hippocrates medicina, qui venientem ab Illyriis pestilentiam praedixit discipulosque ad auxiliandum circa urbes dimisit, quod ob meritum honores illi quos Herculi decrevit Graecia.*

> Innumerable are the men who have excelled in the various arts; we may, however, take a cursory survey of them, by citing the names of the principal ones. Berossus excelled in astrology; and on account of his divinations and predictions, a public statue with a golden tongue was erected in his honor by the Athenians at the gymnasium. Apollodorus, for his skill as a grammarian, had public honors decreed him by the Amphictyonic Council of Greece. Hippocrates excelled in medicine; before its arrival, he predicted the plague, which afterwards came from Illyria, and sent his pupils to various cities, to give their assistance. As an acknowledgement of his merit, Greece decreed him the same honors as to Hercules.[59]

Here and in the following chapter, Pliny reviews the honors bestowed upon men excelling in (a) astrology, grammar, and medicine (37); and (b) geometry, architecture, painting, and sculpture (38). The order of the first three disciplines may seem puzzling at first sight. Berossus and Hippocrates are both honored by the polis of Athens, and the grammarian Apollodorus (from Athens) by the Delphic Amphictyony: the three therefore belong together on account of their relationship with Athens and the classical Greek sphere. However, the insertion of grammar between astrology and medicine is strange. It is likely that astrology was placed first because astronomy was the first topic in Pliny's history. My guess is that

58 The closest sources were perhaps the Persian Zoroaster or the Hebrew Abram. Abram is cited by Valens (in II 29 and 30) and appears in Firmicus Maternus' list of ancient authorities (our F 5). Zoroaster is cited in Valens IX 4.

59 Trans. Bostock with modifications.

grammar came after astrology because Apollodorus was not only a grammarian but also a chronographer like Berossus. Apollodorus was in fact cited by Eusebius as his source for Berossus' count of the number of years of the preflood period of Babylonian history,[60] which was 432,000 years: a number which, as we will see, is related to the 490,000 years of Babylonian observations attributed to Berossos and Critodemus in F 2B.

But where did these stories about the honors for these men come from? Only the story of Hippocrates is well known from other sources, which are worth inspecting now. The oldest witness, the text called *Embassy* (late fourth century BC[61]), is included in the Hippocratic corpus together with the pseudepigraphic letters on Hippocrates' life. This is the purported speech of Hippocrates' son Thessalus to the Athenian assembly on behalf of his Coan fellow citizens, in which he mentions the honor decreed by the Athenians, allowing him and his father to be initiated into the Eleusinian mysteries as a reward for their services to the Hellenes in treating them against a pestilence.[62] Later texts from the corpus like the *Decree* (probably from the third century AD[63]) or the *Life of Hippocrates according to Soranus* repeat the same story, comparing Hippocrates to Heracles, who had received the same award for his twelve labors.[64] The *Decree* is the most specific (*Ep.* 25.18–20):

> στεφανῶσαι αὐτὸν στεφάνῳ χρυσῷ ἀπὸ χρυσῶν χιλίων· ἀναγορεῦσαί τε τὸν στέφανον Παναθηναίοις τοῖς μεγάλοις ἐν τῷ ἀγῶνι τῷ γυμνικῷ.
>
> to crown him with a gold crown worth one thousand gold pieces, and to proclaim the crown at the great Panathenaia at the athletic competition.[65]

The story of the golden crown and its proclamation in the athletic competition (ἐν τῷ ἀγῶνι τῷ γυμνικῷ) is too similar to Pliny's report about Berossus' statue with the golden tongue dedicated at the gymnasium to be a mere coincidence. The anecdote on Berossus was probably fashioned in this tradition of pseudepigraphical accounts of Hippocrates' life.

60 Cf. *FGrHist.* 680 (Berossus) F 3a (p. 377, 27–28); *FGrHist.* 680 F 3b (p. 377, 11).
61 Pinault 1992, 41–42.
62 *Ep.* 27, p. 118 line 1 Smith.
63 Pinault 1992, 43.
64 *Ep.* 25. The extant *Life* by Soranus has been argued to be an epitome of a longer life in Soranus of Ephesus' *Bioi*, and would therefore date back to the second century AD: cf. Pinault 1992, 33.
65 Trans. Smith.

M. Terentius Varro and the pseudepigraphic tradition on Hippocrates' life

M. Terentius Varro was an important source for Pliny. He wrote treatises on grammar (*De lingua latina*), on agriculture (*De re rustica*), and on the classification of the scholarly disciplines (*Disciplinarum libri IX*), which were already influential in antiquity. It is less well known that he was engaged in the study of Roman chronology, where he applied astronomical and astrological methods.[66] Varro could have seen in Berossus a precursor of the discipline.

We know that Varro was involved in the diffusion of a legend about Hippocrates. As a matter of fact, Pliny explicitly cites Varro (*NH* XXIX 2) as a source for the story that Hippocrates gathered his medical knowledge from inscriptions on votive tablets dedicated at the Coan Asclepeion by healed patients, on the basis of which he allegedly founded the new science of medicine. The story is also told by Strabo (XIV 2), but the latter detail about the foundation of the art appears only in Varro's account (in Pliny). Now, Varro is well known for having included medicine, as well as architecture, in his canon of disciplines: it would therefore not seem far-fetched to suggest that Varro could have included this account in his history of the medical discipline. Recounting this kind of anecdote about the greatest figures of each discipline, as we read them in Pliny VII 37, would be a natural fit for his project. We may, then, hypothesize that Pliny possibly took his information about the honors dedicated to the founders of the disciplines from Varro.

Varro's knowledge of the pseudohippocratic letters could also help in the case of the honors decreed for Apollodorus by the Amphictyonic Council in Pliny VII 37, since the *Embassy* contains a similar story. Thessalus reminds the Athenian people not only of the services of his father Hippocrates to the Greek nations but also of those performed by his ancestors. The first story, which occupies as much as half of the entire discourse, concerns the assistance provided by the Asclepiad Nebros to the religious organization of the Amphictyons. The Amphictyonic *poleis* had sworn an oath of mutual assistance against any invasion of their land or that of the Delphic sanctuary, and, according to Thessalus, they launched a military campaign against the Criseans, who had plundered the shrine and enslaved the neighboring people (*Ep.* 27.2 Smith). As a reward for Nebros' services, the Amphictyons decreed a series of honors for the Asclepiads of Cos (*Ep.* 27.4).

Of course, this does not constitute irrefutable proof that Varro was Pliny's source for Berossus' honors at Athens, but we have good grounds to believe that this is a likely scenario, considering the well-established fact that Varro was a frequent source in the *Natural History*. If Varro was interested in reshaping the

---

66 Cf. Grafton and Swerdlow 1985.

existing biographical accounts of Hippocrates in order to explain the foundation of Greek medicine, it does not seem implausible that he pursued a similar project with astrology and grammar, refashioning the figures of Berossus and Apollodorus to endow them with the characteristics expected of disciplinary founders. Furthermore, if the foundational myth of astrology mirrored the biographical tradition of Hippocrates, the story that Berossus settled on the island of Cos, the seat of the Hippocratic school, makes all the more sense.

Varro as a source for Vitruvius and Pliny

If we now turn to the context of Vitruvius book IX, the source for the story of Berossos' foundation of an astrological school at Cos, we detect the same framing as in Pliny VII 37. Vitruvius begins his preface by arguing that sages such as Pythagoras, Democritus, Plato, and Aristotle should not only receive similar honors to athletes, but even be turned into gods for their inestimable contributions to mankind (IX pr.2). He then presents some of these contributions (IX pr.3: *e pluribus singula paucorum uti exempla ponam*), in the hope of demonstrating the validity of his enthusiasm: Plato's procedure for doubling the square (from the *Meno*), a Pythagorean method for the construction of a right triangle, Archimedes' story about Hiero and the golden crown, and Archytas' and Eratosthenes' solution for the Delian problem of doubling the cube. Thereafter, Vitruvius remarks that in the future scholars will still engage in conversation with Lucretius' philosophy, with Cicero's rhetoric, and with Varro's presentation of the Latin language. It is tantalizing to hypothesize that this latter remark could have been a veiled acknowledgment of his source. In any case, it seems natural to assume that Vitruvius also used Varro's work on the disciplines, which he praises in the preface of book VII for dedicating one book to architecture (VII pr.14: *Terentius Varro de novem disciplinis unum de architectura*). One of these disciplines was undoubtedly astronomy/astrology, which Vitruvius could well have employed for his exposition of basic astronomical doctrines in the first half of book IX, a surprisingly long (and, for the most part, unnecessary) introduction before the discussion of sundials, the main topic of the book. Vitruvius' account is remarkably similar to that of Pliny II 8. See for example the following excerpts:

Vitruvius IX 1.16

Esse autem alias stellas temperatas, alias ferventes, etiamque frigidas haec esse causa videtur, quod omnis ignis in superiora loca habet scandentem flammam. ergo sol aethera, qui est supra se, radiis exurens efficit candentem, in quibus locis habet cursum Martis stella; itaque fervens ab ardore solis efficitur. Saturni autem, quod est proxima extremo mundo tangit<que> congelatas caeli regiones, vehementer est frigida. ex eo Iovis, cum inter utriusque circumitiones habeat *cursum, a refrigeratione caloreque earum medio convenientes temperatissimosque habere videtur effectus.*

The reason why some planets are temperate, some hot, and others cold, appears to be this; that all fire has a flame, whose tendency is upward. Hence the Sun warms, by his rays, the air above him, wherein Mars moves, and that planet is therefore heated thereby. Saturn, on the contrary, who is near the extremity of the universe, and comes in contact with the frozen regions of the heavens, is exceedingly cold. Jupiter, however, whose orbit lies between those of the two just mentioned, is tempered by the cold and heat, and has an agreeable and moderate temperature.[67]

Pliny II 8.34

Tertium Martis, quod quidam Herculis vocant, igne ardens solis vicinitate, binis fere annis converti, ideoque huius ardore nimio et rigore Saturni, interiectum ambobus, ex utroque temperari Iovem salutaremque fieri.

The next star, Mars, which some people call Hercules, is of a fiery and burning nature from its nearness to the Sun, and is carried round in little less than two years. In consequence of the excessive heat of this star and the rigidity of Saturn, Jupiter, which is interposed between the two, is tempered by both of them, and is thus rendered salutary.[68]

The parallel is inescapable, in both order and content. The heat of the Sun travels upward to the next planet, which is Mars; Saturn is cold because of its extreme position at the other end; and Jupiter, which lies in between, is tempered by the interaction of the two planets.

Varro is the obvious candidate as a common source. Apart from the geographical, chronological, and linguistic proximity, we can consider Vitruvius' mention of his scholarly merit at the end of the preface, and his status as the first Roman authority for Pliny's book II in the indexes (I 2b). An important consequence follows: if Varro was Pliny's source for the astronomical exposition in II 8, many of the Greek authorities mentioned in the chapter or in its index, such as Critodemus, could have been known to Pliny only through the mediation of Varro.

The same was probably the case with Pliny's knowledge of Berossus. An important piece of evidence implicating Varro as an intermediary source comes from the grammarian and chronographer Censorinus (third century AD), who in his work *De die natali* composed a summary of astronomical lore mostly derived from Varro.[69] Indeed, Censorinus gives exactly the same data that we find in Pliny VII 48, citing

67 Trans. Gwilt.

68 Trans. Bostock.

69 Varro is often cited in the text and is considered his main intellectual model and source: cf. Grafton and Swerdlow 1985.

Epigenes for believing in a maximum life span of 112 years, Berossus for 116, and unspecified others for 120 or even more.[70]

The "school" of Cos and Cicero's *De divinatione*

It is time to return to the question of the 490,000 years of Babylonian observations attributed in F 2B to both Berossus and Critodemus. If the information was trustworthy, we could deduce that Critodemus knew and followed Berossus, giving Varro grounds to believe that there was an astrological school originated with Berossus of which Critodemus formed part. As we have seen, Vitruvius informs us that after Berossus came Antipater, and after him Achinapolus, who, according to the text, did not use the time of birth for his horoscopes but rather the time of conception (*post ei studens Antipater iterumque Achinapolus, qui etiam non e nascentia sed ex conceptione genethlialogiae rationes explicatas reliquit*). However, we do not know any astrologers by these names, nor is there any other attestation of the second of these names for any individual, which is why the name Athenodorus has been proposed by the editor (Krohn).

An astronomer with a similar name, Anchialus, appears in *De divinatione* (II 46), where Cicero reviews Stoic opinions on the possibility and nature of divination. This is also the passage where Cicero gives a number of years (470,000) for Babylonian observations, which is very close (in relative terms!) to the number attributed to Berossus and Critodemus in Pliny. Cicero provides a little more context, informing us that such observations were not only astronomical: life outcomes were continuously compared to the corresponding birth charts to deduce meaningful patterns forming the base of astrological doctrines. Cicero responds throughout this second book to his brother Quintus, who argued in book I, in a Stoic vein, that divination is entirely possible because of the interconnection of all the parts of the world. In his examination, Cicero reviews the opinions of prominent experts. The men concerned here are Eudoxus and three other astronomers (II 42):

> *de quibus Eudoxus, Platonis auditor, in astrologia iudicio doctissimorum hominum facile princeps, sic opinatur, id quod scriptum reliquit, Chaldaeis in praedictione et in notatione cuiusque vitae ex natali die minime esse credendum. Nominat etiam Panaetius, qui unus e Stoicis astrologorum praedicta reiecit, Anchialum et Cassandrum, summos astrologos illius aetatis, qua erat ipse, cum in ceteris astrologiae partibus excellerent, hoc praedictionis genere non usos. Scylax Halicarnassius, familiaris Panaetii, excellens in astrologia idemque in regenda sua civitate princeps, totum hoc Chaldaicum praedicendi genus repudiavit.*

---

70 Censorinus 17.4: *Epigenes in centum duodecim annis longissimam vitam constituit, Berosos autem centum sedecim; alii ad centum viginti annos produci posse, quidam etiam ultra crediderunt.*

> In discussing this, Plato's pupil, Eudoxus, whom the best scholars consider easily the first in astronomy, has left the following opinion in writing: "No reliance whatever is to be placed in Chaldean astrologers when they profess to forecast a man's future from the position of the stars on the day of his birth". Panaetius, too, who was the only one of the Stoics to reject the prophecies of astrologers, mentions Anchialus and Cassander as the greatest astronomers of his day and states that they did not employ their art as a means of divining, though they were eminent in all other branches of astronomy. Scylax of Halicarnassus, an intimate friend of Panaetius, and an eminent astronomer, besides being the head of the government in his own city, utterly repudiated the Chaldean method of foretelling the future.[71]

Anchialus was a real (admittedly little-attested) name, apparently a servile *nomen parlans* (Ἀγχίαλος meaning "near the sea"), appearing mostly in Latin inscriptions.[72] The fact that it was not a common name could have led to the scribal corruption. Most likely, Cicero relied on Posidonius for these astronomers known to Panaetius, since Panaetius is said in book I not to have written anything on divination (I 3). Posidonius was a pupil of Panaetius, and the older contemporary and friend of Cicero who, as the latter recounts in the same passage, wrote as many as five books on the topic of divination. Now, it is possible that Varro also had access to Posidonius' books on this subject, from which he would have known about Anchialus and Cassander, since Cassander is otherwise only mentioned by Censorinus (18.11). Therefore, we have good grounds to deduce that the corrupt Achinapolus in Vitruvius was Cicero's Anchialus.

What is still surprising is that Panaetius, if we believe Cicero's account, said that neither Anchialus nor Cassander used astronomy for divination (*hoc praedictionis genere non usos*), contrary to Vitruvius' specific description of Achinapolus-Anchialus not just as an astrologer, but as one who employed the dubious technique of the time of conception for his horoscopes.[73] Inspection of Censorinus' passage on the Great Years will give us the clues to interpret this inconsistence.

The Great Years in Censorinus

Babylonian astronomical periods for the planets must have become well known in the Greek world by the mid-second century BC, the time of Panaetius. These

---

71 Trans. Falconer.

72 Only once in the PHI database of Greek inscriptions: *Mélanges Daux* 6.5 (a freedman). A robber in Xenophon's *Ephesiaca* (IV 5) also has this name.

73 See on conception horoscopes Frommhold 2004. In the second century AD, the practice was regarded with caution by Ptolemy (*Tetr.* III 2), and rejected outright by Antigonus of Nicaea (T3 Heilen), which may be due to the rationalizing trends of this era; cf. Heilen 2015 I, 59.

periods, used in the so-called Goal-Year Texts, are formed by an integer number of years for each planet, after which the planet roughly repeats the same pattern of phenomena in relation to the Sun and the zodiac. These Babylonian periods revived the Platonic idea of the Great Year (*Timaeus* 39d–e), since, theoretically, after the monstrously long time corresponding to the minimum common multiple of all these periods, all planets would return to the same point and move in the same way. However, all of the numbers reported, although extremely large, are still very far from such a common multiple. Censorinus provides a list of proposals (Censorinus, *De die natali* 18):

> *Hunc Aristarchus putavit esse annorum vertentium IICCCCLXXXIIII, Aretes Dyrrachinus VDLII, Heraclitus et Linus XDCCC, Dion XDCCCLXXXIIII, Orpheus CXX, Cassandrus tricies sexies centum milium.*

> Aristarch considered it [= the Great Year] to consist of 2484 yearly revolutions, Aretes of Dyrrhachium of 5552, Heraclitus and Linus of 10,800, Dion of 10,884, Orpheus of 120,000, and Cassander of 1,800,000.

That this sort of speculation incorporates both myth and phantasy can be easily gleaned from considering some of the authorities in Censorinus' list: apart from Cassander, the only other known astronomer is Aristarchus. Aretes and Dion could be astronomers unknown to us: a Dion of Neapolis appears together with an Adrastus of Cyzicus in another work of Varro quoted by Augustine, dating a strange phenomenon of the planet Venus purportedly occurred at the time of the mythical figure Ogyges.[74] Otherwise, the inclusion of Orpheus and his associated Linus reveals the kind of rationalization that we see, for example, in Pliny's presentation of Endymion as an astronomer (*NH* II 8.43).[75]

The Babylonian time unit *saros*, which equals 3600 years and is used in the Sumerian King List, probably lies behind the number attributed to Heraclitus and Linus, since 10,800 = 3 × 3600. For example, Apollodorus of Athens reported that Berossus counted the number of years of the preflood reigns as 120 *saroi*, that is, 120 × 3600 = 432,000 years.[76] So Orpheus and Linus would have established a Great Year of three *saroi*. Similarly, Cassander's period would be equivalent to 500 *saroi*. Even if Cassander was a real astronomer, his perfectly round number can hardly be the result of a serious astronomical investigation.

The other Great Years in the list seem to be derived from the *saros* as well, although in a more intricate way. A little earlier in the same paragraph, Censori-

74 Augustine *De civ. dei* XXI 8.
75 Orpheus reportedly stopped the Moon with his lyre: Molina Moreno 2009, 42.
76 *FGrHist.* 680 (Berossus) F3a (p. 377, 27–28); *FGrHist.* 680 F3b (p. 377, 11).

nus lists astronomical periods related to calendrical purposes which we need to consider here. There is the well-known Metonic cycle of 19 solar years; another of Hipparchus, of 304 years; and one attributed to Democritus, of 82 years (which should actually be 84).[77] The 5552 years of the Great Year of Aretes can be obtained by adding half the period of 304 years to one and a half *saros*:

$$3600 \times 1.5 + 304/2 = 5552.$$

Similarly, the Great Year attributed to Aristarchus is two thirds of a *saros* plus Democritus' period of 84:

$$3600 \times 2/3 + 84 = 2484$$

As for the Great Year of Dion, we have:

$$3600 \times 3 + 84 = 10{,}884.$$

Number symbolism surely played a role in these manipulations. For example, 84 is singled out by Varro in a passage from his work *Hebdomades* ("groups of seven") because at the time he wrote the work he was entering his 12th hebdomad of age (12 × 7 = 84). This and other remarks on the supposed relevance of groups of seven were considered "trifling" (*frigidiuscula, Noct. Att.* III 10.16) by our source Gellius. For example, according to Gellius, Varro said that the number of circles of the heavens perpendicular to the axis of the Earth was seven, arguing that the poles should be included as minimal circumferences which, because of their size, cannot be accommodated in the armillary sphere.[78] Obviously, there were five circles: the equator, the tropics, and the arctic circles. To include the poles was a stretch. Varro also said that from solstice to solstice, as well as from equinox to equinox, there were seven signs, if one counted inclusively. Using a similar strategy, he adduced that the cycle of the month – presumably the sidereal period of ca. 27 days – takes

77 304 solar years contain a nearly exact number of lunations: 304 × 365.2467 (tropical year according to Hipparchus) divided by 29.53 (Hipparchus' and Babylonian) = 3760.07 lunations. Similarly, 84 solar years contain 1038.97 lunations, whereas 82 solar years are very far from containing an integer number of months.

78 III 10.4: *Circulos quoque ait in caelo circum longitudinem axis septem esse; ex quis duos minimos, qui axem extimum tangunt, πόλους appellari dicit; sed eos in sphaera, quae κρικωτή uocatur, propter breuitatem non inesse.*

28 days, that is, 4 times 7. For this latter statement, he quoted Aristarchus of Samos, even if this was basic knowledge.[79]

As for Orpheus' number of 120,000 years in Censorinus' list, the factor 12 is perhaps related to what is called the "Chaldean Year" in the same chapter, a cycle of 12 years that astrologers did not relate to the Sun or the Moon but to "other observations" (18.7 *huic anno Chaldaico nomen est, quem genethliaci non ad solis lunaeque cursus, sed ad observationes alias habent adcommodatum*). Not by coincidence, this 12-year period was used in astrological texts attributed to Orpheus and known as *Dodecaeterides*, which detailed the weather and other phenomena for a recurring period of 12 consecutive years (cf. Chapter 4).

From the examination of the Censorinus passage it has become clear that Varro engaged in a kind of double play. While he was undoubtedly attracted to chronology, astronomy, and astrology, at the same time he produced pseudoscholarly claims attributing to both historical and mythical figures quite precise numerical lore obtained in a symbolic manner, or consciously reformulating the theory to introduce numerological arguments.

M. Terentius Varro and Menippean Satire

How are we to interpret such obfuscating practice? What bearing does it have on our fragment? Varro is well known for having followed the lead of the cynic philosopher Menippus (third century BC) in his satires, composing, like him, a number of writings featuring a parody of human and philosophical endeavors. Recent studies have looked at Varro's scientific and antiquarian production from this new perspective, suggesting that it bears a strong relation to the persona he assumes in his Menippean satires.[80] Specifically, Kronenberg has brought forward compelling arguments for interpreting many apparent errors and inconsistencies in Varro's

---

79 III 10.5: *in septimo signo fit solstitium a bruma, in septimo bruma a solstitio, in septimo aequinoctium ab aequinoctio* / 10.6: *Praeterea scribit lunae curriculum confici integris quater septenis diebus; "nam die <duo-> detricesimo luna", inquit "ex quo uestigio profecta est, eodem redit", auctoremque opinionis huius Aristidem esse Samium* (in the manuscripts we read the name of Aristides, which the editors are surely right to emend to "Aristarchus"). In I 20 Gellius cites Varro transmitting Pythagoras' statement that the Moon, with its cycle of 27 days, is related to the cube ($3 \times 3 \times 3 = 27$): *Huius numeri cubum Pythagoras uim habere lunaris circuli dixit, quod et luna orbem suum lustret septem et uiginti diebus et numerus ternio.*

80 See Kronenberg 2017 for a Menippean reading of the *Antiquitates rerum divinarum*. For a similar view on *De re rustica*, with arguments in favor of applying the same lens to *De philosophia* and *De lingua latina*, cf. Kronenberg 2009, 76–129. For a similarly ironical interpretation of the *Fasti*, cf. Fantham 2002.

technical writings as the result of self-ironic intellectual parody. It seems, though, that this parodic character was not perceived in the reception of Varro's work. In Kronenberg's words:

> Just as Varro's pedantic farmers lose control of their art of farming in the *De re rustica* by bungling their categories and giving impractical information (and misinformation) about farming, so Varro's pedantic persona in the *ARD* [*Antiquitates rerum divinarum*] creates divisions that make little practical sense for the teaching of Roman *religio*, have nothing to do with cult practice, and are full of blunders or inventions.[81]

This may be what is at stake in the astronomical lore derived from Varro. In the story of the founding of the astrological school at Cos, Varro could have contrived a parody of ancient and contemporary biographical attempts – especially within the field of philosophy – to classify thinkers in schools, aligning them in clusters of master–disciple relationships. But the sketch of Varro's Coan school would be more comical still, since the Antipater who is said in Vitruvius' account to have been the successor of Berossus and the predecessor of Anchialus was likely meant to be the Stoic Antipater who taught Panaetius. The line of three astrologers (Berossus → Antipater → Anchialus) would then mirror the line of the three Stoic philosophers that appears in Cicero (Antipater → Panaetius → Posidonius). Another playful ironic turn would be that, whereas Panaetius used Anchialus precisely as an example of an astronomer who did not practice astrology, according to Varro he would even cast horoscopes using the dubious theory of the time of conception.

A similarly Menippean reading could be made of the series of long periods for the Great Year. It was common for philosophers interested in astronomy like Aristotle or Posidonius to give lists of different values for astronomical periods authored by different astronomers (e.g., lunisolar cycles and length of the year). Varro could have parodied this well-established knowledge-classifying tradition by introducing invented numbers and attributing them to a mixture of well-known astronomers, little-known astronomers, and mythical figures like Linus and Orpheus.

It is true that the parodic tone is more obvious in other ancient works influenced by the Cynic tradition, such as when Lucian's Menippus (*Icar.* 6) mocks the astronomers' pretentiousness in attempting to establish astronomical distances with extreme precision:

---

81 Kronenberg 2017, 325. In support of her thesis, Kronenberg notes that Varro's friend Cicero, who was in a better position than later writers to appreciate the nature of Varro's work, praised it in general terms but never cited Varro in the specific places where he would have been expected to had he given his work greater consideration (ibid. 308–309).

(. . .) πολλάκις, εἰ τύχοι, μηδὲ ὁπόσοι στάδιοι Μεγαρόθεν Ἀθήναζέ εἰσιν ἀκριβῶς ἐπιστάμενοι τὸ μεταξὺ τῆς σελήνης καὶ τοῦ ἡλίου χωρίον ὁπόσων εἴη πηχῶν τὸ μέγεθος ἐτόλμων λέγειν, ἀέρος τε ὕψη καὶ θαλάττης βάθη καὶ γῆς περιόδους ἀναμετροῦντες

(. . .) [they] often could not tell you correctly the number of stades from Megara to Athens, but know exactly the distance in feet from the Sun to the Moon, how high the atmosphere is, how deep the sea, or the measure of the Earth's orbit.

At the opening of *Icaromenippus*, Menippus reviews the large numbers corresponding to the stages of his own astronomical travel (*Icar.* 1):

MEN. Οὐκοῦν τρισχίλιοι μὲν ἦσαν ἀπὸ γῆς στάδιοι μέχρι πρὸς τὴν σελήνην, ὁ πρῶτος ἡμῖν σταθμός· τούντεῦθεν δὲ ἐπὶ τὸν ἥλιον ἄνω παρασάγγαι που πεντακόσιοι· τὸ δὲ ἀπὸ τούτου ἐς αὐτὸν ἤδη τὸν οὐρανὸν καὶ τὴν ἀκρόπολιν τὴν τοῦ Διὸς ἄνοδος καὶ ταῦτα γένοιτ' ἂν εὐζώνῳ ἀετῷ μιᾶς ἡμέρας.

It was three thousand furlongs, then, from the Earth to the Moon, my first stage; and from there up to the Sun perhaps five hundred leagues; and from the Sun to Heaven itself and the citadel of Zeus would be also a day's ascent for an eagle travelling light.[82]

Sometimes, however, the caricature was much subtler, and this is where later tradition became confused, although it is likely that this was part of the game. Lucian's *On astrology* is an appropriate example. This text takes the form of an apparently serious essay on the history of the astrological genre, but as we read it we gradually perceive that it must be a kind of parody. For example, as is typical of historical investigations, the text attempts to trace the origins of the genre, and, plausibly enough, it mentions Orpheus, a pseudepigraphic name which, as we know, had been attached to a number of astrological texts. However, the list of mythological figures with astrological associations is extended to a parodic degree, to include Tiresias, Thyestes and Atreus, Bellerophon, Daedalus, Endymion, Phaethon, and more. Another conspicuous oddity of the text is the polemical assertion that the inventors of astrology were the Aethiopians (*De astr.* 3), no doubt a parody of the historiographical tendency to attribute humanity's most remote inventions to these distant people.[83] Varro's attribution of values for the Great Years to mythological figures probably formed part of the same satiric genre.

A similar argument could be made for another passage of Pliny's *Natural History*. In chapter VII 48 on life span, Pliny gives the maximum human life span according

82 Trans. Harmon.

83 Cf., e.g., Diodorus Siculus, who argues that Ethiopians are in fact "the first of all humans" (πρώτους ἀνθρώπων ἁπάντων, III 2.1). The story of the table of the Sun in Herodotus III 18 may also play a role here, since the text is written in the Ionic dialect and can be understood as a parody of Herodotus: cf. on the character and the authenticity of this essay Tolsa *forthcoming*.a, Berdozzo 2011, 165, McNamara 2013.

to the astrologers: "Epigenes says it cannot reach 112, Berossus that it cannot exceed 116" (*Epigenes CXII annos inpleri negavit posse, Berosus excedi CXVI*). These values correspond to the maximum life spans of individuals born in the latitudes of Alexandria and Babylon, respectively, according to an astrological method called *tetartemorion.*[84] The information may be credible for Epigenes, since he probably knew the work of Nechepsos and Petosiris.[85] However, Berossus was active around 250 BC, roughly a century before the work(s) of Nechepsos and Petosiris in which the theory of the *tetartemorion* for predicting the length of life was first established. It is very doubtful that the doctrine existed by the time of the real Berossus: much more likely is that Varro took Berossus as a Babylonian representative for astrology and used the value for the maximum life span calculated for the latitude of Babylon.

Epigenes, Berossus, and Critodemus on the span of Babylonian observations

Finally, let us tackle the content of F 2B. The same pair, Epigenes and Berossus, accompanies Critodemus here. As with the Great Year periods, Epigenes' number of 720,000 years is striking for its roundness, since it is equivalent to 200 *saroi*. The little we know about Epigenes compels us neither to trust or mistrust the claim, but we should proceed with the same caution we apply to the passage on the maximum life spans. The qualification of Epigenes as "a most serious author" (*gravis auctor in primis*), probably deriving from Varro because Pliny does not seem to have had direct access to this author, could be ironic.

This is not to say that such traditions were completely invented. For example, according to Iamblichus, Hipparchus had reported the claim that the span of Babylonian observations was of 270,000 years, that is, 75 *saroi*,[86] and this is relatively close to ca. 73 *saroi* calculated in the Sumerian king list (of which 67 are from before the flood). We have already seen how the maximum life spans were calculated with legitimate procedures, even if they were attributed, at least in part, to the wrong authors.

In the case of Berossus in F 2B, we do not have an independent witness for the span of Babylonian observations, but his chronology of Mesopotamian history

84 The method consists in counting the time degrees rising in the quarter of the zodiac that yields the greatest number (Leo–Virgo–Libra or Virgo–Libra–Scorpio). The maximum life span attached to Epigenes leads Neugebauer to assume that this was the first evidence for the rising times of the clima of Alexandria, cf. *HAMA*, 721.

85 Seneca *NQ* VII 4.1: *apud Chaldaeos studuisse se*. "Chaldean" usually meant "astrologer" for Roman intellectuals, and Nechepsos and Petosiris wrote about comets, like Epigenes.

86 The information is transmitted in Proc. *Comm. Tim.* I, p. 100 line 29 to p. 101 line 2 Diehl. Cf. *HAMA*, 608.

becomes of course relevant to this question if one equates the years of observation to the whole period of recorded history. Berossus obviously had access to Babylonian sources, which he must certainly have used for his chronology. Eusebius transmits his list of 10 kings for the preflood period, each of them reigning for an unfathomable number of years. The series is comparable to the Sumerian king list, which also separates preflood kings – with several common names – and later regents, assigning to the former similar, extremely long periods:[87]

Berossus' preflood king list		Sumerian preflood king list	
Alorus	10 *saroi*	Alulim	8 *sars*
Alaparus	3 *saroi*	Alalngar	10 *sars*
Amelon	13 *saroi*	En-men-lu-ana	12 *sars*
Ammenon	12 *saroi*	En-men-gal-ana	8 *sars*
Megalarus	18 *saroi*	Dumuzid the shepherd	10 *sars*
Daonus	10 *saroi*	En-sipad-zid-ana	8 *sars*
Euedorachus	18 *saroi*	En-men-dur-ana	5 *sars* 5 *ners*[88]
Amempsinus	10 *saroi*	Ubara-tutu	5 *sars* 1 *ner*
Otiartes	8 *saroi*		
Xisuthrus	18 *saroi*		
10 kings and 120 *saroi*		8 kings and 67 *sars*	

The roundness of Berossus' sum suggests some kind of reworking.[89] Here, it is important to keep in mind that Berossus wrote in Greek for a Greek audience at a time when Babylonian culture was still quite separate from the Greek world (cf. introduction). He could have rounded up the numbers with the purpose of "rationalizing" Babylonia history, making it more attractive to his Greek readers. There could be more examples of the same practice: Steele has pointed out that Berossus' account of the phases of the Moon (in Vitruvius IX 2 and Cleomedes IV 1), despite being utterly un-Babylonian in its description of the Moon as a ball half-painted in blue turning on its axis in the course of its period, shares with Babylonian texts

87 For Berossus' list, cf. *FGrHist.* 680 F3. Syncellus *Chronography* 39 also transmits the list, although it does not specify the regnal periods of the last kings. Eusebius' is less reliable textually, since the original Greek is lost and the text is recovered from an Armenian translation, but the names approximately coincide (and the sum of 120 *saroi*). For the Sumerian King List, see text and translation in Glassner 2004, 117–126.

88 1 *ner* = 600 years.

89 The transmitted Sumerian list ends with the dynasty of Isin, from the beginning of the second millennium BC, which it aims to legitimize. De Breucker 2013, 22, claims that the list of 10 antediluvian kings transmitted by Eusebius and Syncellus was a Hebrew or Christian forgery intended to align Babylonian history with the 10 Biblical patriarchs.

the specification of the day within the Moon cycle, which is not typical of Greek cosmological accounts. By contrast, there are also specifically Greek elements such as the geometrical model, which recalls Heraclitus' image of the phases of the Moon as a bowl containing flames turning slowly around, and the Stoic argument that the luminous side of the bowl is attracted to the Sun as it approaches it (light is attracted to light), turning the dark side toward us at new Moon.[90] In another fragment of Stoic reminiscences (Seneca *NQ* III 29), Berossus is said to have identified a period between a great deflagration and a great flood as half a Great Year in which all the planets moved from Cancer to Capricorn.

We cannot find Pliny's value of 490,000 years in other sources, but Cicero (*De divinatione* II 46) and Diodorus Siculus (II 31) attach periods of 470,000 and 473,000 years, respectively, to the Chaldeans. If both numbers derive from the same place, this is probably Posidonius, who is undoubtedly behind Cicero here.[91] An approximation of the 432,000 years of Berossus' preflood period seems to have been used, since:

$$430{,}000 + 1/10 \times 430000 = 473{,}000\ (\approx 470{,}000)$$

It is possible that Berossus approximated the post-flood time as one-tenth of the preflood time span – again not far from the Sumerian king list, where the proportion between the preflood years and the post-flood period is between 1/8 and 1/9 – but the final number probably does not come from his own pen, since the use of 432,000 would give 475,200 years. In any case, the resulting time span is still far removed from Pliny's 490,000 years.[92] Would it make sense for Varro to give a slightly different number? Possibly, since:

$$490{,}000 = 700 \times 700.$$

90 Another Babylonian detail is a reference to the opposition of the Sun and the Moon at opposite horizons at full Moon: cf. Steele 2013, 109. Lucretius (5.720–730) attributes a very similar model to the Chaldeans. For Heraclitus' view on the Moon, cf. Diogenes Laertius IX 1, 10: "the monthly phases of the Moon are due to the bowl turning around its own place little by little" (τούς τε κατὰ μῆνα τῆς σελήνης σχηματισμοὺς γίνεσθαι στρεφομένης ἐν αὐτῇ κατὰ μικρὸν τῆς σκάφης).

91 It is noteworthy that the review of Chaldean cosmology that precedes Diodorus' discussion of the time span of Babylonian observations is not compatible with Berossus' explanation of the Moon's phases, since, according to Diodorus, Babylonians would conceive the Moon as reflecting the light of the Sun, whereas Berossus said that one half is luminous per se (see above).

92 There is another possibility: a confusion between the alphanumeric character *o micron* (=70) and the *koppa* (=90). However, even if the shape *koppa* in the papyrus writing of the time was more similar to the monumental version than the later cursive version in which the circle is not closed, the opposite mistake (reading *o micron* instead of *koppa*) seems more probable.

We have seen above that Varro, half-mockingly, wrote about the importance of the number 7 and its multiples in many areas, including astronomy. Many of the Varronian values for the Great Years in Censorinus' list use the period of 84 years, which is 12 × 7. It would not be surprising, then, that he 'slightly' modified Berossus' value to make it a multiple of 700.

Critodemus and Berossos

Critodemus' work was an astrological manual, and we rarely find this kind of data in this textual genre. Among the ancient astrologers, only Manilius (*Astronomica* I 25–65) makes a vague reference to the old priestly task of observing the skies to find the patterns matching human affairs that later evolved into astrological theory, but for an astrologer to give a precise number of years for the span of Babylonian observations seems out of place. This was rather the domain of other ancient disciplines such as natural philosophy, chronography, history, or geography. Of course, caution is necessary, but the results of our examination of Varro's astronomical data transmitted by Pliny suggest that the fragment may well be spurious in the sense that Varro just dropped the name of Critodemus to his already half-spurious figure for Berossus.

There might also be a rationale in the attribution of this number to Critodemus. I will argue in Chapter 3 that Critodemus probably wrote his work in seven books, and that he specifically signaled the importance of his seventh book by using oaths of secrecy. Varro was fond of underscoring the significance of the number of books into which his works were divided,[93] so the assignation of a value related to the number 7 such as 490,000 years could have a symbolic connotation here, along the lines of Orpheus' *annus magnus* of 120,000 years (cf. the Orphic *Dodecaeterides*).[94]

93 Square numbers reigned in his *Antiquitates*: Augustine explains the structure of the work in *De civitate Dei* VI 3: the *Antiquitates rerum divinarum* took 16 books (=4 × 4), divided into 1 introductory book and 5 large sections each containing exactly 3 books; and the *Antiquitates rerum humanarum* was 25 books long (=5 × 5), comprising 1 introductory book and 4 sections each containing 6 books.

94 Another possibility which does not exclude the latter is that Varro perceived some sort of affinity between Berossus and Critodemus, and so attributed the same number to both in the same way as the same Great Year is attached to Heraclitus and to Orpheus' brother Linus. For the affinity of this Heraclitus and Orphism, cf. Derveni papyrus frs. B3 and B94, Betegh 2004. Since the narrative of Berossus' astrological school at Cos mirrors existing traditions on the life of Hippocrates, the astrological oath(s) of Critodemus (see Chapter 3) may perhaps have suggested a loose relationship with Cos and Berossus. Note that Cato in the second century BC already knew a version of the medical oath: cf. Tolsa 2019a.

Chapter 2
Critodemus' *Horasis*: the astrologers' opinions (F 3–5)

In F 3, the astrologer Vettius Valens (second century AD) quotes the initial lines of Critodemus' *Horasis*, announcing the completion of a journey. The content of the passage can be compared with the proemium of Dorotheus' astrological manual, where his travels in search of astrological knowledge are described. Critodemus' lines show significant traces of iambic trimeter, the meter of Nechepsos and Petosiris' astrological poem.

The fragment is also important for Valens' observations about the beginning and other aspects of Critodemus' treatise, such as the presence of tables and oaths. Critodemus is praised for the value of his astrological learnings, but he is also criticized on the grounds that his presentation was allegedly unclear, although on this point we can consider Valens unfair in that he appropriated most of these presentational devices for his own work.

The other fragments collected here (F 4–5) are short remarks by Valens and Firmicus Maternus that reveal more dimensions of the reception of Critodemus as well as the prominent place in the history of astrology in which these astrologers situated our author.

F 3. Beginning of the *Horasis*: Valens III 9.1–6 / Valens IX 1.5–10

Type of fragment: Two similar passages from different books of the *Anthologies* in which Valens quotes what he explicitly labels as the beginning of Critodemus' work – a poetic evocation of a voyage – providing us with its title, *Horasis* ("Vision"), and his general impressions of the work, including praise for the author's insights amid criticisms of the fantastical nature of his introduction and general style, as well as his excessive use of tables. The quoted words of Critodemus can be approximately accommodated in three iambic trimeters.

A. Valens III 9.1–6

1 Ἐπεὶ δὲ καθ' ἕκαστον κεφάλαιον τὴν ἐμὴν ἀφθονίαν ἐλέγχων ὑπομιμνήσκω, ἵνα μὴ δόξω διὰ τῦφον ταῦτα ποιεῖν, ἀναδραμὼν εἰς τὰς τῶν παλαιῶν συνταγματογράφων βίβλους κατάμαθε τὸν λόγον μὲν εἶναι κεκαλλωπισμένον καὶ κακόζηλον, ἐκπλῆξαι

https://doi.org/10.1515/9783111329147-003

δυνάμενον τὰς τῶν ἐντυγχανόντων καὶ ἀμαθῶν ψυχάς, ἀληθείας δ' ἀνέφικτον καὶ τοῖς συμφρονοῦσι πολέμιον. **2** πολλοὺς γὰρ τῶν ἀνθρώπων ἐγχρονίσας καὶ βουκολήσας οὓς μὲν τοῦ ζῆν περιέγραψε, οὓς δὲ τελέως ἐκόλασεν.
3 ἐντυχέτω δέ τις τῇ λεγομένῃ **Ὁράσει Κριτοδήμου**, πῶς μὲν τὴν ἀρχὴν εὐφαντασίωτον ἔχει καὶ τὰ λοιπὰ τετερατολογημένα πρὸς τοὺς ἀμαθεῖς· **'πέλαγος προσορμήσας**[a]', φησί, **'καὶ πολλὴν ἔρημον ὁδεύσας ἠξιώθην ὑπὸ θεῶν λιμένος ἀκινδύνου τυχεῖν καὶ μονῆς ἀσφαλεστάτης'**. **4** εἶτα αἱρέσεις ἐκτίθεται καὶ παραδόσεις διὰ φρικωδεστάτων ὅρκων καὶ <κατ'> ἄλλους τινὰς τρόπους ἀπαθανατίζων τοὺς ἐντυγχάνοντας καὶ ὄνησίν τινα ἐνδειξάμενος διὰ τῆσδε τῆς βίβλου τὰ ὅλα συνέχεσθαι δι' ἑτέρων τὰς δυνάμεις ἐσήμαινεν, περικλείσας εἰς ἄπειρον ὕλην τὴν τῶν μαθημάτων ἀλήθειαν. **5** ἄξιον μὲν οὖν ἐπαινεῖν καὶ θαυμάζειν τὸν ἄνδρα διὰ πάσης ἐμπειρίας διεληλυθότα καθηγεμόνα τε τῶν ἐπιζητουμένων γενόμενον. μυστικῶς <γὰρ> καὶ ποικίλως μερίσας αὐτὰ εἰς κανονικὰς ὀργανοθεσίας καὶ πεζικὰς λόγων συντάξεις, πολλοὺς ἐραστὰς ἐπηγάγετο, ὧν οἱ μὲν τὰς ματαιολογίας παραπεμψάμενοι καὶ διεξιχνεύσαντες τὰ δοκοῦντα κεφάλαια μετὰ παντὸς πόνου καὶ ἔπαινον καὶ ἀρετὴν κατὰ τοῦ ἀνδρὸς ἐπεισηνέγκαντο, οἱ δὲ ἀφερέπονον ἦθος ἀναλαβόντες ψόγον τῇ θεωρίᾳ περιεποιήσαντο. **6** ὁ δὲ ἐμὸς λόγος πειθήνιος μὲν καὶ διδασκαλικός, καθὼς οἶμαι, τοῖς ἐντυγχάνουσιν ἔσται καὶ ἀμετανόητος, ὡς καὶ τοὺς ἤδη ἀπεχθήραντας πάλιν ἀγκαλίσασθαι διὰ τὴν τῶν λεχθέντων καὶ ῥηθησομένων ἐνέργειαν.

a πέλαγος προσορμήσας VS, cf. πελαγοδρομήσας IX 1=F 3B

1 Since I am reminding you of my generosity in testing the methods in every chapter, proceed in the following way lest I appear to be doing so to obfuscate: go to the books of the older writers and convince yourself that their text is adorned and affected, and thereby prone to excite the minds of ignorant readers, as well as far from the truth and an enemy of the wise. **2** Delaying many men and herding[1] them, this kind of text has excluded some from life and has punished others thoroughly.
3 Let anyone read the so-called ***Vision* of Critodemus**: how it has a most fantastical beginning and how the rest is so marvelous to the ignorant. "**Having set sail for the sea**", he says, "**and crossed vast deserts, I was deemed worthy by the gods of attaining a safe harbor and a secure resting place**". **4** Then he expounds his

1 βουκολέω, "to herd", metaphorically "to misguide", is a technical term for the leader of second-century-AD Dionysiac groups in Western Asia Minor which appears in the collection of *Orphic hymns*: cf. Morand 2001, 255–275. This, along with καθηγεμών below (see note), could suggest that Valens alluded to the Orphic/Dionysiac presentation of Critodemus' work.

methods and transmissions[2] with frightful oaths and, deifying his readers in other ways and pointing out a profit throughout the book, he discloses the many powers to control everything, locking up in infinite material the truth of his teaching. **5** This man should be praised and admired, since he has gone through the whole endeavor and has become a guide[3] for seekers. Indeed, by arranging his material mystically and complexly in tabular settings and regular compilations of doctrines,[4] he has attracted many lovers, some of whom have rejected his futile material, have tracked down the relevant chapters with great toil, and have brought praise and fame to the man. **6** Others, adopting a less tenacious spirit, have forged a contempt for this art. My text will, I think, be persuasive and educational for my readers, and not to be regretted, so as to bring back those who have come to detest it, through the force of what has been said and what will be said.

B. Valens IX 1.5–10

5 Ὁ δὲ σοφώτατος **Κριτόδημος** ἐν τῇ ἐπιγραφομένῃ αὐτοῦ **Ὁράσει** συνεκτικωτάτῃ πολλῶν μυστηρίων[a] ἀρχὴν τοιαύτην ἐποιήσατο· **‘ἤδη ποτὲ πελαγοδρομήσας καὶ πολλὴν ἔρημον διοδεύσας ἠξιώθην ἀπὸ θεῶν λιμένος ἀκινδύνου τυχεῖν καὶ μονῆς ἀσφαλεστάτης’**. **6** ἄλλως τε καὶ ὁ Τίμαιος καὶ Ἀσκλατίων καὶ ἕτεροι πλεῖστοι· οὗτοι μὲν οὖν καλλονῇ λόγων ἐνεχθέντες καὶ τερατολογίᾳ οὐ κατὰ τὰ ἐπαγγέλματα τὰ ἔργα ἐπέδειξαν οὐδὲ τὰς συντάξεις πλήρεις καὶ ἐπιλελυμένας, ἀλλὰ λειπομένας ἐν πολλοῖς τισιν ἀεὶ τῶν ἐντυγχανόντων, ἐν πᾶσι δὲ τὸ σκολιὸν καὶ ἐφθονημένον καὶ ἀναδυόμενον[b] καὶ ἐπιπλεκόμενον· καὶ μηδεμιᾷ ὁδῷ διευθύνοντες ἀλλὰ προσεισφέροντες αἵρεσιν αἱρέσει καὶ ἀναπομπίμους βίβλους, πλάνης μᾶλλον ἢ ἀληθείας τεκμήρια.

7 ὁ μὲν οὖν Κριτόδημος, πλῆθος θεωρημάτων κεκτημένος καὶ τοῖς ἄλλοις συγκεκοσμημένος καὶ δυνάμενος σαφῶς διερμηνεῦσαι, διὰ τῆς τούτων τῶν ὀργάνων φαντασίας ἠμαύρωσε τὴν ἐπιστήμην. **8** ἐγὼ δὲ ἐν ταῖς προσυντεταγμέναις μοι βίβλοις οὔτε κενῶν οὔτε ματαίων ὕθλων πόνους διήνυσα οὔ<τε> μὴν κατὰ τὴν δόκησιν τινῶν ἀμφιβόλους ἐπιλύσεις καὶ γραφὰς ἀναρίθμους περιττῶν[c] συντάξεων,

2 παραδόσεις and ἐπιμερισμοί (“distributions”) are technical terms for the partition of human lifespan set up in the fragments of Chapter 4.

3 The term could allude to Dionysus Kathegemon, the main divine figure worshipped in the *boukoloi* groups of Asia Minor in the time of Vettius Valens. Cf. the collection of inscriptions related to the Dionysiac associations in Jaccottet 2003.

4 In the original, πεζικὰς λόγων συντάξεις. The adjective πεζός (“on foot”), when applied to text means prosaic, often in contrast to poetic: cf. DL III 37: φησὶ δ’ Ἀριστοτέλης τὴν τῶν λόγων ἰδέαν αὐτοῦ (Πλάτωνος) μεταξὺ ποιήματος εἶναι καὶ πεζοῦ λόγου (see discussion below).

διὸ καὶ τὰ δοκοῦντα ἀληθείας ἐφικνεῖσθαι εἰς ἄπειρον ζήτησιν καὶ ψόγον ἐχώρησεν. **9** χρὴ δέ γε τὸν βουλόμενον συγγράφειν ὡς μὴ χρήζειν ἑτέρου· εἰ δ' οὖν, διὰ τὸ ἀγνοεῖν καὶ φθονεῖν τὴν πλοκὴν ἐπεισφέρει. **10** πελαγοδρομήσας οὖν καὶ πολλὴν γῆν διοδεύσας, κλιμάτων τε καὶ ἐθνῶν κατόπτης γενόμενος, πολυχρονίᾳ πείρᾳ καὶ πόνοις συνεμφυρείς, ἠξιώθην ὑπὸ θεοῦ καὶ τῆς προνοίας βεβαίου καὶ ἀσφαλοῦς λιμένος τυχεῖν.

a μυστηρίων: μαρτυριῶν **S** corr. Kroll
b ἀναδυόμενον: ἀναλυόμενον **S** corr. Kroll
c περιττῶν: ὑπὲρ τῶν **S** sugg. Wendland

5 The great Critodemus, in the most important[5] *Vision*, attributed to him, made such a beginning of many mysteries: "**Already having traversed the seas and crossed vast deserts, I was deemed worthy by the gods of attaining a safe harbor and a secure resting place**". **6** Timaeus, Asclation, and many others have said other things.[6] These men were carried away by the beauty of words and by marvel-telling, and they did not present works which accorded with their promises, nor were these compilations complete and self-explanatory,[7] but they left their readers abandoned in many places, and were always warped, envious, withdrawn, and deceptive. They never travelled a straight road, but piled method over method, and books which could be prosecuted as tokens of deceit rather than truth.
7 Critodemus then, although he had inherited a mass of theorems, had developed others himself, and was able to interpret clearly, still obscured the science because of the fantasy[8] of these tables. **8** I, on the other hand, in my above-compiled books, have accomplished labors not of vain and empty babble; nor dubious, questionable solutions following someone's opinion and infinite writings of useless compilations;

5 In the original συνεκτικός. It has the sense of "fit for holding together". Valens only uses the term in the sense of "important" and "indispensable", e.g., in IV 11.9: ἐπειδὴ δοκεῖ συνεκτικώτατον κεφάλαιον παρὰ τὰ λοιπὰ ὑπάρχειν. τούτου γὰρ ἄνευθεν οὐδὲν οὔτε ἔστιν οὔτε ἔσται ("this is a most essential topic well above the others; without this there is nothing nor will be anything").
6 The almost incomplete appearance of this sentence is typical of astrological manuals, where ἄλλως is typically used to introduce a different doctrine. Here the text probably means something like "different things in the same vein". Cf. Dorotheus' preface to book I, where he claims that he has traveled widely in order to learn from the sages in these regions; Manilius I 13–15 claims that he is traversing the air and touring the sky. We know nothing about Timaeus and Asclation, except for a doctrine about the parents in Val. II 32 attributed to the former.
7 Rarely used to describe writings, ἐπιλύω means "untie", and, in the closer context of teachings, "solve" or "explain".
8 The term is also used metaphorically by Valens elsewhere, to convey the idea of ostentation, as in I 19.10: "rich and living with much spectacle" (πλουσίους καὶ μετὰ πλείστης φαντασίας διεξάγοντας).

this is why that which seemed to arrive at the truth passed into an endless search and argument. **9** One who wishes to write must proceed as if he needed no one else;[9] for if he does, he will bring in deceit because of his ignorance and spite. **10** Therefore I, having traversed the sea and crossed many lands, surveyed many climes and nations, and combined long practice and labors, have finally been deemed worthy by God and Providence of attaining a secure and safe harbor.[10]

Context in Valens

F 3A is found toward the middle of book III of the *Anthologies*, where Valens tackles for the first time the most frequently discussed topic in his work, the doctrines on determination of the length of life. A little earlier in the book is Critodemus' F 20, containing an example of one such doctrine. However, Valens' discussion of Critodemus' work in F 3A is only tangentially related to these topics, since Critodemus is cited as a counterexample to Valens' self-proclaimed generosity in examining in detail the doctrines he presents "in every chapter" (καθ' ἕκαστον κεφάλαιον τὴν ἐμὴν ἀφθονίαν ἐλέγχων ὑπομιμνήσκω), in contrast to what he finds in the old compilations (τὰς τῶν παλαιῶν συνταγματογράφων βίβλους). Indeed, in both the previous chapters (5–8) and the following chapters (10 and 11), Valens makes frequent use of concrete examples that are useful to understanding the procedures. Valens certainly stands out as the astrological writer who provides by far the most examples, including a vast number of real horoscopes. This is not to say that the older authors did not provide any,[11] but many doctrines were merely outlined in their work. Nechepsos and Petosiris, for example, despite their consideration as *the* undisputed authorities of Greek astrology, also received Valens' criticism on this account (see F 4).

F 3B forms part of Valens' preface to the last book of his compilation, which begins with a salutation to his dedicatee Marcus and with a general reflection on his own work.[12] This is the expected place for these kinds of positive observations on the achievements and final form of the work: see, for example, the end of the

9 That is, having mastered astrology to the point of being autonomous. Valens is apparently criticizing Critodemus for relying too much on other astrologers' opinions. For his part, Valens claims to have perfected his art so as to be able to use only his own, tested methods.

10 A clear imitation of Critodemus' beginning. Providence (πρόνοια) was an important entity for Valens; it is one of the witnesses of his oaths (Chapter 3), e.g., F 6A (Val. IV 11.11). Cf. Plut. *Pyth. Or.* 406B οὐ μὴν ἀλλὰ καὶ τὸ τοῦ θεοῦ καὶ τῆς προνοίας σκοποῦντες.

11 A group of more than 100 idealized (nonreal) horoscopes in Firmicus VI 29–31 could go back to the early Egyptian pseudepigraphs: cf. Heilen 2010, 137.

12 See Appendix III for a characterization of book IX as a relatively independent closure of the work.

Almagest (XIII 13), where Ptolemy addresses Syrus, commenting on the successful completion of the topics and on the general character of the work. Valens cites Nechepsos as the authoritative text that has formed the basis of his discussion in the previous books:[13] the king must be taken as a model for confessing his previous errors and for abandoning his royal duties in favor of astrology.[14] This is followed by the example of Critodemus, who is a model only in part. Alongside similar criticisms to the ones expressed in F 3A, which are again used to present his own astrology as a significant improvement, Valens acknowledges many positive qualities of Critodemus, calling him "most wise" (σοφότατος), and crediting with collecting a mass of doctrines, with producing other theories, and with being able to interpret clearly (πλῆθος θεωρημάτων κεκτημένος καὶ τοῖς ἄλλοις συγκεκοσμημένος καὶ δυνάμενος σαφῶς διερμηνεῦσαι). Valens then imitates Critodemus' words on arduous travel and safe arrival to make a similar claim for himself (ἠξιώθην ὑπὸ θεοῦ καὶ τῆς προνοίας βεβαίου καὶ ἀσφαλοῦς λιμένος τυχεῖν). Developing further this narrative on divine inspiration, he closes the preface with reflections on the divine nature of the soul, which he illustrates by quoting four Orphic fragments (*OF* 436, 422, 426, 425).

Valens' criticisms and his own contradictions

F 3A and B offer, in only slightly different ways, similar views on Critodemus' work. What is immediately apparent is that Valens criticizes Critodemus on various accounts, contrasting what he claims are the features of his own project with what he says are those of Critodemus and others. Let us summarize Valens' criticisms:

(a) Critodemus and others, such as Timaeus and Asclation, used an affected and bombastic style to attract the attention of the ignorant, and recounted fantastic tales supposedly of the discovery of astrological tenets. These treatises therefore lacked a straightforward educational progression through the astrological

13 Valens seems to indicate that Nechepsos' work begins with book XIII (Ὅσα μὲν ὁ θειότατος βασιλεὺς εἴρηκε Νεχεψὼ ὁ τὴν ἀρχὴν ποιησάμενος τῆς ιγ' βίβλου. . .). Indeed, it would be odd if Valens implied that it was just the beginning of Nechepsos' book XIII which he had covered with his treatise; it is more reasonable to assume that he is noting the curious beginning of Nechepsos' work at the 13th book, and that he has followed the work as a whole. This would explain why only books XIII, XIV, and XV of this ancient Graeco-Egyptian manual are cited. Perhaps, as Heilen 2011, 24, suggests, a previous writing by Petosiris addressed to king Nechepsos is assumed.

14 Val. IX 1.3. Valens qualifies the King's reign as a tyranny (τυραννίς), so his errors should perhaps be considered ethical and amended for through his abandonment of royal duties.

doctrines, which made it very difficult for their followers to track down the relevant information.
(b) Critodemus' work was incomplete and partly unsound, including dubious techniques from other writers (only in F 3B).
(c) Two devices in Critodemus' work contributed to its obscurity: frightful oaths and ostentatious tables (see Chapters 3, and 7 and 8, respectively).

It is puzzling that, despite his harsh criticisms, the very features of Critodemus' treatise with which he finds fault can also be found in Valens' own manual. Most clearly of all, at several points, Valens exacts an oath from the readers, asking them to preserve the astrological learnings that he is imparting to them (F 6A–C). We will leave discussion of such oaths until Chapter 3, but it is worth noting that they were a rarity in astrological writing, used elsewhere only in the work of Firmicus Maternus, who was also influenced by Critodemus.

Valens refers in F 3B to the "fantasy of these tables" (τῆς τούτων τῶν ὀργάνων φαντασίας), and in F 3A to the varied mixture (ποικίλως μερίσας) of tables and text. Despite modern assumptions and expectations, tables were not common in Greek astrological writing. Ancient bookrolls showed a striking uniformity in and between themselves, and tables and diagrams appeared only in textual genres in which they were felt to be indispensable.[15] Lettered diagrams formed an essential part of the text of mathematical demonstrations,[16] and astronomy was the obvious domain of tables. Otherwise, we only find tables in the arithmetical treatise of Nicomachus of Gerasa: it therefore seems that, outside of mathematical disciplines, such departures from monotonous columns full of text were rare.[17] As for astrology, even if Graeco-Roman astrologers were frequently known as μαθηματικοί/mathematici, as a general rule astrological treatises and compilations contain no diagrams or tables.[18]

This is a genre-related question. Bookrolls containing oratory are frequently arranged in narrow columns, literary commentaries in wide columns, and history and philosophy somewhere between the two.[19] Like mathematical treatises, astro-

15 See Johnson 2004 for an analysis of the formatting of literary bookrolls.
16 For a detailed analysis of the conventions of mathematical writing, cf. Netz 1999.
17 An interesting exception is the case of illustrated herbals, but the practice was not without its detractors: cf. the arguments of Pliny XXIV 91.142 adducing the changes in plants throughout the seasons (what time of year do we choose for the illustration?) and the difficulty of copying the illustrations. Cf. Harvin and Totelin 2016, 168. From the number of surviving papyri, we can deduce that illustrated herbals were much more common than the illustrated medical books.
18 See Tolsa 2018 for the tables of terms in Ptolemy *Tetrabiblos* I 21, which were probably not original.
19 Johnson 2004, 34.

logical texts frequently required complex visualizations to illustrate arguments about the relationships between points in the chart: it is possible that such astrological operations were often performed outside the text, on an astrological board.[20]

This is why Valens finds the presence of tables in Critodemus' work striking and obfuscating: not because they made no sense, but because they were unsuited to the genre. Nevertheless, Valens reproduces the tables for the length of life which he found in Critodemus' work (F 16) and dedicates half a book (VIII) to their discussion. At the end of book IX, Valens similarly explains the construction and use of two other large tables for the correction of the hourly position of the Moon, but the tables themselves did not survive (see Appendix III).

Valens also frequently displays what can be qualified as a bombastic and fantastic expositional style, for example, when he describes his own learning as divinely inspired, following the example of king Nechepsos.[21] But we need not even look that far: his emulation of Critodemus' words at the end of F 3B is also an appropriate example.

Finally, if one wished to describe Valens' astrological work from a general viewpoint, the keywords would hardly be "clear", "complete", or "educational". Aside from the frequent self-aggrandizing passages in which he comments on his own astrological project, breaking the continuity of the explanation, his work is evidently not structured as a whole according to a predefined plan. Rather, we have the impression that the author worked on it at very separate times, which is confirmed by the chronologically distinct sets of horoscopes that he inserts as examples.[22] The title *Anthologies*, whether original or not, is a very apt description of the treatise.

It is impossible that Valens was unaware of his own ambivalence. In fact, at one point he acknowledges that the work is intentionally obscure in places, and that this makes it more interesting for discussion in learning contexts.[23] Interestingly, numerous contradictions have also been identified in the astrological poem of Manilius (first century AD): the main type is formed by contradictory ideas stemming from different approaches: for example, Manilius states variously that the study of the heavens is open to everyone (e.g., II 105–125) and that it is only for a few selected initiates (II 136–149).[24] These two levels of readership seem to be present in Valens' mind, too, when he claims that Critodemus' bombastic passages are designed to attract new adepts, whereas those already initiated have to skip

---

20 See the discussion of this topic in relation to Egypt in the Introduction.

21 Val. VI 1.9 = Nech. et Pet. fr. 1 Riess. For a discussion of this fragment, cf. Heilen 2011, 37–56.

22 Neugebauer 1954. Cf. Pingree 1986, xviii–xx.

23 Val. III 10.19.

24 Volk 2011.

these chapters in order to find the specific theories (F 3A). We can establish the following chart to differentiate the characteristics appropriate for these two kinds of readers, roughly according to Valens' interpretation:

	Uninitiated readers	**Initiated readers**
Style	Bombastic, marvel-telling, oaths	Matter of fact
Astrological doctrines	Obscurity in the details of procedures	Clarity in all the steps of the method
General structure	Mysterious (=chaotic)	Clear
Tables	Inclusion	Exclusion

Valens interprets that new learners of astrology were frequently lured not by well-structured, dry, and monotonous manuals, but by writings that attempted to be original in presenting their doctrines in a convoluted, mysterious, and unexpected fashion. This can be attributed in part to the pseudepigraphic nature of the first Hellenistic Egyptian manuals like the poem of Nechepsos and Petosiris, in which the king told his own story of discovery and enlightenment; the *topos* was later inherited and somewhat rationalized in later treatises like those of Critodemus and Dorotheus, which opened with the account of the astrologer's journey in search of knowledge. Another factor is the very nature of astrological investigation: unless one is very cautious when assessing the methods, compilations often tend to present allegedly infallible doctrines in an accumulative fashion that inevitably leads to contradictory results.[25] In any case, both approaches (obscure and contradictory vs. rational or rationalizing) were probably integral to Graeco-Roman astrological writing. With time, writers gradually prioritized the rationalizing approach, one aspect of which is the use of real names for authors. For example, Ptolemy's *Tetrabiblos* falls almost completely within this second approach.[26] Vettius Valens, on the other hand, while criticizing the obscurity of his predecessors' work, cannot resist the temptation to consciously use such devices in his own manual. Even if some of the inconsistencies can be ascribed to carelessness, the possibility of allowing oneself such carelessness may have formed part of the astrological genre.

25 See, for example, the experiment of Barton 2004, chapter 5, applying the methods of Firmicus Maternus and Dorotheus to the chart of Prince Charles.

26 Cf. Tolsa 2018 for one exception: the obfuscating explanation of Ptolemy's own terms and the strange tale of their discovery. Firmicus Maternus in Critodemus F 5 seems to allude to this tendency in the history of Graeco-Roman astrology (see discussion below).

Two slightly different witnesses

As we have seen, the two attestations of Critodemus' text differ slightly in the first part:

F 3A

> πέλαγος προσορμήσας, φησί, καὶ πολλὴν ἔρημον ὁδεύσας ἠξιώθην ὑπὸ θεῶν . . .
>
> Having set sail for the sea – he says – and crossed vast deserts, I was deemed worthy by the gods . . .

F 3B

> ἤδη ποτὲ πελαγοδρομήσας καὶ πολλὴν ἔρημον διοδεύσας ἠξιώθην ἀπὸ θεῶν . . .
>
> Already having traversed the seas and crossed vast deserts, I was deemed worthy from the gods . . .

Similarly, small differences are found in the two quotations of the same poem by Cleanthes in Valens' books VI and VII.[27] Valens may have quoted from memory in at least one of the passages,[28] and we are probably seeing the same phenomenon here, with the version in F 3B being somewhat less reliable, for the following reasons:

(1) Beginning of a work with ἤδη ποτὲ (Already then. . .) is odd, since this connector generally presupposes a previous narrative.[29]

(2) The term πελαγοδρομήσας and its cognates are only attested in late sources: the *Orphic hymn* to Leucothea (74.5) from the collection dated between the second and the fourth centuries AD;[30] a magical hymn to the Moon (11a line

27 VI 9.13: ἄγου δέ μ', ὦ Ζεῦ, καὶ σύ γ' ἡ πεπρωμένη, / ὅποι ποθ' ὑμῖν εἰμὶ διατεταγμένος, / ὡς ἕψομαί **γ' ἄοκνος· ἢν δέ γε** μὴ θέλω / κακὸς γενόμενος, **οὐδὲν ἧττον ἕψομαι**. VII 3.53: ἄγου δέ μ' ὦ Ζεῦ καὶ σύ γ' ἡ πεπρωμένη, / ὅποι ποθ' ὑμῖν εἰμι διατεταγμένος, / ὡς ἕψομαί **γε κἂν ὀκνῶ· κἂν** μὴ θέλω, / κακὸς γενόμενος **αὐτὸ τοῦτο πείσομαι**.

28 Both versions of Cleanthes' poem might in fact be wrong, since they differ slightly from the same quotation in Epictetus, *Ench.* 53.1.

29 In the *TLG* it is not even found at the beginning of a chapter, except for Valens IX 16, which is in fact an imitation of Critodemus' own beginning as quoted in the same book (F 3B).

30 For the date of the *Orphic hymns*, see West 1983, 63, who argues that they represent a mature Stoicism, comparable to Mesomedes in the age of Hadrian. See also Morand 2001, 255–275, who links the hymns to the preserved inscriptions of the *boukoloi* in Asia Minor, most of them also from the age of Hadrian. The *boukolos*, a leader of Dionysiac cultic groups in Asia Minor (cf. Jaccottet 2003), is alluded to in two of the hymns (1.10 and 31.7). It might seem a noteworthy coincidence

12 Heitsch) of uncertain but roughly similar age;[31] late ancient lexicographers explaining the characteristics of vessels;[32] late ancient paroemiographers;[33] and two Christian sermons.[34]

(3) ἠξιώθην more naturally calls for the preposition ὑπό, as in F 3A.

In conclusion, the first version is probably preferable to the second one. Interestingly, it also contains a rare word, προσορμήσας, which was nevertheless already used by Polybius (X 42.1) in the second century BC. An imitation of Critodemus' fragment echoing this verb is found at the beginning of the Byzantine astronomical poem of Kamateros (*In astr.* 1–2):

> **Πρὸς** ὕψος τέχνης **ὥρμησα** μέγιστόν τε καὶ βάθος
> καὶ **πρὸς πέλαγος ἄπειρον**, μικρὸν ἔχων τὸ πλοῖον (. . .)
>
> Having **set sail to** the height and depth of this art
> and **to the infinite sea**, with a small vessel (. . .)

Kamateros followed Critodemus very closely in the description of the astrological doctrine of the terms (see Chapter 5), but this does not imply that Kamateros had

that among these few attestations we find one from the *Orphic hymns*, given that Critodemus was influenced by Orphism in many ways. However, the poems in this collection are basically built from long lists of epithets, so the appearance of this term may not in fact be of particular relevance. Alternatively it is possible that Valens knows these hymns and inadvertently changes Critodemus' expression for the epithet in the hymn to Leucothea, which also depicts a situation of salvation.

31 The two collections of hymns are probably directly related. Selene receives τόδ' ἄρωμα in the first line of the next magical hymn, similarly to what we find in the *Orphic hymns* (cf. *H. O.* 9, Εἰς Σελήνην θυμίαμα ἀρώματα). But there are other quite specific common features: if the *Orphic hymn* to Selene describes her as "female and male" (9.4, θῆλύς τε καὶ ἄρσην), the same magical hymn calls her "male-female" (11a.26 ἀρσενόθηλυν). Another hymn calls her "Physis mother of all" (10.33, Φύσι παμμήτωρ), and the next poem in the *Orphic hymns* after Selene is dedicated to Physis, who is invoked with the same epithet (*H.O.* 10.1, Ὦ Φύσι, παμμήτειρα θεά). However, the direction of the influence is difficult to establish (should we instead consider it to be bidirectional?) because the preferred dating of both types of compositions is the same (second/third century AD): cf. Blanco Cesteros 2020.

32 As a synonym of ποντόπορος (Hesychius) and to remark that a certain small vessel (πορθμίς) does not traverse the sea (οὐ πελαγόδρομος) but is only used as a ferry (ps-Zonaras and others).

33 Diogenianus, Zenobius, and Apostolius, interpreting an expression in Chrysippus (οὐ νυκτιπλοεῖς) as meaning that someone is not precise. They say that those who traverse the sea (πελαγοδρομοῦντες) use the stars and are therefore very precise, but the word only appears in their explanation.

34 Hesychius and Ephraem Syrus, qualifying sailors.

direct access to any version of Critodemus. Indeed, from the analysis of Kamateros' sources it will become apparent that he read Critodemus through the work of Valens. However, the fact that he chose to follow Valens' first version is perhaps indicative of the less credible appearance of the second quotation.

Rhythmic traces

Roughly one third of F 3A can be accommodated in iambic trimeter with the more flexible structure typical of comedy. It seems plausible that Critodemus' astrological manual was originally written in this metrical form. This kind of verse was used in Critodemus' age for the poem of Nechepsos and Petosiris, but also by the geography writers Apollodorus of Athens and his follower, conventionally known as pseudo-Scymnus. Later, in the first of second century AD, the author Damocrates wrote a treatise on antidotes using this type of verse.[35] Judging from the scientific texts written in verse form, one can deduce that an important factor in the choice of meter was its function in facilitating memorization, which was especially challenging in genres containing long, unstructured lists, such as astrology, geography, and antidotes. With particular regard to the iambic form, Aristotle remarks on the similarity between this verse form and spoken language (*Poet.* 1449a: μάλιστα γὰρ λεκτικὸν τῶν μέτρων τὸ ἰαμβεῖόν ἐστιν), and pseudo-Scymnus shows himself to be proud of the accuracy in the content that the flexibility of the comic trimeter allowed (33–35):

> μέτρῳ δὲ ταύτην ἐκτιθέναι προείλετο,
> τῷ κωμικῷ δέ, τῆς σαφηνίας χάριν,
> εὐμνημόνευτον ἐσομένην οὕτως ὁρῶν.
>
> this [abridgment] has been chosen to be set up in verse,
> namely in the comic one, for the sake of clarity,
> seeing that it will be memorable in this form.

Other astrological writers chose the hexametric form (Manilius, Dorotheus, and Anubio wrote in elegiac couplets). This represented a tour de force due to the nature of the rhythm and was at the same time perhaps considered fit for a doctrine like astrology that involved the heavens: after all, the hexameter was the rhythm of Hesiod, Orpheus, and the oracles. Galen's comparison of Damocrates' iambic trimeters with Andromachos' elegiac couplets (containing an hexameter and a hemie-

---

35 He is quoted at length by Galen: cf. Vogt 2005. See West 1982, 160 for this kind of verse used in didactic works.

pes) is illustrative of the challenges and possibilities observed in iambs and dactylic verse, respectively, since he praises the technical accuracy of Damocrates while criticizing the obscurity of Andromachos' lines.[36]

The clearest iambic trimeter in F 3A begins with ἔρημον and ends in θεῶν, after which a line end feels most appropriate. Of course, in order to calibrate the plausibility of this analysis, one should also estimate the probability of finding a sequence corresponding to an iambic trimeter in a prose text. This may initially seem likely because of the flexible structure of the meter, but in fact the probability is rather small. For example, looking at the first 24 prose lines of Aelius Aristides' *Hymn to Zeus*, there is only one cluster that conforms to the required sequence of long and short syllables of the comic iambic trimeter, even without taking into account caesurae. Finding sequences that correspond to two iambic meters is much easier: in these 24 lines there are 6 examples, but the probability decreases greatly if we looking for trimeters. It is possible that iambic trimeter was consciously avoided in prose texts. Therefore, it seems enormously coincidental to find one such cluster in the short passage from Critodemus quoted by Valens, unless it was, at least originally, written in this meter.

Furthermore, only minimal changes in the rest of the fragment give us another two and a half iambic trimeters. It is admittedly true that the verse lines are odd in comparison with the iambs of Greek canonical literature, especially because of the enjambment between the first two lines, but the trimeters of pseudo-Scymnus and Damocrates also present special characteristics which would be difficult to accept were it not for the evidence of the text.[37]

πέλαγος προσορμήσας <τε> καὶ πολλὴν <χθόνα>
ἔρημον ὁδεύσας ἠξιώθην ὑπὸ θεῶν
λιμένος ἀκινδύνου <τε> καὶ μονῆς τυχεῖν
<τῆσδ'> ἀσφαλεστάτης < . . . >

1 τε: an emphatic coordination between the objects here is plausible, since earth and sea are frequently treated as a strongly marked pair: cf. e.g. *Hymn. Orph.* 75.5 κατά τε χθόνα καὶ κατὰ πόντον, Herodotus V 17.5 γῆν τε καὶ ὕδωρ. In the third verse, the emphasis is again welcome for the double parallel sea-desert / harbor-resting place. | χθόνα: hiatus at the end of line seems permissible in the case of apposition (see the

36 Gal. *De comp. med. per gen.* V 10 (XIII 820.15–17 Kühn), cf. Vogt 2005, 71. Indeed, the numerous quotations of the latter show a very matter-of-fact, prosaic style, while Andromachos' text (*De ant.* I 6 [XIV 32–42 Kühn]) has a predominantly literary character.

37 For example, consider the placing of articles at the end of line in ps.-Scymn. *Perieg.* τόδε / σύνταγμα (8–9), εἶτ' Ἴβηρες οἱ / προσεχεῖς (199–200).

counts in the classical tragedians in Stinton 1977). A good parallel for the wording at the beginning of a canonical literary work could be Soph. *Phil.* 1–2: Ἀκτὴ μὲν ἥδε τῆς περιρρύτου χθονὸς / Λήμνου, βροτοῖς ἄστιπτος οὐδ' οἰκουμένη (another description of a deserted land). Another similar beginning is that of Eur. *Suppl.* 1–2: Δήμητερ ἑστιοῦχ' Ἐλευσῖνος χθονὸς / τῆσδ'. This offers a parallel for the conjectural addition in the fourth line too (see below), and it could have been significant to Critodemus for its mysteric context. On the other hand, even if it can be argued that χθόνα ἔρημον is less economic than just ἔρημον, one must consider that the latter does not appear as a substantive in the classical dramas. | **3** τυχεῖν: I have moved τυχεῖν to the end of the line to create a hepthemimeres caesura, considering that in a prose version it would plausibly have been moved back after the first object (λιμένος ἀκινδύνου), whereas in iambic trimeter the infinitive quite frequently occurs at the end of the line (e.g. Eur. *Heracl.* 351, 525; *Hipp.* 1326, 1455). | **4** τῆσδ': I propose this conjectural restoration to present Critodemus' statement as a lively reference to his own final destination, in agreement with Valens' imitation (F 3B). This kind of demonstrative frequently accompanies similar terms indicative of a welcoming place in the tragic authors (e.g. χθών in Eur. *Cyc.* 468, *Med.* 1237, *Heracl.* 397, 591; στέγη in *Med.* 1293; ἕδρα in *Heracl.* 631).

Having set sail for the sea and traveled across a vast desert, I was deemed worthy by the gods of attaining a safe harbor and a secure resting place < . . . >

Astrological prosifications and other textual transformations

Valens seems to be unaware of the metrical traces outlined above. When he remarks on the mixture of tables and text in Critodemus' *Horasis* (F 3A), he uses the expression πεζικὰς λόγων συντάξεις, "regular compilations of doctrines", for the non-tabular sections. This is in striking contrast with the double quotation of Cleanthes' iambs. Even if those lines were in part wrongly recalled from memory (see above), in both cases, the quotations maintain the exact structure of the iambic trimeters, which demonstrates that he was entirely conscious of their use and attempted to adhere to it. This is not the case with Critodemus' words. The situation is similar to Valens' own quotation of the beginning of the poem of Nechepsos and Petosiris:

ἔδοξε δή μοι πάννυχον πρὸς ἀέρα
<. . . . .>
καὶ μοί τις ἐξήχησεν οὐρανοῦ βοή,
τῇ σάρκας [μὲν] ἀμφέκειτο πέπλος κυάνεος
κνέφας προτείνων

1 δή: δέ MSS, coni. Usener | **3** μέν del. Riess | κυάνεος: κυανόχρα MSS

I decided, then, (to gaze in prayer) all night long (up) to the sky, (. . .) and a shout sounded forth from heaven. Around its flesh a mantle of dark blue color was wrapped, stretching out darkness before itself.[38]

In this fragment, the metrical structure is more apparent than in Critodemus' fragment, but we can nevertheless see that some particles were changed (δή) or added (μέν), and that the resulting text did not exactly conform to iambic trimeter either. It is therefore reasonable to assume that Valens did not recognize the metrical nature of the text here either. Stephan Heilen poses the question as to why Nechepsos and Petosiris are only once acknowledged as poets in the ancient sources.[39] This must be regarded in light of the fact that none of their surviving metrical fragments accurately respects a metrical sequence.[40] The more natural conclusion is that both Critodemus and Nechepsos/Petosiris were originally in verse form, but were turned into prose texts at an early stage of the textual transmission. Indeed, our longer fragments of Critodemus point in this direction, since they appear to be standardized abridgments of a lost original work (cf. Chapters 4–6).[41]

This kind of transformation of ancient metrical compositions with technical content is not infrequent. For example, in the case of the medical work of Andromachos in elegiac couplets, it seems that his own son – also named Andromachos – wrote a prose version which, according to Galen, was still, unfortunately, inaccurate.[42] Transformation of a previous authoritative work by paraphrase, epitome, or expansion (or any combination thereof), with or without attribution, was a common practice in ancient astrology, judging from the surviving material. To list just a few instances, we have the so-called Proclus paraphrase of Ptolemy's *Tetrabiblos*, four epitomes of Hephaestio's *Apotelesmatica*, and several paraphrases of astrological texts attributed to Orpheus.[43] There are also partial paraphrases of

38 Trans. Heilen. Cf. reconstruction of the fragment with app. crit. and extensive discussion in Heilen 2011, 37–52.

39 Heilen 2011, 82.

40 They are edited and commented in Heilen 2011.

41 On the other hand, the simplified form of the text beyond the first lines makes it impossible to find metrical structures.

42 Gal. *De ant.* I 7 (XIV 44.10–17 Kühn). Cf. Vogt 2005, 70–71.

43 *OF* 779 is about ingresses; *CCAG* IV, 43 (attributed by Tzetzes to Amon, *OF* 781a–b) on catarchic astrology. Cf. Martín Hernández 2006, 115–116.

Dorotheus.[44] Sometimes authors incorporated practically unaltered sections of older authorities, as Hephaestio did with Ptolemy and Dorotheus, while unattributed chapters of Valens' *Anthologies* were translated into Latin and included in the Hermetic corpus.[45]

The progress represented by new manuals throughout the history of ancient astrology consisted to a degree in rearranging and selecting the material rather than looking for new theories. This was clearly not the case of Valens, but he is illustrative of the perils of the latter approach. Valens frequently sought to develop new doctrines, but the reader is left with the impression that his project exploded into a myriad of incompatible doctrines presented in a chaotic way, in part due to his wish to improve the theory without discarding the current or older doctrines. Ptolemy represents another, sounder solution: a middle course consisting in systematizing the astrological system by selecting, organizing, and simplifying the major doctrines that were, at least to a certain degree, compatible with each other. Naturally, manuals often displayed a mixture of approaches, including rearrangement, selection, collection, and rephrasing of older doctrines, interspersed with idiosyncratic theories.[46]

The fluidity of the astrological genre also entailed possible enlargements over time. Pseudepigraphic names like Nechepsos and Petosiris, or Hermes Trismegistus, allowed for the accretion of the corpus at different times and by different individuals. Additions are also found within the works: the example of an astral chart for a person born in AD 381 in Dorotheus III 1.28 ("the ninety-sixth year of the year of Darinūs [possibly Diocletian]") is an obvious insertion. Sometimes, chapters representing evident additions with varying degrees of relevance to the author's doctrine, and more or less marked as such, were appended at the end of the work. Valens' text, for example, is supplemented with material of this kind, which is edited under the title of *additamenta antiqua*.[47] In the Arabic Dorotheus, the heading of V 41 explicitly attributes the chapter to a certain Qīṭrnūs, which could in fact be a corruption of the name "Critodemus" (see Chapter 7). As for Critodemus, large tables for calculating the length of life (Chapter 8) were probably added at the end of the work, too.

Reuse and reworking to varying degrees were typical of the astrological field, in which readers were often practicing astrologers themselves, who copied and

44 For example, the one wrongly attributed to Anubio, covering II 14–19 and 27–33 of the Arabic version, and edited in Pingree 1976, 344–367; cf. Heilen 2010.

45 Pingree 1978, 432–433.

46 See, e.g., Chapter 6 for the collection attributed to Rhetorius and the inclusion of doctrines from Critodemus and Dorotheus.

47 Pingree 1986, 349–368.

adapted the texts to their own needs. Vettius Valens was perfectly aware of the propensity of astrological texts to undergo such transformations. It is for this reason that, as we will see in the next chapter, he includes among the clauses of his astrological oaths the promise from his readers that they will not copy the treatise under another name and that they will not cut anything out (F 6A).

Metaphor and reality in the voyage motif

In F 3B, Valens remarks, after quoting the beginning of Critodemus, that "Timaeus and Asclation, as well as many others, have said other things". He is likely referring to similar awe-inspiring beginnings of astrological manuals: we have seen the example of the poem of Nechepsos and Petosiris, where the king reveals that he has been awake all night looking at the sky awaiting inspiration, until finally a shout (!) came forth therefrom, incarnated in a dark-blue robe. Dorotheus (I 1) is more similar to Critodemus:

> I have traveled, oh my son, through many places, and I have seen the wondrous things which are in Egypt and Babylon, which is in the mouth of the Euphrates. I collected the best of their sayings from the best [authorities] who were before me like the bees which gather from the trees and all kinds of plants; for from it there is good honey.[48]

Both Critodemus and Dorotheus might be interpreted to use an appropriate *topos* of the genre of astrological poems when referring to their own knowledge-gathering journeys.[49] However, considering that these authors belong to the early period of astrological practice in the Greek-speaking world, it is not implausible that such trips may also have been real. By this early time in the history of Hellenistic astrology, travel to Egypt could have been the easiest way to gain access to a wide array of astrological doctrines and learnings. After all, Valens himself appropriates Critodemus' words to recount his own real experience of emigrating to Egypt to learn

---

48 Trans. Pingree with modifications. There is an important mistake at the end of the passage, where Pingree writes "the honey of medicine" instead of "good honey": *ṭayyib* means "good" (in flavor), which was probably confused with *ṭabīb*, doctor. There is another mistake, noted in Cottrell 2015, 365: Pingree translates *maṣabb* ("mouth") as "direction" (reading *muṣīb*). Furthermore, some of the elements in this proemium might have been altered through the transmission of the work. In particular, the text claims that the book is addressed to his son, who is said to be Hermes (I 2), which seems to frame the treatise as a sort of Hermetic dialogue. Such a mixture of authorial and pseudepigraphical presentation seems suspicious. However, the bee metaphor is undeniably Graeco-Roman, as is the voyage motif. Cf. on these issues Tolsa *forthcoming*.c.

49 Cf. Ferella 2018 for the use of the metaphor by early Greek thinkers.

astrology.[50] Thus, it does not seem far-fetched to hypothesize that Critodemus really traveled to Egypt to learn and practice astrology, and that he had thereafter moved again to a place which he considered to be a safe port, perhaps returning to Rhodes (see Introduction).

Another question is the raison d'être of these veiled references. Why did Critodemus not simply name Egypt and Rhodes, rather than alluding to them vaguely? There are many factors at play here. One reason might be the aura of mystery that was a hallmark of early astrological works. Another factor is that providing concrete names precludes the possibility of multiple allusion. Critodemus' expressions ἀκίνδυνος λιμήν ("safe harbor") and μονὴ ἀσφαλεστάτη ("most secure resting place") carry strong connotations of salvation after a dangerous journey.[51] Even if a real context may be alluded to, the literary metaphor is also at work: the end of the epistemic journey has finally arrived, and the author can finally transmit his learnings. One may compare this with the ambitious cosmological discourse of Timaeus' character in Plato's dialogue, which is brought to a close at the beginning of the *Critias* (106a):

> {ΤΙ.} Ὡς ἅσμενος, ὦ Σώκρατες, οἷον ἐκ μακρᾶς ἀναπεπαυμένος ὁδοῦ, νῦν οὕτως ἐκ τῆς τοῦ λόγου διαπορείας ἀγαπητῶς ἀπήλλαγμαι.
>
> TIMAEUS: How gladly, Socrates, I have now happily escaped from the path of the discourse, as if resting from a long journey!

The difference is that, whereas Timaeus transmits his speculations *live* during his journey, Critodemus does so only after his Egyptian voyage, in the same way as Dorotheus.

An Orphic background may also be playing a role here. Like many Greek sages, Orpheus is said to have traveled to Egypt for his initiation (Diod. Sic. IV 25.3). Also, he was part of the naval expedition of the Argonauts narrated by Apollonius of Rhodes. Orpheus was the spiritual leader of this expedition, the vigilant discoverer of portents and of the proper rituals crucial for securing a safe journey.[52] Critodemus could be alluding to this kind of famous episode in his emphatic description

---

50 Valens, who was from Antioch in Syria, gives credible details of his negative experiences with teachers in Egypt (IV 11.4).

51 Cf., e.g., *LXX 1 Macch.* 7.39; Philo Alex. *De sobr.* 49.6 (ἡσυχίαν καὶ μονὴν), *De leg.* III 36.5 (σωτηρίαν καὶ μονὴν).

52 For the divine portents, cf. *Argonautica* IV 1411–1421; for the offerings to the indigenous gods in IV 1547–1555, see Karanika 2010. In the fifth century AD, these religious implications were amplified by an author thoroughly familiar with the Orphic tradition in the reworking known as the *Argonautica Orphica*.

of arrival to that safe port granted by the gods, and, in connection with this, his astrological vision would recall the portents which Orpheus was able to decipher.

The title of Critodemus' work

If we are familiar with the beginning of the poem of Nechepsos and Petosiris (see above), the dark-blue robe (πέπλος) in which the heavenly shout is incarnated immediately comes to mind. This provides the astrological inspiration which enables Nechepsos to write his astrological work. Valens does not tell us what concrete vision Critodemus refers to, but it must ultimately have revealed to him the learnings of the astrological art.

Again, this may be related to the Orphic material. There was a tradition of Orphic texts carrying titles corresponding to aspects of mystery rituals – *Katharmoi, Thronismoi, Hierostolika, Katazostikon* (*OF* 601–608)[53] – which could have influenced the choice of Critodemus for the title. The theories contained in the *Vision* could be understood as emanating from religious mysteries, and they would have to be protected against impure individuals, which is precisely the aim of the Orphic-style oaths (F 6).

It is tempting to hypothesize that Critodemus alluded (even if only in the title) to the robe of Nechepsos and Petosiris. He could have found this image attractive as the robe was one of the objects the Orphics had used to allegorize their explanation of the cosmos: in the Orphic poem *Robe* (Πέπλος), dated before Hellenistic times, Persephone is weaving and is captured by Pluto at the point of embroidering the figure of Scorpio, a constellation signifying winter.[54]

---

53 Cf. discussion in Herrero de Jáuregui 2010, 37.

54 For the dating, it is relevant that Epigenes wrote about the poem: cf. Herrero de Jáuregui 2010, 208. The fragments related to the *Peplos* are *OF* 286–290; cf. Bernabé 2009a, 318–319. Epigenes disclosed several *mots-clefs* used by the Orphic poet, among which were the "parts" (μοῖραι), called "white-robed" (λευκόστολος) on account of their correspondence with three phases of the Moon (new, 15th day, and 30th day): cf. Clem. *Strom.* V 8.49. In the *Orphic hymns* we also find the characterization of the parts of the world as three μοῖραι, from which Poseidon receives the sea (17.7: ὃς τριτάτης ἔλαχες μοίρης βαθὺ χεῦμα θαλάσσης) and Pluto the earth (18.6: ὃς τριτάτης μοίρης ἔλαχες χθόνα παμβασίλειαν).

F 4. Valens VIII 5.19

Type of fragment: A brief passage in Valens, where Critodemus, Nechepsos, and Petosiris are listed as obscure astrological authorities of the past who need correction and clarification.

Οἱ οὖν ἐντυγχάνοντες ταῖς ὑφ' ἡμῶν συντεταγμέναις βίβλοις πάσας αἱρέσεις διελεγχούσαις μὴ λεγέτωσαν· αὕτη μέν ἐστιν τοῦ βασιλέως, ἑτέρα δὲ Πετοσίρεως, ἄλλη δὲ **Κριτοδήμου** καὶ τῶν λοιπῶν, ἀλλ' ἰδέτωσαν ὅτι ἐκεῖνοι μὲν προθέμενοι κωφῶς καὶ κατεζητημένως ἀνυπόστατον τὴν ἐπιστήμην κατέδειξαν, ἡμεῖς δὲ τὰς ἐπιλύσεις ποιησάμενοι οὐ μόνον θνήσκουσαν τὴν αἵρεσιν ἀνερρώσαμεν, ἀλλὰ καὶ ἑαυτῶν δόξαν κατεψηφισάμεθα καὶ ἑτέρους δὲ ἀξίους ἐμυσταγωγήσαμεν, οὐ χρημάτων προτροπῇ θελχθέντες ἀλλὰ φιλομαθεῖς καὶ ἐπιθυμητὰς ἐπιγνόντες· καὶ γὰρ αὐτοὶ τοιούτῳ χαρακτῆρι Νεμέσεως ἐχαλιναγωγήθημεν.

Let the readers of our compilations, when testing all procedures, not say: "This procedure is from the King, this other is from Petosiris, that one is from **Critodemus**, etc". Instead let them know that these men showed their science to be groundless by presenting it in an obtuse and involved way. We on the other hand provided solutions, and not only revived this dying art, but also voted fame for ourselves[55] and initiated other worthy men, seducing them not with the impulse toward money, but by recognizing them as lovers of knowledge and enthusiasts. We too have been guided with bit and bridle by this type of Nemesis.[56]

Against the old authorities and their bare compilation

F 4 provides more context to understand Valens' self-positioning against Critodemus and Nechepsos/Petosiris. Valens criticizes these authors for their obscurity, rejects the idea that his own work is a mere compilation of their doctrines, and declares to have provided "solutions" (ἐπιλύσεις) to these problems of the old trea-

55 A strange expression, taken from judicial discourse, where it normally indicates a vote of condemnation. But elsewhere in Valens it can be both positive and negative: ἑαυτῶν δόξαν κατεψηφισάμεθα (V 6.13).

56 Allusion to the classical picture of Nemesis riding a chariot. As the divine distributor of fortune, Nemesis is an appropriate tutelar divinity of astrology. Valens playfully portrays himself as the horse ridden by Nemesis, who guides it through the path of astrological learning. He uses the same image elsewhere, again in reference to astrology, among other arts: Νέμεσις χαλιναγωγὸς ἔπεστι (VI 9.14).

tises. In the same spirit, in the paragraph preceding the fragment, he signals as his greatest achievement the work to go through and explain the doctrines that the older authorities have just outlined.[57]

Indeed, Valens is entirely justified in his claim that his treatise cannot in any way be considered a slavish compilation of older doctrines. The dispositions in his oaths concerning the intact transmission of his work (Chapter 3) are another expression of this pride. A different consideration is in what sense he provided solutions, and whether it was always the case that the older treatises left the theory half-explained. In fact, in many instances we find that Valens simply produced a variation of an older, simpler doctrine.

The immediate context of the fragment is relevant to the question: it is found between the detailed explanation of the construction of the tables for calculating the length of life (F 17) and Valens' own tabular method for determining the length of life, which he calls "three factors" (VIII 6 title: τρεῖς ὅροι). Valens seems to have considered this latter method to be his own improvement of the tables in Critodemus, so his remark on his improvement of previous theories is consistent with the content of the present book.

If Valens' work may be described as chaotic and obscure, these qualities arise in part from the author's constant search for new doctrines, frequently departing from the older ones to the point that it can be difficult to distinguish the new developments from the old doctrines. This is the case of Critodemus F 9 (Chapter 4) and 20 (Chapter 8), in which a very specific doctrine is apparently attributed to the old authority but close inspection reveals that we are in fact reading Valens' particular version of it.

F 5. Firmicus IV pr.5

Type of fragment: Firmicus lists astrological authorities which he claims to have used for his astrological manual. Like Valens in F 4, Firmicus insists on clarity, which he deems necessary in order to present the astrological art to the Romans in an adequate manner, and he appears to interpret the history of the astrological genre as the result of a continuous effort of clarification undertaken by his predecessors. Critodemus is listed after Orpheus.

Omnia enim, quae Aesculapio Mercurius et Cmifis[a] *uix tradiderunt, quae Petosiris explicavit et Nechepso et quae Abram, Orfeus et* ***Critodemus*** *ediderunt [et] ceteri-*

57 Valens VIII 5.15: τοῦτο δέ μοι δοκεῖ μέγιστον, τὸ ἐξελέγξαι ἀλλοτρίας ἐνθυμήσεις μυστικῶς κατακεχωσμένας.

que omnes huius artis scii, perlecta pariter atque collecta et contrariis sententiarum diversitatibus comparata illis perscripsimus libris divinam scientiam Romanis omnibus intimantes, ut hoc, quod quibusdam difficillimum videbatur propter Latini sermonis angustias, ostensa Romani sermonis licentia veris ac manifestis interpretationibus explicarem.

a *et Cmifis* coni. Quack; *einhnus* MSS; *et Chnubis* Reitzenstein; *et Hanubius* Teuffel

Everything which Hermes and Kneph managed to hand down to Asclepius;[58] which Petosiris and Nechepso explained;[59] which Abram,[60] Orpheus, and **Critodemus** wrote, and all the others knowledgeable in this art; such theories, well-studied as well as collected and collated from the contrary diversity of methods, we have written in these books, presenting this divine science to all the Romans, in such a way that I have explained from a manifest freedom of the Latin language, with

58 Compare the similar setting in the dialogues of the *corpus hermeticum*, in which Asclepius (who was indeed an astrological pseudepigraphic author, cf. Valens IX 3.5) is Hermes' interlocutor. The Egyptian god Kneph or Chnubis was probably equivalent to the Greek Agathodaimon, since in the magical papyri Agathodaimon sometimes receives an epithet connected to Kneph or Chnubis: HARPENKNOUPHI (*PGM* XXXVI 220, VII 1025). Also, there were texts of the Hermetic tradition in which Agathodaimon was mentioned as a teaching authority: cf. Cyrill. Alex. *Contra Iulianum* II 30: "And Osiris said – says [Hermes] – : Oh greatest Agathos Daimon, how did the whole Earth appear? And Agathos Daimon replied. . ." (διαμεμνήσομαι δὲ τῶν Ἑρμοῦ τοῦ τρισμεγίστου λόγων. Ἔφη γὰρ οὕτως ἐν τῷ Πρὸς Ἀσκληπιόν· Καὶ εἶπε, φησίν, Ὄσιρις· Εἶτα, ὦ μέγιστε ἀγαθὸς δαίμων, πῶς ὅλη ἡ γῆ ἐφάνη; Καὶ εἶπεν ὁ μέγας ἀγαθὸς δαίμων. . .). Heilen 2011, 51, argues that Kneph is alluded to as transmitter in the preface of Nechepsos' poem (cf. above), although the identity of the divine inspiration, if any particular god was meant at all, might never have been disclosed in the poem. Hermes and Asclepius were perhaps also not at all alluded to, but the fact that they were divine, and therefore ancient, astrological authorities, probably led later authors to infer their influence. Ryholt 2011, 71, argues, on the basis of P. Louvre 2342 bis and from the frequent coupling of Imhotep (=Asclepius explicitly in the papyrus) with Amenhotep, that the Hermes in both the papyrus and Firmicus is Amenhotep son of Hapu. Quack 2002, 90, however proposes the more obvious Thoth, but the two possibilities are not mutually exclusive.

59 A similar sequence is found in P. Louvre 2342 bis = *GH* 137c, 2–6: σκεψάμενος ἀπὸ πολλῶν βιβλίων ὡς παρεδόθη ἡμεῖν ἀπὸ σοφῶν ἀρχαίων τουτέστιν χαλδαικῶν καὶ [Π]ετοσίρις μάλιστα δὲ καὶ ὁ βασιλεὺς Νεχεύς ὥσπερ καὶ αὐτοὶ συνίδρυσαν ἀπὸ τοῦ κυρίου ἡμῶν Ἑρμοῦ καὶ Ἀσκληπιοῦ ὁ ἐστιν Ἰμούθου υἱὸς Ἡφήστου. ("Having examined many books as it has been handed down to us from ancient wise men, that is the Chaldeans, and Petosiris and King Necheus, just as they themselves took counsel from our Lord Hermes and Asclepius, that is Imouthes, son of Hephaestus").

60 Abram is mentioned as a source for the predictions on travel in the horoscopes of Valens II 29 and 30.

true and clear explanations, what appeared most difficult to some because of the limitations of the Latin language.[61]

An organized list of authorities

F 5 is the list of authorities that Firmicus provides at the beginning of book IV, from which he claims to have gathered the astrological teachings he is now presenting in Latin. In this preface (pr. 1–3), Firmicus explains that he has abandoned his legal duties, presenting himself to the reader as freed from such vexatious obligations, which recalls King Nechepsos' abandonment of his royal persona in favor of astrology (Valens IX 1.1–3).[62]

If Firmicus is consciously emulating Nechepsos, his self-presentation here can be compared to the strategy adopted by Valens in his treatment of Critodemus. The astrologer declares his admiration for his predecessor as a worthy transmitter of doctrines, but at the same time he feels the need to justify the value of his own contribution by criticizing matters of presentation in his predecessor's work. Similarly to Valens, Firmicus alludes to the obscurity in the presentation of astrological knowledge by the first authorities (*uix tradiderunt*), and to its further clarification (*explicauit*). He then introduces the new problems of translating the theory into Latin and of making sense of the myriad of contradictory theories (*contrariis sententiarum diversitatibus*). This is indeed the character of Firmicus' treatise: while it mainly gathers together the doctrines of other authors without introducing new theories, it integrates these doctrines in a consistent whole.

The order of Firmicus' list underlines the chain-like character of the transmission of astrological knowledge and the fact that, at each step, less divine and more human authorities explain and clarify the previous doctrines, bringing them closer to their readers.

61 It is indisputable that Firmicus used the first-century-AD didactic astrological poem of Manilius (Volk 2009, 106), who in turn calls himself the first to bring these teachings to the Latin-speaking world, noting that he does so in Latin verse form (*Astr.* I 3–5).

62 The manuscripts call Firmicus *V. C.*, that is, *vir clarissimus* or *consularis*, implying senatorial rank. Cf. Chapter 3 on the purity associated with astrologers and physicians, and the related mechanism of the oath.

Different traditions and Critodemus' place

Firmicus apparently organizes his list in three branches, comprising Egyptian authorities (Hermes, Kneph, Asclepius, Nechepsos, and Petosiris), Hebrew (Abram), and Greek (Orpheus and Critodemus). Valens remarks once on the fact that Egyptians, Babylonians, and Greeks all agree on one particular doctrine (Val. IV 30.21), and he also uses Abram as an astrological authority (Val. II 29–30). This is most likely a Greek pseudepigraphical author, exploiting the Biblical association of Isaac's father with the city of Ur (*Gen.* 11.31). As early as the beginning of the first century AD, Philo of Alexandria had interpreted the name Abram to mean "sublime father", describing him as knowledgeable of the astronomical theories of the Chaldeans (*On Abr.* 82, *Gig.* 62).[63]

The inclusion of Orpheus just before Critodemus is surely not coincidental. Here, Firmicus had an ample array of "Greek" sources to choose from. It is, then, reasonable to think that he was underscoring the association between Critodemus and Orpheus. This would identify Orpheus as a Greek version of the Egyptian Hermes Trismegistus. Orpheus was then interpreted as a divine transmitter of astrology whose work had been continued by Critodemus in the same way that the work of Hermes had been taken up by Nechepsos and Petosiris.

63 It is striking that Abram acts as the closest representative of the Babylonian branch in Firmicus' list. This is probably due to the fact that Mesopotamian culture was not integrated into Greek culture to the same degree as Egyptian culture (cf. Introduction).

Chapter 3
Orphic oaths (F 6)

F 6A–D are oaths in Valens' and Firmicus' manuals by which their readers must swear not to transmit the astrological knowledge the manuals to impure or uninitiated individuals. They exhibit typical elements of ancient Greek oaths, such as divine witnesses acting as warrant, and rewards and curses for the oath-keepers and the oath-breakers, respectively.

Oaths were not a usual feature of ancient astrological texts; indeed, the ones in Valens and Firmicus are the only preserved examples. Otherwise, as far as we know, Critodemus is the only other astrological writer to include oaths in his treatise (F 3A). In fact, even if the oaths in Valens and Firmicus are not expressly related to Critodemus, there is enough circumstantial evidence to deduce that they took inspiration from our author.

Both Valens and Firmicus allude to Orpheus immediately after their oaths. As we shall see, these present significant affinities with oaths of silence from the Orphic tradition and, more specifically, with an oath found in an Orphic poem called *Oaths*. The ideal function of Orphic oaths was to prevent external access so as to avoid misunderstandings from inexperienced individuals, who could easily fail to observe the allegorical dimension of the teachings and thus distort the tenets of the sect. The oaths in Valens and Firmicus clearly aim to exercise this kind of control over the transmission of their science.

There is another feature of Critodemus' treatise that brings it close to the texts of the Orphic tradition. Critodemus' oath (or oaths) seems to have been introduced in the seventh and last book of his poem, which was marked as an especially important book. It is well known that Orphic texts often showed structural similarities with religious rituals. For example, the collection of *Orphic Hymns* is arranged by first presenting hymns dedicated to gods associated with birth and beginnings, such as Hecate and Prothyraia, and ends with sleep (Hypnos), the dreams (Oneiroi), and death (Thanatos), thus dramatizing the path of life and mimicking the spatial conceptualization of the stages of life in mysteric rituals, in which the initiate gradually proceeded from the outer to the inner parts of the sacred enclosure, the *adyta* ("impenetrable"). Thus, the presence of the oath(s) in the last book of an astrological manual reveals a conception of the process of reading and learning through the treatise as an initiation in itself, ending in the innermost chamber, whose knowledge is protected.

We may compare the astrological oaths to the well-known Hippocratic *Oath*, which, according to its own text, had the object of preventing the teaching of medicine outside of the family, to students who had not taken that very oath. The

https://doi.org/10.1515/9783111329147-004

medical oath stipulates conditions for the practice of medicine such as the prohibition of using medicine for putting persons to death (including abortion), of divulging secrets, or of sexually abusing patients by taking advantage of home visits. Astrology can be likened to medicine here in the sense that the authors aim to safeguard the integrity of astrological discourse and practice, but the conditions of the astrological oaths are simpler. Valens stresses the need to have progressed through the basic astrological learnings (F 6B), which is the basis of the repeated warnings against passing on the work to uneducated or impure individuals.

Another question is whether such oaths were ever really taken by practitioners, in either medicine or astrology. It is even possible that a first version of the Hippocratic *Oath*, not including the ethical stipulations, was the invention of an ancient historian, intended to explain the fact that at the beginning of the Hippocratic tradition medical learning and teaching was apparently limited to the Asclepiad family, as well as the subsequent dissemination of such knowledge.[1] In Roman imperial times and late antiquity, when the oath was introduced to the medical curriculum, there is no evidence that students actually swore by it.[2] It is perhaps also naïve to imagine a reader of Critodemus, Valens, or Firmicus taking an oath before continuing with the text. Nevertheless, even if no one took such oaths, their very presence in the curriculum – in the case of medicine – or in an astrological treatise probably had an impact on the way in which readers approached the disciplines.

F6. Valens IV 11.11–13 / Valens VII 1.1–4 / Valens VII 6.230–234 / Firmicus VII pr.

Type of fragment: Four passages present similar astrological oaths of silence, three in Valens and one in Firmicus. Oaths B and C open and close, respectively, book VII of Valens, and D opens book VII of Firmicus, which suggests that Critodemus also presented his oath(s) in his book VII, probably the last one, at least originally. The three oaths in Valens are related to the so-called distributions, a type of time rulerships plausibly originating from Critodemus that will be discussed in the next chapter, and Firmicus deals with this doctrine just before his oath, at the end of book VI. It is therefore likely that the oath(s) in Critodemus were also related to the distributions. As for the oaths themselves, it can be shown that they are modeled on oaths of the Orphic tradition.

1 Cf. Tolsa 2019a. The ethical statements in the second part of the Hippocratic *Oath* could have been inspired by conditions of purity for participation in the mysteries of Cybele.

2 Some medical contracts have been found, but nothing similar to the ethical commitments of the oath: cf. Anagnostou-Canas 2017 and Straus 2017.

A. Valens IV 11.11–13

11 Ὁρκίζω σε, ἀδελφέ μου τιμιώτατε, καὶ τοὺς μυσταγωγουμένους ταύτῃ τῇ συντάξει οὐρανοῦ μὲν ἀστέριον κύτος καὶ κύκλον δυοκαιδεκάζῳδον, Ἥλιόν τε καὶ Σελήνην καὶ τοὺς ε′ πλανήτας ἀστέρας δι᾽ ὧν ὁ πᾶς βίος ἡνιοχεῖται, αὐτήν τε τὴν πρόνοιαν καὶ τὴν ἱερὰν ἀνάγκην, ἐν ἀποκρύφοις ταῦτα συντηρῆσαι καὶ μὴ μεταδοῦναι τοῖς ἀπαιδεύτοις εἰ μὴ τοῖς ἀξίοις καὶ δυναμένοις διαφυλάσσειν καὶ ἀμείβεσθαι δικαίως, αὐτῷ τε ἐμοὶ Οὐάλεντι τῷ εἰσηγησαμένῳ ἀείμνηστον καὶ ἀγαθὴν φήμην ἀπονέμειν, καὶ μάλιστα ἐπιγνόντας τὸ ἄφθονον καὶ τὸ τῆς ἀληθείας μέρος ὡς ὑπὸ οὐδενὸς ἀνδρὸς ἐπιλελυμένον αὐτὸς ἐφώτισα, μηδὲ παρέντας τὸ ἐμὸν ὄνομα ἑτέρους ἐπεισφέρειν ταύτῃ τῇ συντάξει μηδὲ λωβῆσαί τι τῶν προγεγραμμένων ἢ μελλόντων λέγεσθαι πρὸς τὸ ἀθετῆσαι τοὺς ἐντυγχάνοντας καί μοι ψόγον ἐπενέγκαι.
12 καὶ ταῦτα μὲν διαφυλάσσουσιν οἱ προειρημένοι θεοὶ πάντες εὐμενεῖς ἔσοιντο καὶ βίος εὐσταθὴς καὶ καταθύμιος λογισμῶν συντέλεια, ἐπιορκοῦσι δὲ τὰ ἐναντία, μήτε γῆ βατὴ μήτε θάλασσα πλωτὴ μήτε τέκνων σπορά, τυφλός τε νοῦς καὶ πεπεδημένος ὑπάρχων ἀσχήμονα βίον καὶ ἀνεπίτευκτον ἀγαθῶν ἐπάγοι. ἐὰν δέ τις καὶ μετὰ θάνατόν ἐστι κακῶν τε καὶ καλῶν ἀμοιβή, κἀκεῖ τῶν ὁμοίων μεταλάβοιεν.
13 ὅθεν <εἰ> καὶ μετὰ τὴν ἐπίγνωσίν τις τῆς ἐνθάδε διδασκαλίας ἐν ἑτέρῳ συντάγματι αἰνιγματωδῶς εὕροι προκειμένην τὴν εὕρεσιν, οὐκ ἐκείνῳ δεῖ ἐγκώμιον ἀπονέμειν, ἀλλὰ ἡμῖν χάριν ὁμολογεῖν ὡς οὐ μόνον προμηνυταῖς ἀλλὰ καὶ εὑρεταῖς πολλῶν γενομένοις καὶ συγκοσμήσασι τὴν αἵρεσιν· καὶ γὰρ πολλοὶ ἀφθόνως τινὰ παραλαμβάνοντες μετὰ φθόνου συνέταξαν.

11 I put you under oath, my dearest brother, and the initiates into this compilation, by the starry vault of heaven and by the twelve-sign circle, by the Sun and the Moon and the 5 planets by which all life is guided, and by Providence itself and Holy Necessity, to keep these matters in secret and not to distribute them to the uneducated; only to those worthy of them and able to preserve them and respond to them with justice, and to bestow eternal and noble fame on me, Valens, the one who introduced you, especially recognizing that it is myself who generously illuminated this part of the truth never before released by anyone, and not omitting my name to add others' to this compilation, nor cutting out anything from what has been written or what will be said in order to put out of place the readers and bring discredit to me.
12 May the previously mentioned gods be well-disposed to those who keep these promises and may their life be steady, and may the goal of their calculations[3] be as they wish, and the opposite to those who foreswear the oath: may the earth not be

3 Cf. the very similar expression τὰς τῶν λογισμῶν ἐπιθυμίας in Val. I 48.2.

passable, the sea not be sailable, may there be no offspring of children, may their mind be blind and bound, bringing on a shameful life not attaining the good. If after death there is a retribution for good or evil deeds, may they receive the same there afterwards.
13 Therefore, if after learning the present instruction someone mysteriously finds these findings in another treatise, he must not bestow praise to that, but he must concede gratitude to us as being not only the first informant but also the discoverer of many theories as well as the one who adjusted the method, for many borrow a doctrine without problem, but arrange them in a treatise with envy.

B. Valens VII 1.1–4 (=*OF* 618 T) (beginning of book VII)

1 Χρὴ μὲν οὖν πρὸ πάντων καὶ περὶ ταύτης τῆς βίβλου ὅρκον προτάξαι τοῖς ἐντυγχάνουσιν, ὅπως πεφυλαγμένως καὶ μυστικῶς ἔχωσι τὰ λεγόμενα· ποικίλως γὰρ καὶ ἀκολούθως τὰς προκειμένας αἱρέσεις ἐθέμην εἰς προτροπὴν καὶ ἀγωγὴν φιλόκαλον πολλὴν δύναμιν ἐχούσας, ὅπως ἀπὸ τῶν ἐλαχίστων ἐπὶ τὰ μείζονα τὴν ἐπιθυμίαν ἐπεκτείνωσι καὶ μὴ διαλιπόντες ψόγον καθ’ ἡμῶν οἴσονται. **2** βέλτιον οὖν ἐστι προγυμνασθέντας ταῖς συντεταγμέναις ὑφ’ ἡμῶν βίβλοις ἐπὶ ταύτας ἐλθεῖν· ἐκ γὰρ τῶν καθολικῶν καὶ μερικῶν συγκρίσεων ἄπταιστος ἡ διαίρεσις καὶ σεβάσμιος κριθήσεται. **3** καὶ καθάπερ ἂν εἴς τινα ἀκρώρειαν διὰ βαθμῶν καὶ περικαμπῶν τόπων ἀνελθών τις μετὰ πολλοῦ πόνου καὶ εὑρὼν ναοῦ κατασκευὴν καὶ πολυτέλειαν ἀγαλμάτων

χρυσοῦ τε καὶ ἀργύρου ἠδ’ ἐλέφαντος[a]

ἤ τινα ἁλουργήματα ἀμετανόητον καὶ ἀκοπίατον ἡγεῖται τὴν ἄνοδον καὶ μεθ’ ἡδονῆς θρησκεύει φαντασιούμενος οὐρανίοις θεοῖς προσομιλεῖν, τὸν αὐτὸν τρόπον καὶ οἱ ταῖς παραγγελίαις ἡμῶν πειθόμενοι· οὓς ὁρκίζω Ἡλίου μὲν ἱερὸν κύκλον καὶ Σελήνης ἀνωμάλους δρόμους, τῶν τε λοιπῶν ἀστέρων δυνάμεις καὶ κύκλον δυοκαίδεκα ζῳδίων, ἐν ἀποκρύφοις ταῦτα ἔχειν καὶ τοῖς ἀπαιδεύτοις ἢ ἀμυήτοις μὴ μεταδιδόναι τιμήν τε καὶ μνήμην τῷ εἰσηγησαμένῳ ἀπονέμειν. **4** εὐορκοῦσι μὲν εὖ εἴη καὶ καταθύμιοι οἱ προκείμενοι θεοί, ἐπιορκοῦσι δὲ τὰ ἐναντία.

a χρυσοῦ τ’ ἠλέκτρου τε καὶ ἀργύρου ἠδ’ ἐλέφαντος Hom. Od. IV 73

1 First of all, and especially regarding this book, it is necessary to present an oath to the readers, so that they keep the things said well-guarded and in mystery, since I have set out complexly and sequentially the present methods, which have a great power encouraging and guiding toward the good, in that they enlarge desire from the lesser things to the greater and do not have gaps that bring discredit to me. **2** Thus it is best to come to these books after having practiced with the ones that we have previously composed, for the division[4] will be judged to be infallible and venerable from the general and particular combinations.[5] **3** And just as someone climbing to a mountain top by steps and zigzagging places, with lots of effort, and finding the furniture of a temple and the costliness of statues of gold and silver and ivory [Hom. *Od.* IV 73] or some purple clothing, does not regret nor finds the ascension tiring, and worships with pleasure imagining that he is talking to the gods in heaven, it is the same with those who are persuaded by our rules. Them I put under oath by the sacred circle of the Sun and the irregular paths of the Moon, by the powers of the other stars and the circle of twelve signs, to keep these things in secret and not pass it over to the uneducated and uninitiated, and to bestow honor and remembrance to their guide. **4** May things be good, and the present gods propitious, for those who keep the oath, and the opposite for those who forswear.

C. Valens VII 6.230–234 (end of book VII)

230 Καὶ ταῦτα μέν, ὦ Μάρκε, μετὰ πολλοῦ πόνου καὶ ἐγκρατείας ζητήσας καὶ ἀνευρὼν συνέταξα καὶ τὰς αἱρέσεις ἐξεδόμην. **231** τοιγαροῦν ὁρκίζω σε Ἥλιον καὶ Σελήνην καὶ τῶν ε´ ἀστέρων τοὺς δρόμους, φύσιν τε καὶ πρόνοιαν καὶ τὰ δ´ στοιχεῖα μὴ ταχέως τινὶ μεταδοῦναι καὶ ταῦτα ἀμαθεῖ μηδ' ὅσ<τις> ἔτυχεν, λογισάμενον τὸν πόνον καὶ τὸν πόθον καὶ τὴν πολυχρονίαν εἰς τὰ τοιαῦτα διατριβήν τε καὶ ζήτησιν. **232** τοῦ γὰρ χρόνου τὴν πρᾶξιν εἰκάσας ἀντὶ πολλῶν σοι χρημάτων κατέλιπον· τὸ γὰρ ἀργύριον εὐανάλωτον καὶ ἐπίφθονον καὶ εὐεπιβούλευτον ὑπάρχει, τὰ δὲ συντάγματά μου καὶ βίον σοι καὶ δόξαν καὶ τιμὴν καὶ ἡδονὴν καὶ ὠφέλειαν παρέξει ἐάνπερ κοσμίως καὶ ἀσφαλῶς ἐνεχθῇς, καθὼς προγέγραπται, καὶ μὴ ἐγκληματικῶς μηδ' ἀκροθιγῶς. **233** καὶ αὐτὸς οὖν ὁμοίαν τὴν κακοπάθειαν ἡγησάμενος, ὥσπερ ἂν εἰ αὐτὸς συνέταξας – ἀλλ' ὅμως ἐκακοπάθησας παραλαμβάνων καὶ ἀνεκρίθης ἀξίαν ἀμοιβὴν κομισάμενος – μετάδος τοῖς δυναμένοις· τοῦτο δὲ ποιήσας ἐμὲ μὲν καὶ τὸ μάθημα δοξάσεις, σαυτὸν δὲ ὠφελήσεις καὶ φιλόπονον καὶ φιλόκαλον ἀποδείξεις. **234** εὐορκοῦντι μέν σοι καταθύμιος συντέλεια.

---

4 This surely means the division of times, as in the method of Critodemus: cf. Val. VI 6.1: καὶ ταύτην δὲ τὴν διαίρεσιν. Critodemus proposed a system of chronocratorships, that is, a method for dividing the length of life and assigning ruling stars to the stars, which will be dealt with in the next chapter.

5 In Critodemus' system, there is a "general" sequence of stars governing the successive periods of the individual's life, as well as a subdivision, meaning that each star in the general division subdivides its time assigning it to all the stars in sequence, which are then constituted as subrulers (cf. next chapter).

230 And all these methods, Marcus, which I have researched and discovered thanks to great toil and mastery, I have collected and published. **231** Thus I put you under oath by the Sun and the Moon and the paths of the 5 stars, by Nature and Providence and the 4 elements, not to pass this over rashly to any uneducated or casual person, taking into account the toil and the longing and the extended time of this study and research. **232** I estimate the work of so much time to cost a lot of money, but I have left it to you, since money is easy to spend, attracts envy and treachery, but my collections will offer you a means of living, fame, honor, pleasure, and profit, if you behave orderly and securely, as has been written before, and not in a criminal and superficial way. **233** And yourself, after becoming aware of a similar strain – in any case you strained yourself in your learning and were examined giving a good response – pass it over to the worthy ones. If you do this, you will honor me and the science, you will be useful to yourself and will appear as devout of toiling and beauty. **234** If you keep this oath, may your life goal be as you wish.

D. Firmicus VII 1 (=*OF* 617 T)

1 *Cum incognitis hominibus Orpheus sacrorum caerimonias* <intimaret>, *nihil aliud ab his quos initiabat in primo vestibulo nisi iurisiurandi necessitatem [et] cum terribili auctoritate religionis exegit, ne profanis auribus inventae ac conpositae religiones proderentur. Sed et Platonici deum*[a] *perpetuum †a se eum frequenter convenit†*[b], *nec secretarum disputationum veneranda commenta inperitis aliquando auribus intimari.*[c] *Patiuntur enim haec omnia iacturam, cum perditis ac desperatis animis ingeruntur. Pythagoras etiam et noster Porphyrius religioso putat animum nostrum silentio consecrari.*
2 *Vnde et ego horum virorum legem insecutus convenio te iureiurando, Mavorti decus nostrum, per fabricatorem mundi deum, qui omnia necessitate perpetuitatis excoluit, qui Solem formavit et Lunam, qui omnium siderum cursus ordinesque disposuit, qui maris fluctus intra certos terrae terminos coartavit, qui ignem ad sempiternam substantiam divinae perpetuitatis inflammat, qui terram in medio collocatam aequata moderatione sustentat, qui omnes homines feras alites et omnia animantium genera divina artificii maiestate conposuit, qui terram perennibus rigat fontibus, qui ventorum flatus cum quadam facit necessitatis moderatione variari, qui ad fabricationem omnium quattuor elementorum diversitate conposita, ex contrariis et repugnantibus cuncta perfecit, et ortum occasumque terraemotum omnium*[d] <*decretat*>; *per descensum ascensumque animae; per immortalem aeternae perpetuitatis ordinem.*
3 *Ne haec veneranda commenta profanis vel inperitis auribus intimentur, sed his quos animus incorruptus ad rectum vivendi ordinem casto ac pudico praesidio mentis ordinavit, quorum illibata fides, quorum manus ab omni sunt facinorum scelere separatae,*

integris pudicis sobriis ac modestis, ut puro mentis splendore decoratis integra se scientia divinationis insinuet. Accipe itaque omnia partili ratione collecta, et quia te iurisiurandi religione convenimus, accipe quod tibi cum maxima animi trepidatione misimus.

a *deum: meum* codd.
b *Plato Timaeum perpetuum apud se frequenter continuit* ed. Aldina
c *ne . . . intimaret* ed. Aldina
d in cruc. Kroll. I think that the received text works well here as it is, and that the editor is confused assuming that *ortus* and *occasus* must refer to star risings and settings, while it can also simply mean beginning and end (by transference), which would apply well to earthquakes. Cf. Lucretius VI 587: *terrae motus obortus*.

1 When Orpheus brought strangers into the ceremonies of his mysteries, he required nothing of those whom he initiated in the first stage but the necessity of an oath with the fearful authority of religion, that the rites learned and collected would not be betrayed to profane ears. And the followers of Plato . . . the eternal god . . . *[corrupt text]*, and not to reveal to unlearned ears the revered words of the secret dialogues. For these things suffer damage, when they are presented to lost and desperate minds. Pythagoras, and our Porphyry too, believes in consecrating our mind to religious silence.
2 Therefore, following the law of these men I come to you with an oath, dear Mavortius, by the god maker of the world, who has perfected everything with the necessity of eternity, who has formed the Sun and the Moon, who has placed the paths and the arrangements of all the stars, who has compressed between definite limits the flow of the sea, who kindles the fire of the eternal substance of divine eternity, who maintains the Earth placed in the middle with equal balance, who composed all humans, beasts, birds, and all kinds of animals with divine ingenuity, who waters the Earth with perpetual fountains, who makes the blowing of the winds change with a certain balance of necessity, who for the construction of everything made all out of contrary and diverging things, composed with a variety of four elements, and [controls] the beginning and end of all earthquakes; by the descent and ascent of the soul; by the immortal order of eternal perpetuity:
3 That these sacred words be not revealed to profane and uneducated ears, but to those whom an unspoiled mind has driven toward a correct mode of life under the chaste and humble lead of reason, whose unharmed faith, whose hands are apart from the sin of crimes; those virtuous, humble, moderate and modest, so that the virtuous science of divination enters those embellished with the pure splendor of reason. Receive, therefore, all things collected here with a single condition and, because we came to you with the religiousness of an oath, receive what we sent you with great trepidation of mind.

Context in Valens and Firmicus, and relationship with Critodemus

In F 3A, Valens says of Critodemus: "with frightful oaths he expounds his methods and his transmissions" (αἱρέσεις ἐκτίθεται καὶ παραδόσεις διὰ φρικωδεστάτων ὅρκων). Since the only oaths attested in astrological writings are found in Valens and Firmicus, and both authors are familiar with the work of Critodemus, it is a plausible assumption that these extant oaths were at least inspired by Critodemus. Furthermore, as we shall see, the oaths in both writers' work share specific features that confirm a common source, despite some important differences.

Crucially, there are parallels in the places the oaths appear in Valens and Firmicus that can hardly be attributed to chance. At the beginning of his seventh book (F 6B), Valens is quite explicit about the fact that an oath of secrecy is required of his readers (and, in particular, his dedicatee Marcus) precisely because they have reached this book ("it is necessary to establish an oath in relation to this book", περὶ ταύτης τῆς βίβλου ὅρκον προτάξαι). This is further reinforced by the fact that at the very end of the same book Valens again addresses Marcus, demanding that he not share the details of the imparted astrological doctrine too quickly (F 6C). It is surely, then, no coincidence that Firmicus also places his oath (F 6D) at the beginning of his book VII.

It is worth noting that the content of book VII in Valens and Firmicus differs substantially. Firmicus VII consists of the exposition of different paradigmatic outcomes of the stellar configurations at birth. This is a typical section of astrological manuals. Ptolemy, for example, distributes the treatment of these questions between his last two books, presenting in book III the astrological configurations typical of various life circumstances around the time of birth, and leaving for book IV the astrological features related to events occurring later in life (as he explains in his preface to book IV). Among the first group we find topics such as the existence of siblings, information about the parents, or the temperament of the person concerned. As still occurs in astrological practice today, the astrologer would prove himself to be proficient by revealing this kind of information to their client. In the second group of topics, Ptolemy classifies future predictions such as marriage, children, travel, or the kind of death.

In contrast with Firmicus' seventh book, Valens' book VII addresses matters related to the distribution of the times that he has already advanced in previous books. As is often the case in his treatise, Valens seems to lack a clearly defined plan: he jumps from one technique to another without notice and returns to the same point several times. The overall impression from the many different approaches related to the distribution of times that he tries and explains, and from the great amount of space it occupies in this and other books, is that it was one of Valens' favorite topics. In the oath that forms the preface of the seventh book (F 6B), Valens seems to assume that the teaching is now about this topic, making it clear that his readers are progressing from lesser to greater matters (ἀπὸ τῶν ἐλαχίστων ἐπὶ τὰ μείζονα) and that it is preferable they have practiced the previously instructed methods, as this ensures "the distribution will be infallible and sacred" (ἄπταιστος ἡ διαίρεσις καὶ σεβάσμιος); the word διαίρεσις ("distribution") is frequently used before in the technical sense of the distribution of times.[6]

It may be significant that Valens qualifies as "sacred" (σεβάσμιος) the topic of the distribution of times that forms the content of book VII. However, what more clearly ties the distribution of times to the oath is the fact that this same topic is already developed in book IV after the first oath (F 6A). Unlike book VII, in book IV the oath is found half-way through the text (IV 11). After some preliminary material, Valens presents a long list of the effects of the so-called transmissions or distributions (παραδόσεις or ἐπιμερισμοί) in IV 17–25, which has a clear parallel in the last part of Firmicus' book VI (33–39), just before his own oath. These lists seem to go back to Critodemus F 7, extant in Valens' *Additamenta antiqua* (cf. next chapter).[7]

Significance of the number 7 and the number of books of the Horasis

From what has been said above, it can be deduced that both Valens and Firmicus remained faithful to Critodemus not only in reproducing the doctrine of the distributions but also in signaling book VII as one of especially sensitive content, by having their readers swear an oath of secrecy at its beginning. The fact that Firmicus expounds the distributions not in book VII but at the end of book VI suggests that both aspects were respected to a certain degree: a connection with the doctrine and the book number. Critodemus, then, probably presented his doctrine of the distributions in his seventh book, introducing it with an oath and perhaps ending it with another, like Valens.

The earliest astrological writers whose treatises are extant, Dorotheus and Manilius, wrote their works in five books.[8] Hübner is probably right in suggesting

6 For example, IV 5.5 (τὴν διαίρεσιν τῶν χρόνων), 9 tit. (τὴν διαίρεσιν ἐνιαυτοῦ), 11.2 (τὴν τῶν χρόνων διαίρεσιν), etc.

7 Also, both Valens and Firmicus introduce the meanings of the 12 astrological places – the "houses" in modern terminology – just before the list of distributions (Val. IV 12 / Firm. VI 32), which could imply that Critodemus treated them in this sequence, too. Nevertheless, Firmicus already explains the meaning of the houses in II 19.

8 Cottrell 2015, 362, casts doubt on the authenticity of book V of Dorotheus (extant only in Arabic translation) on the grounds that it mainly presents the theory of interrogations, influenced by Indian astrology. However, in this book we also find other material which is reproduced (and thus validated) by Hephaestio (fourth century AD), such as V 31 (cf. Heph. III 31).

that this was an allusion to the five wandering stars, the planets.[9] It is not far-fetched to suppose, then, that Critodemus wrote his treatise in seven books in allusion to the five planets plus the Sun and the Moon. This choice could have been reinforced by the Orphic associations of the number, especially the association between the strings of Orpheus' lyre and the seven planetary bodies.[10]

The marked character of Critodemus' seventh book was probably comparable to the case of Dorotheus, whose last book is conceived as a departure from the other four, in that it is concerned not with birth charts but with the so-called initiatives: that is, charts taken at the beginning of events such as diseases with the aim of making predictions about their evolution and outcome. Thus, in these cases, the last book probably represents the most original contribution of each astrologer, in contrast to the first books, which contained the more basic lore.[11]

Valens' oath witnesses

The table below lists the witnesses invoked in Valens' oaths:

F 6A	B	C
Starry vault of heaven	Circle of the Sun	Sun
Twelve-sign circle	Paths of the Moon	Moon
Sun	Powers of the other stars	Paths of the five stars
Moon	Circle of the 12 signs	Nature and Providence
The five planets		The four elements
Providence and Fate		

9 He adduces as supporting evidence the fact that in one of the main manuscripts containing the text of Ptolemy's *Tetrabiblos*, the title is given as τετράβιβλος καὶ ἕτερον βιβλίον τοῦ Ἀντιόχου θησαυρῶν, which presents as essentially the same treatise Ptolemy's four books plus a fifth book formed by the following part of the manuscript, which is ascribed to Antiochus: Hübner 1998, xxxvi–xxxvii.

10 Cf. schol. Verg. *Aen.* VI 119 *apud* Par. Lat. 7930 = *OF* 417: *si potuit Manes] dicunt tamen quidam liram Orphei cum VII cordis fuisse, et celum habet VII zonas, unde teologia assignatur. Varro autem dicit librum Orfei de vocanda anima Liram nominari, et negantur animae sine cithara posse ascendere.* Cf. also Lucian *De astr.* 10: ἡ δὲ λύρη ἑπτάμιτος ἐοῦσα τὴν τῶν κινεομένων ἀστέρων ἁρμονίην συνεβάλλετο. There were other Orphic associations of the number 7: several authors reveal that in the Orphic tradition Dionysus was dismembered in seven parts by the Titans (*OF* 34–35, 210), which could be related to an Orphic lamella from Olbia whose reverse shows a rectangle divided into seven parts: cf. *IOlb.* 94b Dubois (*OF* 464). Cf. Bernabé 2009b, 542.

11 The situation in Manilius is similar, in that he devotes his fifth and last book to the so-called *paranatellonta*, zodiacal and non-zodiacal stars that rise and set at the same time, after three books dealing with the zodiac. Manilius in fact begins the book by claiming that another astrologer would

The Sun and the Moon appear together in all three lists of witnesses. The five planets also appear in the three oaths. This does not seem surprising, since they represent the basic material from which astrological predictions are made. It seems natural that divine entities directly relevant to the matter at hand are the ones chosen as witnesses. For example, the *Hippocratic Oath* invokes four divine figures traditionally associated with medical practice: Apollo, Asclepius, Hygieia, and Panacea. Similarly, the Pythagoreans were known to swear by the *tetraktys* – the number 10 distributed in a progressive series of 1 + 2 + 3 + 4 tokens forming a triangle (*DK* 58 B 15).

It is noteworthy that the divine entities in Valens' oaths are sometimes substituted by their paths in the sky. This is the case of the Sun and the Moon in B, and of the planets in B and C. This is, however, easy to explain through the principle of literary variation. The fact that the Sun and the Moon appear in their simple form before and after (i.e., in A and C), and that the planets appear in their simple form first (in A), then substituted by their powers (B), and finally by their orbits (C), suggests that Valens wanted to avoid repetition. We should therefore not suppose that he intended to downplay their divine character.[12] In fact, Valens calls his witnesses "gods" when he points out the rewards that the oath-taker can expect if they do not break the oath (A: οἱ προειρημένοι θεοὶ πάντες εὐμενεῖς). The other essential object of astrological practice that Valens uses as a witness is the circle of the zodiac (A and B); also fundamental to astrology is Providence (A and C), which is first accompanied by Fate and then by Nature.

More striking is the presence of the four elements in C, which play no evident role in astrological theory or practice. However, they also appear in Firmicus' oath, which might indicate that they were also included by Critodemus. The four elements, together with the Sun and the Moon, also appear as oath witnesses in a quotation by the first-century Platonist Theon of Smyrna from the Orphic poem *Oaths* (*De utilitate* 104):

> ἔνιοι δέ φασιν ὀκτὼ τοὺς πάντων κρατοῦντας εἶναι θεούς, ὡς καὶ ἐν τοῖς Ὀρφικοῖς ὅρκοις ἔστιν εὑρεῖν·
>
> ναὶ μὴν ἀθανάτων γεννήτορας αἰὲν ἐόντων

have ended the treatise at that point (*hic alius finisset iter*, 1): see Hübner 2010, 9–12, for a summary of the book and *passim* for a detailed commentary thereof.

12 Cf. Murphy 2016 for the Socratic oaths "by the dog" and similar examples. According to Philostratus (*Vit. Ap.* VI 19), when Socrates swore by animals or the plane tree, he was following Rhadamantys, who was said to have recommended saving the gods for serious, judicial oaths. Murphy reviews other ancient interpretations of Socrates' unusual oaths, including Augustine's thesis that he sought to undermine Greek religion (*De ver. rel.* 2), Josephus' that he was joking (*Contra Ap.* II 263), and Porphyry's (*De abst.* III 16) and Lucian's (*Vit. auct.* 16) portraying Socrates as a believer in the holiness of animals.

> πῦρ καὶ ὕδωρ γαῖάν τε καὶ οὐρανὸν ἠδὲ σελήνην
> ἠέλιόν τε Φανῆ τε μέγαν καὶ νύκτα μέλαιναν.

> Some say that the number of gods governing everything is eight, as can be found in the Orphic *Oaths*:
> the parents of all immortal, ever-lasting beings
> fire and water and earth and heaven, as well as the Moon
> and the Sun and great Phanes and black Night.

These eight entities and their paired arrangement recall the four Hermopolitan cosmogonic pairs, each of which is formed of two consorts representative of the same primeval concept: Nu/Naunet (water), Huh/Hauhet (unlimitedness), Kek/Kauket (darkness), and Amun/Amaunet (hiddenness). Note that Phanes – an important god in the Orphic creation myths, also called Protogonus – is the consort of the Night. It is an open question whether the Egyptian myth influenced the Orphic poem, but, considering another three lines of the same poem quoted by Pseudo-Justin and their invocation of the world-creating voice of a father (*Cohortatio* p. 16 Morel = *OF* 620), which is believed to be foreign to traditional Greek religion and instead reminiscent of Egyptian or Egyptianizing texts such as the proem of Nechepsos' astrological poem, a plausible case can be made:[13]

> οὐρανὸν ὁρκίζω σε, θεοῦ μεγάλου σοφὸν ἔργον,
> αὐδὴν ὁρκίζω σε πατρός, τὴν φθέγξατο πρῶτον,
> ἡνίκα κόσμον ἅπαντα ἑαῖς στηρίξατο βουλαῖς.
>
> I swear upon you, heaven, wise work of the great God.
> I swear upon you, father's voice, which he first sang
> when he fixed the whole cosmos with his will.[14]

A creator god accompanied by the Hermopolitan Ogdoad is invoked in a magical papyrus;[15] an ensemble of 1 + 8 divine entities like this seems to be what the Orphic *Oaths* propounded. Was this text, then, an Egyptian creation? If this were the case,

---

13 Cf. Chapter 2 for the creative shout in Nechepsos-Petosiris fr. 1 Riess. Scholars have expressed doubts about the authenticity of the second fragment of the Orphic *Oaths*, adducing that it could have been a Jewish forgery. However, as Herrero de Jáuregui notes, the fact that the word αὐδή is never used for the divine creative voice in the Jewish-Christian literature, as well as the Hermetic parallels, suggests an Egyptian influence and that the fragment is not a forgery. Cf. Herrero de Jáuregui 2010, 197, with references.

14 Trans. Herrero de Jáuregui.

15 The Ogdoad appears by this name (Ὀγδοάς) in a section of *PMag.* XIII: cf. Merkelbach and Totti 1990 I, 208; Herrero de Jáuregui 2010, 59.

and if Critodemus had visited Egypt (cf. Chapter 2), he would perhaps have read it there.

Firmicus' witnesses

Let us now discuss the oath of Firmicus (F 6D), which, although it mentions most of Valens' witnesses, presents a wholly different structure. Firmicus demands that his readers take an oath by three entities, the first of which is a demiurge who is praised on account of his many creations and powers. The sheer length of this invocation transforms the first part of the oath into a sort of aretalogical hymn. Below is a schematic representation:

God, creator of the Universe, who:
- made and regulated everything according to Necessity
- shaped the Sun and the Moon
- ordered courses of all the stars
- limited the flux of the sea with land
- kindles the fire of divine eternity
- maintains the earth balanced in the middle of the universe
- created men, beasts, birds, and life in general
- moistens the earth with fountains
- controls the winds
- created all things out of the 4 elements
- causes rising and setting (of the stars)
- causes all earthquakes

the descent and ascent of the soul
the immortal order of eternal perpetuity

In Firmicus' oath, many elements relevant to astrological practice that Valens treats directly as witnesses appear as particular creations of the demiurge. There is significant overlap: Sun and Moon, the other stars, the four elements, and perhaps Nature, which is not named directly but rather alluded to as "men, beasts, birds, and life". As in Valens' last oath, the natural world appears just before the four elements, probably on account of their role as founding blocks of nature, as underlined by Firmicus (*ad fabricationem omnium quattuor elementorum diversitate conposita, ex contrariis et repugnantibus cuncta perfecit*), who makes a smooth transition mentioning the earth with its fountains (water) and the winds (air) in between.

Whereas Valens' oaths have a general Stoic flavor, especially palpable in his invocation of Providence (Πρόνοια), Fate (Μοῖρα), and Nature (Φύσις), the three separately invoked entities in Firmicus – the creator, the ascent and descent of the soul, and the immortal order – correspond loosely to principles operative in late Platonic metaphysics. It is true that the kindling of the eternal fire by the creator still recalls the Stoic creative fire, but fire performs a similar function in the *Chaldean Oracles*,[16] a collection of hexameter verses purportedly handed down through revelation to a certain Julian the Chaldean in the late second century AD, which was fundamental in the metaphysical theorization of Neoplatonists such as Porphyry or Proclus.

Porphyry is referred to by Firmicus in the preface of his book VII as one of the ancient authorities for whom, Firmicus claims, secrecy was important. It is remarkable that Firmicus calls him *Porphyrius noster*, revealing himself to be a follower of the Tyrian philosopher. Firmicus later converted to Christianism and disparaged Greek religion and Greek philosophers, writing the anti-pagan treatise *On the Error of Pagan Religions*, in which he did not even spare Porphyry, citing his work *Philosophy from the Oracles* (13.4 = Porph. fr. 306 Smith). Porphyry's *Philosophy from the Oracles* contained warnings against wide distribution because of the numerous oracles it contained (Porphyry fr. 304 Smith). It is perhaps on the basis of Porphyry's admonition in this work that Firmicus included him in his list of authorities who demanded religious secret.[17]

It is easy to imagine that Firmicus perceived a strong link between the practice of astrology, of well-known Babylonian origin, and the *Chaldean Oracles* while he was still a Porphyrian Neoplatonist. Indeed, his invocations seem to reveal a combination of first principles related to astrology (Necessity, Sun and Moon, stars, and the Earth as a central astronomical point) and a triadic metaphysical framework which to some extent reflects that of the *Chaldean Oracles*. If the *Chaldean Oracles* posit a supreme, completely transcendent, and self-contemplating God, the so-called Father, Firmicus' god is a direct creator of Nature, like the god of the Orphic poem *Oaths* in the Pseudo-Justin fragment. The second entity that Firmicus invokes is "the ascent and descent of the soul" (*per descensum ascensumque animae*), which recalls the dynamic character of Hecate in the *Chaldean Oracles*: she is located between the intelligible and sensible realms, identified with the world-soul, and sometimes called a membrane that facilitates travel from one realm to the other.[18] The other basic metaphysical principle in the *Chaldean Oracles* is a demiurgic *Nous* ("Intellect") that imposes the Platonic Forms upon matter,[19] which is comparable to

---

16 Cf., e.g., frs. 32, 34–37 Majercik.

17 The reference may also be more general, since Porphyry also used a well-known Orphic formula demanding secrecy at the beginning of his *Imagenes* (ap. Euseb. *Praep. ev.* III 7.1 = Porph. fr. 44 Smith): φθέγξομαι οἷς θέμις ἐστί, θύρας δ' ἐπίθεσθε βέβηλοι (cf. below in this chapter on this formula).

18 Cf. frs. 6, 51–53, 56, and Majecik 1989, 7.

19 Cf. frs. 5, 33, 37, and Majecik 1989, 6.

Firmicus' "immortal order of eternal perpetuity" (*immortalem aeternae perpetuitatis ordinem*).

It can appear puzzling that Firmicus' second witness is defined as a movement rather than as a stable entity, but this has a clear parallel in an oath of clear Orphic resonances that is found with minor variations in two third-century papyri (*OF* 621) (I print the most complete version, in *Pap. Soc. Ital.* X 1162):[20]

> [Ὀμνύω κατὰ τοῦ διχάσ]αντος γῆν ἀπ' οὐρανοῦ
> [καὶ σκότος ἀπὸ φωτὸς κ]αὶ ἡμέραν ἐγ νυκτὸς
> [καὶ ἀνατολὴν ἀπὸ δύσ]εως καὶ ζωὴν ἀπὸ θα-
> [νάτου καὶ γένεσιν ἀπὸ] φθορᾶς· ἐπόμνυμαι
> [δὲ καὶ οὓς προσκυνῶ θεούς] συντηρήσειν
> [καὶ φυλάξειν τὰ παραδ]εδομένα μοι μυστή-
> [ρια (. . .)

> I swear by the separation of the earth from heaven, and of darkness from light, and day from night, and rising from setting, and life from death, and birth from death; I swear by the gods which I obey to observe and guard the mysteries that have been given to me (. . .)

Again, the coincidences with Theon of Smyrna's hexameters from the Orphic *Oaths* are inescapable: the oath-taker swears by the same type of pairs of primeval deities, and in particular the pairs earth/heaven and day/night are found in both texts. It differs from Theon's lines only in the fact that it is not the divine entities themselves or their powers that bear witness to the oath, but the very separation of these entities. A similar dynamic character can be observed in some of the creations of Firmicus' demiurge, such as the limit that the motion of the seawaters imposes on land, and the separation of the contrary elements. A striking parallel with the text of this oath on papyrus is the pair rising/setting (of the stars).

Petitions in Valens' oaths

Valens' principal demand is that his text be treated carefully, but this takes different forms. Invariably, in the three oaths, Valens first asks his reader to keep the astrological doctrines secret from the ignorant, and then to honor their teacher for transmitting them. In the first oath (F 6A), Valens concentrates on the copyright conditions:

(1) Do not put another's name to this compendium.

---

20 Cf. Herrero de Jáuregui 2010, 59 n.69. Herrero de Jáuregui argues that the existence of the two slightly different versions seems to imply that the text enjoyed a relatively wide distribution.

(2) Do not blot out anything, which would result in confusion for the readers and damage to Valens' reputation.

(3) (As an afterthought, after the specification of the rewards and punishments in relation with keeping or breaking the oath.) If you find these doctrines attributed to someone else somewhere, do not praise the alleged author, but Valens.

From our perspective, we can see that Valens was right to fear the anonymous distribution of his doctrines, since many chapters of his treatise were indeed translated into Latin and attributed to Hermes Trismegistus.[21] In the second oath (F 6B), Valens claims that his reader will experience the same joy as someone who has climbed a mountain with great toil, but who is happy to have found treasure at its peak. In speaking about this treasure, Valens misquotes part of a line from the *Odyssey* (IV 73), a very apt erudite illustration of his words which he no doubt expected his readers to recognize ("gold and silver and ivory", χρυσοῦ τε καὶ ἀργύρου ἠδ' ἐλέφαντος). The Homeric context is relevant: as part of the search for his father, Telemachus, accompanied by his friend Peisistratus, visits the palace of Menelaus in Sparta, where a luxurious banquet is taking place to celebrate the sending off of Menelaus' daughter as a bride to Achilles' son. First, the poet says that Telemachus and Peisistratus admired the shining beauty of the palace, which he describes as "a ray of light from the Sun or the Moon" (45: ὥς τε γὰρ ἠελίου αἴγλη πέλεν ἠὲ σελήνης). This astronomical reference had probably caught Valens' attention. Then Telemachus, talking to his friend, remarks on the beauty of the bronze in the hall, as well as the amber, the gold, the silver, and the ivory, and compares it to the palaces of Zeus (71–74). Menelaus overhears him and replies that he has no pleasure in such things; that he has acquired them through hardship in his wanderings, but that he would have preferred to possess only one third of these treasures and instead have been in Sparta to preserve the life of his brother Agamemnon, who was killed by his wife Clytemnestra in his absence (78–112). Valens clearly identifies himself with Menelaus, establishing a sharp contrast between material richness and astrological knowledge. Like the Homeric hero, Valens has traveled in many lands and experienced hardship, as he himself has mentioned, imitating the opening lines of Critodemus' poem (cf. Chapter 2), but instead of precious objects, he has gathered the valuable astrological doctrines which he now reveals to his audience.

In his last oath (F 6C), Valens repeats the same idea in a more direct way, explaining that he has toiled for a long time; this could have been spent accumulating riches, but he chose instead to compose this precious work. He adds one last jus-

21 Pingree 1978, 432–433.

tification for the demand of secrecy, involving his dedicatee in the project: Valens encourages Marcus, who is now addressed by name, to think of the treatise as if he were the author, and even corrects himself, arguing that Marcus is in fact an essential piece of it, since he has also studied so much as he has made his way through the treatise. It will therefore be easy for him, Valens argues, to understand that such an important treatise cannot merely be shared with any casual acquaintance and must only be passed on to trusted friends.

Content of Firmicus' oath

If in Valens we find detailed specifications regarding the handling of the text (F 6A), as well as protracted justifications of the need for such careful treatment, adducing hard work on the part of the astrologer (B and C) and even of his reader (C), Firmicus simply begs his dedicatee Mavortius to swear that his treatise will not be revealed to profane ears (*profanis vel inperitis auribus*). This is an echo of an old Orphic formula in which the uninitiated are asked to stop reading, invoking the image of shutting the doors, with the ears coming to substitute the doors in the metaphor at some point, as in this passage of Plato's *Symposium* (218b):

> πάντες γὰρ κεκοινωνήκατε τῆς φιλοσόφου μανίας τε καὶ βακχείας – διὸ πάντες ἀκούσεσθε· συγγνώσεσθε γὰρ τοῖς τε τότε πραχθεῖσι καὶ τοῖς νῦν λεγομένοις. οἱ δὲ οἰκέται, καὶ εἴ τις ἄλλος ἐστὶν βέβηλός τε καὶ ἄγροικος, πύλας πάνυ μεγάλας τοῖς ὠσὶν ἐπίθεσθε.
>
> and all the rest, have all shared in the madness, the Bacchic frenzy of philosophy. And that's why you will hear the rest of my story; you will understand and forgive both what I did then and what I say now. As for the house slaves and for anyone else who is not an initiate, my story's not for you: block your ears![22]

Many similar instances survive in the ancient literature, invoking the Orphic formula to signal especially sensitive content.[23] Although the formula ultimately became a sort of cliché, the Orphic connection did not fall into oblivion, as is clear in both Plato's and Firmicus' reproductions.

Firmicus is somewhat more prolix than Valens in his specifications regarding the appropriate receivers of his doctrines. Whereas Valens simply indicates that

---

22 Trans. Nehamas and Woodruff.

23 See the collection of passages in Bernabé 1996, 19–31; I cite only those explicitly referring to profane ears: Dionys. Hal. *De comp. verb.* 6, 25, Ael. Arist. *Or.* 3.50, Galen *De usu part.* XII 6, Euseb. *Laud. Const.* proem 4, Greg. Naz. *Carm. de se ipso* Patr. gr. XXXVII 1367 Migne. Orphic verses mentioning ears are quoted in Elias *in Aristot. Cat.* proem 2.

the work should not be shared with a chance acquaintance, and this only in the last oath (F 6C), Firmicus establishes three conditions of purity for these ideal persons:

(1) An uncorrupted reason (*animus*) which has led them into the right path of life (*ordo vivendi*) with the chaste and pure assistance of their mind (*mens*)
(2) A loyalty (*fides*) beyond reproach
(3) Hands free of all crime

As in his list of witnesses, we can perceive the influence of Neoplatonism. If the third point is a well-known traditional condition of purity for participation in cultic activities, the first condition expresses a general Neoplatonic ideal of salvation through contemplative life, and the second (*fides*, πίστις) is the first element of the triad of virtues in the *Chaldean Oracles*.[24]

Rewards and punishments in Valens' oaths

Ancient Greek and Roman oaths in their most developed form often incorporated at the end a brief account of the rewards that oath-keepers could expect, and of the punishments that would befall oath-breakers. That oath-keeping and oath-breaking were deeply-ingrained categories in ancient society is witnessed by their very treatment in the astrological treatises, in which, for example, the influence of certain planets at birth is said to produce oath-breakers.[25] Punishments for oath-breakers are attested from the most ancient times in the oath ceremonies described in the Homeric epic.[26]

It is remarkable that Firmicus does not make any reference to rewards or punishments. He obviously chooses to focus on the philosophical questions related to the nature of the witnesses and on the ethical qualities of the reader. By contrast, Valens does not shy away from expressing his good wishes toward those who keep the oath and, most significantly, from listing a series of evils that will befall oath-breakers:

24 Fr. 46 Majercik. Cf. Brisson 2015, 122.

25 For example, Mars indicates false oaths in Val. I 1.22; Saturn, Mars, and Mercury together produce oath-breakers in Val. I 20.9; beginning an activity when the Moon is at a node (i.e., passing through the ecliptic) indicates that the oaths will not be fulfilled in Val. V 3.25. Cf. similarly Ptol. *Tetr.* III 14.15, 30, 35.

26 See Berti 2006.

F 6A	B	C
Rewards		
May gods be well disposed: - Prosperous life - Fulfillment of wishes	May it go well: - Gods according to wishes	Fulfillment of wishes
Punishments		
The opposite: - Earth not passable - Sea not sailable - No offspring - Blind minds - Shameful/unsuccessful lives	The opposite	

Probably to avoid repetition, Valens summarizes the rewards of the first oath in the subsequent oaths, using more general expressions. Thus, in the second oath he seems to combine the blessings concerning divine disposition, and he paraphrases the first oath's desire for a prosperous life (βίος εὐσταθής) with the common formula εὖ εἴη ("may it go well"). A similar pattern is observed in the punishments, which are summed up in the common formula τὰ ἐναντία ("the opposite") in the second oath and disappear completely in the third. Both abbreviated formulas for rewards and punishments were also often used to substitute concrete lists of rewards and punishments in civic oaths,[27] and appear, for example, in the Orphic oath *OF* 621 (I 10–11).

Valens qualifies Critodemus' oaths as "frightful" (φρικωδέστατοι) in F 3A, above. We should therefore probably expect that Critodemus threatened his readers in a similar way to Valens in his section on punishments. Comparison with the punishments announced in Greek civic oaths reveals common use of the curse of infertility, as well as the godly induced *mania*, but the civic oaths usually go much further, threatening oath-breakers with utter destruction. The other punishments in F 6A, and especially the first two concerning the transitability of the Earth and sea, were perhaps taken from Critodemus. As we have seen, Valens often touches upon the topic of travel, taking his cue from the first lines of Critodemus, which refer to travel by sea and by land (F 3).

As for the blessings for the oath-keeper, it is noteworthy that in all three oaths Valens mentions the fulfillment of wishes, using the word καταθύμιος ("according to

27 Cf. Berti 2006, and the Cretan inscriptions *IC* III 4, 8 (Itanos) and *IC* I 9, 1 (Dreros), featuring lists of rewards and curses. They are good illustrations of the meaning of the formula "the opposite", since in both cases the blessings (offspring of children, crops, and farm animals) are explicitly negated in the section of the curses (lines 44–49/78–89).

one's wishes") in all three and συντέλεια ("fulfillment") in the first and third. This specific blessing, at least expressed in these terms, is not typical of other oaths. It derives in all likelihood from Critodemus himself, since Valens uses both words, either together or separately, mostly in passages related in one way or another to the *Horasis*.[28]

Relationship with the Hippocratic oath

The date and nature of the Hippocratic *Oath* document are controversial, but a *terminus ante quem* can be posited from an allusion by Cato the Censor (*apud* Pliny XXIX 7.14 = Cato *Libri ad M. filium*, fr. 1 Jordan).[29] Since Cato wrote in the first half of the second century BC, it is pertinent to inquire whether Critodemus took into consideration the Hippocratic *Oath* when composing his own. The probability that Critodemus knew about the medical oath would in theory seem to be higher if Critodemus was active on the island of Rhodes (cf. Introduction). It might also be tempting to argue, in light of the general affinity between astrology and medicine, that such an influence would be natural. However, on close inspection we find that there is no evidence to support this possibility. It is difficult to find any specific similarities between the Hippocratic *Oath* and the astrological oaths, in either content or vocabulary, that go beyond general characteristics of ancient Greek oaths. To begin with, the main content of astrological oaths – that is, that which is sworn – is basically self-restraint with regard to the dissemination of the imparted doctrines, which is an exceedingly common aim in ancient oaths.[30] The Hippocratic *Oath* reproduces this structure in a more convoluted form and accompanied by other pledges. After the invocation of the witnesses, the Hippocratic oath-taker swears to consider their teacher as a father, and to impart the medical doctrines to their teacher's sons, if they wish to learn them, in the same way as they would do with their own sons, but to no one else (ἄλλῳ δὲ οὐδενί) except students who have already taken the oath. A little further on in the text, the issue of secrecy appears again, when the oath-taker

28 συντέλεια: IV 18.6; IV 21.9; IV 25.1 VI 6.9 (all passages relative to the doctrine of the transmissions or distributions, cf. above and Chapter 4) / καταθύμιος: III 10.19 (Valens confesses to write mystically); IV 12.5 (on the 12 places in relation to the transmissions); IV 16.19 (on the transmissions); IV 25.3 (on the transmissions of the four lots).

29 Cf. Tolsa 2019a, *pace* the list of ancient witnesses of the *Oath* in Jouanna 2017, 30, where he classifies the passage under the heading "Élimination de faux témoignages", adducing Plutarch's explanation in *Vit. Cat.* 23.4. Plutarch suggests that the Roman censor knew about the famous story according to which Hippocrates refused to help the Persian king for patriotic reasons, but Cato's wording implying treachery rather points to the narrative in the pseudo-Hippocratic *Embassy*.

30 Like, e.g., the Isiac oath: *PSI* 1290 (first century AD) and 1162 (third century AD). Cf. Alvar 2008, 177.

swears not to divulge anything that they hear in the house of their patients, which they will safeguard as "sacred mysteries" (ἄῤῥητα), in a clear allusion to religious oaths of secrecy.

This is related to another important difference between the Hippocratic *Oath* and the astrological oaths deriving from Critodemus: the former is a separate document that purportedly establishes rules for medical practice, whereas the astrological oaths concern the very text in which they are inserted, and are therefore akin in this respect to the Orphic formulas which simply entreat impure readers to stop reading. The Hippocratic *Oath* is, at least at face value, addressed to prospective physicians in a context of direct, oral teaching, hence the recursivity in the condition that outside of the teacher's and the oath-taker's families the oath-taker can only impart instruction to indentured students (through this very oath). In this way, the regulations contained within the Hippocratic *Oath* would be imposed on an entire medical community, composed of an aggregation of medical families. In contrast with this, the astrological texts in which these oaths are inserted were written for independent circulation outside of an oral community, even if Valens sometimes refers to his own students. This situation is comparable to the context of Greek mathematical treatises, a genre which developed a very specific device in order to function independently of teaching, namely, the formulaic language referring to lettered diagrams typical of this kind of work.[31]

The distinct contexts of medicine and astrology affect the wording of the oaths, in that the Hippocratic oath-taker is, so to speak, already conceptualized as a physician integrated into a community, whereas the astrological oaths generally regard the readers as distant laymen, even if they were probably astrologers themselves. We may note the coinciding reference to reputation in the Hippocratic *Oath* (δοξαζομένῳ παρὰ πᾶσιν ἀνθρώποις ἐς τὸν αἰεὶ χρόνον) and in Valens' astrological oaths, but Valens refers mostly to his own renown, which he predicates on the reader's correct handling of the material, whereas in the case of the Hippocratic *Oath* the oath-takers wish an excellent reputation for themselves as a reward for keeping the oath. Thus, in the rewards section of the astrological oaths, Valens' emphasis is not on professional success but more generally on good life, similarly to what we find in civic oaths. It is only in the third oath (F 6C), which is explicitly requested of his student Marcus, that Valens refers to the oath-taker's career as an astrologer.

We should, then, probably rule out any particular influence of the Hippocratic *Oath* on the astrological oaths of Critodemus and his followers.

---

31 For a thorough analysis of the language of Greek mathematics, see Netz 1999. For a comparison of the contexts and characteristics of Greek mathematical treatises with the non-elite mathematical practices, see Asper 2009.

Second part: **Technical fragments**

Chapter 4
The “distributions”: Critodemus’ innovation? (F 7–9)

A doctrine called “distributions” (ἐπιμερισμοί) or “transmissions” (παραδόσεις) by Valens was probably introduced by an oath of secrecy in book VII of Critodemus (cf. Chapter 3). Similarly to Dorotheus and Manilius, who divided their astrological treatises into five books in allusion to the five wandering stars, Critodemus probably alluded to the greater set including the two luminaries. Like these early astrologer-poets, he seems to have reserved the last book for a more personal contribution, conceiving the treatise as a symbolic ritual in which the most sacred part comes at the end.

Critodemus’ doctrine purports to give meaning to the periods of time in a person’s life, which are conceived as ruled in succession by the seven celestial bodies. It is possible that the doctrine was influenced by the so-called *Dodecaeterides* attributed to Orpheus, astrological prognostications related to weather phenomena and agricultural productivity associated with a 12-year period.

The lengthier expositions of the distributions are in Valens IV 17–18/20–24, Firmicus VI 33–36, and Hephaestio II 29–36, but they do not mention Critodemus. The brief summary of the so-called *Epitome Parisina* and a version of the doctrine from Valens’ *additamenta antiqua* do mention Critodemus as the source, but only the latter will be included as a fragment (F 7), since, as we shall see, the summary was likely produced from the expositions in Valens and in the *additamenta*.

I include two other fragments from Valens: one (F 8) briefly mentioning the doctrine and ascribing it to our author, and the other (F 9) alluding to a theory of Critodemus on the transits of the stars (their positions in horoscope recastings) as enhancers of the effects of the changing rulerships in the distributions, which Valens seems to develop in his own way. The explanation of this theory includes a horoscope corresponding to Nero, including his death chart from AD 68, which for reasons of date cannot be the original work of Critodemus.

Astrological significations of the planets and the distributions

In Critodemus’ system, the time of life is divided so that each star (I will use this term to refer to any of the five planets or the luminaries) is allotted 10 years and 9 months. In turn, when each of them is the ruler of time, it “distributes” its rulership to all the stars in sequence, thus forming two levels of time rulerships or *chronocrators*. So, for example, for a person’s first 20 years we might have:

https://doi.org/10.1515/9783111329147-005

Birth →							10 years 9 months		
Sun rules for 129 months (=10 years 9 months)							Saturn rules for 129 months		
Sun 19	Saturn 30	Jupiter 12	Mars 15	Venus 8	Mercury 20	Moon 25	Saturn 30	Jupiter 12	. . .

The 10 years and 9 months of all general rulerships correspond to the sum of the subrulerships. The number of months corresponding to the subrulers is always the same for each star, as in this example. Also, the order of the general rulerships is always the same as that of the subrulerships, and the sequence of subrulerships always begins with the star holding the general rulership. The order of the general sequence is established as follows: individuals born by day begin with the Sun, and those born by night with the Moon; thereafter, the sequence follows the stars found along the zodiac in the birth chart, in the order of the signs, after the Sun or the Moon, respectively.

In the main fragment (F 7), the effects of the stars are listed, often consisting of an amalgam of the properties of the general and the particular ruler. As Hephaestio puts it at the end of his own exposition of the transmissions (II 36.25):

> Ἔστι διὰ μνήμης πως καὶ ταῦτα τὰ ἀποτελέσματα ἔχειν σκοποῦντα ἐκ τῆς φύσεως τοῦ τε μερίζοντος ἀστέρος καὶ τοῦ λαμβάνοντος, καὶ κατὰ τὸν αὐτὸν τρόπον ταῖς ἐπεμβάσεσι σχεδὸν εὑρήσεις τὰ αὐτὰ ἀποτελέσματα.
>
> It is possible, for the sake of memorization, to obtain these effects by considering the nature of the distributing star and the receiving, and in this way you will find almost the same effects in the ingresses.

A long, complete list of such attributes for each star is found at the very beginning of Valens' treatise (I 1), briefly summarized below (see also Appendix II):

Saturn: Deceit, lawsuits, imprisonment, high ranks, injuries from cold and moisture
Jupiter: Ranks, offices, friendship with great men, childbearing
Mars: Harm, violence, diseases
Sun: Light, the father[1]
Venus: Generally favorable, women affairs, friendship
Mercury: Brothers and younger children[2], writing and documents
Moon: The mother

1 Ptolemy *Tetr.* III 5 says that the Sun and Jupiter are naturally associated with the father, in the same way as the Moon and Venus with the mother.

2 This field of action of Mercury could be quite specific to Critodemus, since it only appears in Valens, who, as we know, frequently resorted to our author.

Thus, for example, we read that when Venus distributes the rulership to Mars, she is "bad; anxiety from women, attacks from women, adulteries, hatreds, hemorrhages, unpleasantness" (χαλεπή· διὰ γυναῖκας μερίμνας, ἐκ γυναικῶν ἐπιθέσεις, μοιχείας, ἐχθρασμούς, αἱμορροίας, ἀηδίας). The effects apparently mix up the negative character of Mars with the feminine characteristics of Venus. Similarly, when Mars transmits the rulership to Venus, he is "good and favorable, but blaming for female-related things" (ἀγαθὸς μὲν καὶ πρακτικός, ἐπίψογος δὲ εἰς τὰ περὶ γυναῖκα). Nevertheless, at some point, Hephaestio implies that the distributer is more influential at the beginning of the rulership, its effects disappearing later on, as in the distribution from Mars to Jupiter (II 32.13):

> ἐπὶ δὲ νυκτερινῆς γενέσεως ἐν ἀρχῇ ἔχθρας καὶ θορύβους καὶ ἀπραγίας καὶ ψύξεις, ὕστερον δὲ περιγενήσεται τῶν φαύλων καὶ συστάσεις καὶ κέρδη καὶ φιλίας ἕξουσιν.
>
> In a nocturnal chart at the beginning [of the rulership]: hatreds, tumults, inaction, chillings; then the bad effects will be superseded and there will be associations and gain and friendships.

Often enough, rather than a mixture of both characters, the effect of a particular distribution is defined as the effect of one of the two stars in the field of action represented by the other: thus, the effect of the distribution of Saturn to the Sun is harm (Saturn is negative) to the father (a field of action of the Sun). Similarly, Mars to Venus implies violence in relation to females, Mars to Mercury indicates attacks using legal documents, etc.

The only significant exception to the traditional attributes of the stars concerns the luminaries. Whereas their fields of action (the father and the mother, respectively) are used as expected, they generally have negative effects. Is it possible that Critodemus pictured them in this way because they govern the first stage of life, which was especially dangerous?

F 7. Valens *Additamenta* 4

Type of fragment: List of effects of the distributions. The two stars in each of the predictions in the list should be interpreted as the general ruler and the subruler (see discussion below). This highly standardized text is found among the materials appended to the manual of Valens, the *Additamenta antiqua.*

Ἀστέρων μεταπαραδόσεις

1 Κρόνου ἐπιμερισμός. Κρόνος ἑαυτῷ παραδιδοὺς βλαβερὸς καὶ ἄπρακτος. **2** Κρόνος Διὶ παραδιδοὺς ἀγαθός· ἐξ ἐγγαίων καὶ θεμελίων κτήσεις καὶ κληρονομίας, καὶ ὑπερεχόντων φιλίας. **3** Κρόνος Ἄρει ἂν οἰκείως σχηματισθῇ, ἀγαθός· εἰ δὲ μή, πένθη, νόσους, ζημίας. ὅλος ὁ ἐνιαυτὸς καταράσιμος. **4** Κρόνος Ἡλίῳ δίκας ἐγγαίων χάριν καὶ πατρὸς κινδύνους καὶ ζημίας καὶ τοὺς ἐνιαυτοὺς ἐπιβλαβεῖς μᾶλλον. **5** Κρόνος Ἀφροδίτῃ ἐὰν μὲν οἰκείως, γυναικῶν ψυγμοὺς ἢ ἀηδίας ἢ καὶ πένθη θηλυκῶν προσώπων· οὐ μέντοι γε ἐν τοῖς κατὰ βίον πονηρὸς ὑπάρχει. **6** Κρόνος Ἑρμῇ πράγματα χρόνια καὶ ἐκ παλαιοῦ καὶ καθύγρους ἀσθενείας καὶ ἐξ ὑποτεταγμένων βλάβας· οὐ μέντοι κατὰ πᾶν ἀδικουμένους. 7 Κρόνος Σελήνῃ νόσους καὶ ὑπὸ ὑγρῶν[a] ὀχλήσεις καὶ μητέρων κινδύνους καὶ πτώσεις ἀπεργάζεται· πάντα δὲ πρὸς τὸ κατὰ γένεσιν σκόπει σχῆμα.

8 Διὸς ἐπιμερισμός. Ζεὺς ἑαυτῷ ἀγαθός. **9** Ζεὺς Κρόνῳ παραδιδοὺς ἐπιβλαβής, ἄπρακτος, ὁτὲ δὲ καὶ τέκνων κινδύνους. **10** Ζεὺς Ἄρει βλάβας, ἀναστασίας, νόσους, ἔσθ' ὅτε καὶ συνοχῆς πεῖραν καὶ τέκνων θανατηφόρους κινδύνους. **11** Ζεὺς Ἡλίῳ· τὰ μὲν κρύφιμα πάντα φωτίζει· πρακτικὸς δὲ καὶ περὶ πατέρα δοξαστικὸς καὶ πρὸς ὑπερέχοντας συστατικός· ὁτὲ καὶ τέκνων σπορὰς ποιεῖ. **12** Ζεὺς Ἀφροδίτῃ νυκτὸς μὲν ἀγαθός, ἡμέρας δὲ μέτριος· ἀγαθοὺς μέντοι γε τοὺς ἐνιαυτοὺς καὶ ἐπικερδεῖς καὶ καταγραφὰς ἔχοντας· ποτὲ δὲ καὶ αἱ γυναῖκες ἔγκυοι γίνονται. **13** Ζεὺς Ἑρμῇ πρακτικός, ἐπικερδής, πρὸς φίλους χρηματιστικός. **14** Ζεὺς Σελήνῃ εὔδιος, ἐκ πραγμάτων ῥύεται καὶ κινδύνων, πρακτικούς τε ποιεῖ καὶ ἐπικερδεῖς καὶ ἐκ θηλυκῶν εἰδῶν προβιβασμούς, καὶ μητρὸς δοξαστικός τε καὶ τέκνων σπορὰς ποιῶν.

15 Ἄρεως ἐπιμερισμός. Ἄρης ἑαυτῷ ἐπιμερίσας θορύβους καὶ πράξεις μετὰ ταραχῆς. **16** Ἄρης Κρόνῳ ἀπραξίας, δυσαρεστησίας, καταψύχεται, τῶν πράξεων ζημίας, πένθη, νόσους. **17** Ἄρης Διὶ πρακτικός, ἐπικερδής, πρὸς ὑπεροχὰς καλός, δυναμικός. **19** Ἄρης Ἡλίῳ οὐκ ἀγαθός, κλιμακτηρικός, ταρακτικός, θορυβώδης, ὀφθαλμίας, σκυλμοὺς ποιῶν ἐπισινεῖς ἢ καὶ πυρὶ ἢ σιδήρῳ πειραζομένους καὶ πατρὸς κινδύνους ἰσοθανάτους, χόλους τε ὑπερεχόντων ἢ καὶ νεύρων ἀσθενείας ἢ ὀστέων κατεάξεις. **19** Ἄρης Ἀφροδίτῃ ἀγαθὸς μὲν καὶ πρακτικός, ἐπίψογος δὲ εἰς τὰ περὶ γυναῖκα· ποιεῖ γὰρ καὶ χωρισμοὺς καὶ μάχας καὶ εἰσπρώσεις καὶ αἱμαγμοὺς καὶ ἐκ γυναικῶν ἐπιθέσεις καὶ μοιχείας· οὐ μέντοι γε ἄπρακτος ὁ ἐνιαυτός. **20** Ἄρης Ἑρμῇ συκοφαντίας, ἐκ γραπτῶν δόλους καὶ ἐπιθέσεις, κλοπάς· ἐν μέντοι κριτηρίοις οὐ κακός· πάντα δὲ πρὸς τὸ κατὰ γένεσιν ὅρα σχῆμα. **21** Ἄρης Σελήνῃ χαλεπὸς ἐν πᾶσιν, συνοχάς, κρυβάς, φυγάς, βλάβας, νόσους ἐπιφέρων καὶ μητράσι κινδύνους καὶ ἐπὶ θηλυκοῖς εἴδεσι λύπας.

22 Ἡλίου ἐπιμερισμός. Ἥλιος ἑαυτῷ παραδιδοὺς οὐκ ἀγαθός, σκυλτικός, μεριμνητικός, νυκτὸς δὲ χείρων. **23** Ἥλιος Κρόνῳ χαλεπὸς ἐπὶ νυκτερινῆς γενέσεως, ἐπιβλαβής· καὶ πατρὸς ζημίας· ἐπὶ δὲ ἡμερινῆς γενέσεως μέτριος, ἐκτὸς εἰ μὴ πρὸς ἀλλήλους ἀνοικείως σχηματίζοιντο. **24** Ἥλιος Διὶ ἀγαθός, πρακτικός, δοξαστικός, πρὸς ὑπερέχοντας συστατικός, προσλαμπής, τέκνων σπορὰς διδούς. **25** Ἥλιος Ἄρει δίκας, ἐκπτώσεις, κλιμακτῆρας, ὀφθαλμοῖς πυρίκαυτα, νόσους, κινδύνους, ὑπερεχόντων χόλους, νεύρων ἀσθενείας, ὀστέων κλάσεις. **26** Ἥλιος Ἀφροδίτῃ

a **S**: ὑγρανὸς **V** ὑγράνσεως Pingree

πρακτικὸς μὲν ὑπάρχει, γυναικῶν δὲ ποιεῖ ἐπαναστάσεις πονηρὰς καὶ θορύβους ἐν ὄχλοις καὶ κρίσεις· ἐὰν καλῶς ἥ τε Ἀφροδίτη καὶ ὁ Ἥλιος σχηματίζωνται τῇ γενέσει, πρακτικοὺς ἐνιαυτοὺς καὶ ἐπαφροδίτους· καὶ γὰρ ἐπιπλέκονται γάμοις ἢ γυναιξίν· τὸ δὲ ὅλον καλοὶ οἱ ἐνιαυτοὶ καὶ ἐκ θηλυκῶν προβεβιβασμένοι· ἢ διὰ φίλων εὐτυχίας. **27** Ἥλιος Ἑρμῇ πράξεις καὶ κέρδη ποιεῖ, σὺν μέντοι φόβοις καὶ μόχθοις· οὐ μετρίους, ἐπινοηματικοὺς δὲ καὶ ποριστικοὺς τοὺς ἐνιαυτούς. **28** Ἥλιος Σελήνῃ ἀνωμαλίας, ἀστασίας, μετεωρισμούς, σκυλμούς, ἀλλ' ἐὰν μὲν αὔξῃ δι' ὑπερεχόντων προκοπὰς καὶ συστάσεις, ἐὰν δὲ λείπῃ τούτων ἀτονία.
29 Ἀφροδίτης ἐπιμερισμός. Ἀφροδίτη ἑαυτῇ ἀγαθὴ πρὸς γυναῖκας καὶ πάντα τὰ θηλυκὰ πρόσωπα, πρὸς φίλους καὶ πάντα, εὐφροσύνας, ἱλαρίας, ἡδονάς. **30** Ἀφροδίτη Κρόνῳ γυναικῶν χωρισμοὺς ἢ θανάτους καὶ ἀηδίας ἐκ θηλυκῶν εἰδῶν, ἀλλ' ἐπικερδὴς ὁ χρόνος καὶ προσφιλὴς καὶ καταγραφὰς ἔχων, καὶ γυναῖκες ἔγκυοι γίνονται. **31** Ἀφροδίτη Ἄρει χαλεπή· διὰ γυναῖκας μερίμνας, ἐκ γυναικῶν ἐπιθέσεις, μοιχείας, ἐχθρασμούς, αἱμορροίας, ἀηδίας. **32** Ἀφροδίτη Ἡλίῳ γυναικῶν χωρισμούς, δίκας, μάχας, ἀηδίας·ἄλλως δὲ προσφιλεῖς, εὐσχήμονας, ἐπιτευκτικοὺς τοὺς ἐνιαυτούς. **33** Ἀφροδίτη Ἑρμῇ πρακτικοὺς τοὺς ἐνιαυτούς, ἐπιχαρεῖς, ἡδεῖς, φίλους τέκνοις, γυναιξὶν ἀρεστούς. **34** Ἀφροδίτη Σελήνῃ· τὰ πρὸς γυναῖκα ἀνώμαλα, ζηλοτυπίας, μάχας, ἀνωμαλίας, πρακτικοὺς τοὺς ἐνιαυτοὺς καὶ ἱλαροὺς καὶ εὐσχήμονας· μεμνῆσθαι μέντοι τοῦ κατὰ γένεσιν σχήματος ἵνα πρὸς τοῦτο συγκρίνῃς.
35 Ἑρμοῦ ἐπιμερισμός. Ἑρμῆς ἑαυτῷ πρακτικός, ἐπινοηματικός, εὔπορος, ἐπικερδὴς ἐὰν μὴ πρὸς τοὺς κακοποιοὺς ἀνοικείως σχηματίζηται, ἀλλὰ πρὸς τοὺς ἀγαθοποιοὺς οἰκείως· πάντα γὰρ πρὸς τὸ κατὰ γένεσιν σχῆμα δεῖ λογίζεσθαι. **36** Ἑρμῆς Κρόνῳ κακός· οἰκείων θανάτους ἐπιφέρει καὶ ὑποτεταγμένων καὶ ἀδελφῶν νεωτέρων, καὶ αὐτῷ νόσους καθύγρους καὶ παλαιῶν πραγμάτων ἐπεγέρσεις καὶ δίκας. **37** Ἑρμῆς Διὶ καλωνυμίας, εὐμνημίας ἐπιφέρει, ἱλαρίας, εὐσχημονίας, χρηματισμοὺς διὰ χειρός. **38** Ἑρμῆς Ἄρει ἐχθρῶν ἐπεγέρσεις, δίκας, ἐγκλήματα, κατηγορίας, ἐλαττώσεις, νεωτέρου ἀδελφοῦ ἢ τέκνου πολλὰ κακὰ ἐπόψεται, ὁτὲ δὲ καὶ θανάτους. **39** Ἑρμῆς Ἡλίῳ κατὰ πᾶν ὠφέλιμος, μᾶλλον δὲ βλαβερός. **40** Ἑρμῆς Ἀφροδίτῃ καλός, εὔπρακτος, ἐπιχαρής, προσφιλής, ἡδὺς καὶ φίλοις καὶ γυναιξίν, καθ' ὅλον εὐάρεστος. **41** Ἑρμῆς Σελήνῃ ἀγαθοὺς τοὺς ἐνιαυτοὺς καὶ πρακτικούς, ἐπικερδεῖς τε καὶ οἰκονομικούς· πιστευομένους τινὰς χρησμούς.
42 Σελήνης ἐπιμερισμός. Σελήνη ἑαυτῇ παραδιδοῦσα ἐὰν κακοποιοῖς συσχηματίζηται, οὖσα δὲ καὶ ἀφαιρετική, κακή· πολλάκις καὶ θανάτους ἐπιφέρει· εἰ δὲ πρὸς ἀγαθοποιοὺς συσχηματίζοιτο καὶ αὐξίφως ὑπάρχει, μετὰ πολλῶν κόπων ἐλαχίστην παρέχει κτῆσιν· εἰσὶ γὰρ οἱ ἐνιαυτοὶ Σελήνης καὶ Ἡλίου ἀλλότριοι, μάλιστα ἂν κακοποιὸς συμπαρῇ· κινδύνους γὰρ ἐν ὑγροῖς καὶ βλάβας ἐπιφέρει. **43** Σελήνη Κρόνῳ μητέρων νόσους καὶ αὐτῶν δὲ ἀσθενείας καθύγρους, ναρκώδεις, ναυάγια, πτώσεις ἀπὸ τετραπόδων, κινδύνους ἐν ὁδοῖς, ἐναντιώματα. **44** Σελήνη Διὶ λαμπροὺς τοὺς ἐνιαυτούς, φαντασιωτικούς, ἐπικερδεῖς, συστάσεις ὑπερεχόντων προσώπων,

τέκνων σποράς, θηλυκῶν εἰδῶν εὐνοίας, μητέρων δόξας. **45** Σελήνη Ἄρει θορύβους ἐπιφέρει, βλάβας, κινδύνους αἰφνιδίους, μητέρων κινδύνους, ἐπὶ θηλυκοῖς εἴδεσι λύπας. **46** Σελήνη Ἡλίῳ κακή· πυρετούς, ἐνιαυσιαίας ἀσθενείας. **47** Σελήνη Ἀφροδίτῃ νυκτερινῇ γενέσει εὐσχήμων, ἀγαθή, εὐεπίβολος, ἐπαφρόδιτος· αἰεὶ μέντοι τοῦ θέματος ἐναντιουμένη ἢ καθυπερτεροῦσα τὴν Ἀφροδίτην διὰ γυναῖκας ἀδικίας ποιεῖ· ἡμερινῆς γενέσεως ποιεῖ ζηλοτυπίας. **48** Σελήνη Ἑρμῇ δυσαρεστήσεις, ψυχρίας· ἄλλως δὲ πρακτικὸς ὁ χρόνος· πρόβλεπε δὲ μή πως ἐναντιοῦνται πρὸς ἀλλήλους ἢ καθυπερτερεῖται ἡ Σελήνη· ἕνεκεν γραπτῶν ἢ χαλκῶν ἐπιφέρουσι ταραχὰς καὶ ἐναντίαις γνώμαις περιπίπτουσιν.
Παράδοσις χρόνων ἀστέρων β̄· ἐπιμερισμοὶ ἐκ τῶν Κριτοδήμου.

Transmissions of the stars

1 Distribution of Saturn. Saturn transmitting to himself, harmful and unprofitable. **2** Saturn transmitting to Jupiter, good: acquisitions and inheritances of lands and buildings, and love from superiors. **3** Saturn to Mars, if well-positioned,[3] good; if not, grief, diseases, losses: the whole period accursed. **4** Saturn to the Sun, lawsuits related to lands and dangers to the father and losses and unfavourable periods in general. **5** Saturn to Venus, if well positioned, chilliness[4] of women or discomfort or grief of female persons; but it is not harmful in the affairs of life.[5] **6** Saturn to Mercury: old troubles from the past and humid diseases and harm from subordinates: but not wholly wrongful. **7** Saturn to the Moon: it carries diseases and disturbances from humidity and dangers to the mother and falls; you must consider everything in relation to the birth chart.
8 Distribution of Jupiter. Jupiter to himself, good. **9** Jupiter transmitting to Saturn, harmful, unprofitable, and sometimes dangers to children. **10** Jupiter to Mars: harms, uprisings, diseases, and sometimes attempts of detention and deathly dangers to children. **11** Jupiter to the Sun: it illuminates everything hidden; favorable, it causes reputation for the father and agreements with superiors; it also makes offspring of children. **12** Jupiter to Venus: good by night,[6] moderate by

---

3 The word is οἰκείως, which alludes to "familiar" places in the zodiac. Valens clarifies this in his own section on the distributions of Saturn (IV 20), where he says that things are not especially bad if the concerned stars are in operative signs (the signs where they rule, or where they have an exaltation, or belonging to a triangle which they rule) or aspected by benefics.

4 ψυγμός: a rare word, but appearing as early as the Septuagint, e.g., *Num.* 11.32.

5 τὰ κατὰ βίον: a very rare expression before late antiquity, found in Philo Alexandrinus, e.g., *Spec.* II 195.

6 Here, the astrological division of the planets in diurnal and nocturnal (cf. Appendix II) plays a role, since Venus is nocturnal (with the Moon and Mars), and is thus more familiar to a night birth.

day; nevertheless, good and profitable periods, bringing acquisitions; sometimes women get pregnant too. **13** Jupiter to Mercury: favorable, advantageous, profitable for friends. **14** Jupiter to the Moon: calm, it saves from affairs and dangers, and makes good and advantageous periods, and advancements regarding females, and brings honor to the mother and offspring of children.

15 Distribution of Mars. Mars distributing to himself, disturbances and tumultuous affairs. **16** Mars to Saturn: inaction, malaise, it chills, damages of businesses, grief, diseases. **17** Mars to Jupiter: favorable, profitable, good for authority, powerful. **18** Mars to the Sun: not good, crisis-inducer, turbulent, frightful, tumultuous, provoking diseases of the eyes, infected irritations,[7] or also deathly dangers threatening the father by fire or sword, anger of superiors as well as feebleness of nerves and breaking of bones. **19** Mars to Venus: good and favorable, but blaming for female affairs, for it creates distancing and disputes and exactions and bloodsheds and deceptions and adultery from women; but it is not an unprofitable period. **20** Mars to Mercury: blackmail, treachery with legal documents and deceptions, thefts; but not bad in tribunals;[8] nevertheless inspect everything against the birth chart.[9] **21** Mars to the Moon, bad for everything: it brings detentions, secrets, escapes, damages, diseases, and dangers to mothers and sorrow to female persons.

22 Distribution of the Sun. Sun transmitting to himself, not good, vexating, anxiety-producing; worse by night.[10] **23** The Sun to Saturn: bad in night birth,[11] damaging, and loss of the father; in day birth moderate, except if they are not badly positioned to one another. **24** The Sun to Jupiter: good, favorable, good reputation, consolidating with regard to superiors, shining, producing offspring of children. **25** The Sun to Mars: accusations, losses, crises, burning eyes, diseases, dangers, anger of superiors, feeble nerves, breakings of bones. **26** The Sun to Venus: it is favorable, but it makes worthless uprisings of women and armed turbulences and crises; if

---

7 σκυλμός: a very rare word, first appearing in the Septuagint (e.g., 3 *Macch.* 3.25) and the epigrammatist Hedylus from the so-called Alexandrian school of poets (Callimachus, Posidippus, etc.).

8 κριτήρια: not attested before the third century BC in this sense. Again, first appearing in the Septuagint (e.g., *Ex.* 21.6, *Jud.* 5.10).

9 Frequent advice in this version of the doctrine, also found in Firmicus (cf. below). In order to prognosticate correctly, astrologers need to consider the often contradictory effects predicted by other theories.

10 This negative effect of the Sun transmitting to himself, even by day, is quite surprising, given the exclusively positive qualities attributed to the Sun (e.g., in Valens I 1). A possible reason is that all individuals born by day would have the Sun distributing to itself as the first period of their life, a period of high mortality. While Firmicus gives the same verdict that nothing good can be expected from the Sun distributing to itself (VI 36), it is noteworthy that Valens says this is a time of brilliant prospects and activities (IV 17), as does Hephaestio for day births (II 33.1).

11 Saturn is of the day sect, with the Sun and Jupiter, so its effect will be worse in a night birth.

Venus and the Sun are well positioned in the birth chart, favorable periods and full of love, since they are interwoven[12] with marriages or women; in general good periods and productive in women affairs, or happiness from friendships. **27** The Sun to Mercury: it makes actions and gains, but mixed with fears and troubles; not moderate periods, but reflective and solution-bringing. **28** The Sun to the Moon: anomalies, unsteadiness, turbulences, vexations; but if she is waxing, advancements and unions through superiors; if she is waning, weakness.

29 Distribution of Venus. Venus to herself, good in relation to women and everything related to feminine persons, for friends and everything: good cheer, merryment, pleasures. **30** Venus to Saturn: separation from women or deaths and disgust coming from female individuals, but period of gains, agreeable and with contracts, and women become pregnant. **31** Venus to Mars, bad: anxiety from women, attacks from women, adulteries, hatreds, hemorrhages, unpleasantness. **32** Venus to the Sun: separation from women, accusations, disputes, unpleasantness; otherwise agreeable, becoming, successful periods. **33** Venus to Mercury: favorable periods, cheerful, pleasant, good for children, pleasing for women. **34** Venus to the Moon: anomalous in relation to the woman: jealousy, disputes, anomalies, but favorable, cheerful and becoming periods; remember though to compare the birth chart against this.

35 Distribution of Mercury. Mercury to himself favorable, reflective, resolutive, profitable if not inappropriately positioned toward malefics, and if well-positioned toward benefics;[13] everything must be calculated with the birth chart. **36** Mercury to Saturn, bad: it brings deaths of relatives and of inferiors and of younger brothers, and to oneself humid diseases and attacks from old affairs and accusations. **37** Mercury to Jupiter: it brings good reputation, honorable remembrance, cheerfulness, dignity, businesses under control. **38** Mercury to Mars: attacks from enemies, accusations, charges, judicial processes, defeats, many bad things for the younger brother or child will be observed, sometimes death. **39** Mercury to the Sun: beneficial in general, but sometimes harmful. **40** Mercury to Venus: good, favorable, rejoicing, agreeable, pleasant for friends and women, in general satisfying. **41** Mercury to the Moon: good and favorable periods, profitable and frugal; some reliable omens.

42 Distribution of the Moon. The Moon transmitting to herself, if configurated with malefics, and when she is waning, bad; she often brings deaths, too; and if positioned with benefics and waxing, she provides a minimal gain after many toils; the

12 ἐπιπλέκω: not attested before the second century BC (Posidonius fr. 28b, Chrysippus fr. 986, *PTebt* 6).

13 Here, we see the influence of the doctrine of the aspects (cf. Appendix II).

periods of the Moon and the Sun are unfavorable, especially if a malefic is present; she brings dangers in humid places and damages.[14] **43** The Moon to Saturn: diseases and humid feebleness of mothers, sluggishness, shipwrecks, falls from animals, dangers in paths, obstacles. **44** The Moon to Jupiter: brilliant periods, ostentatious, profitable, associations with superiors, offspring of children, favor from women, reputation for mothers. **45** The Moon to Mars: it brings disturbances, harm, unforeseen dangers, dangers to the mother, grief for female persons. **46** The Moon to the Sun, bad: fevers, periodical weaknesses. **47** The Moon to Venus in a night birth: agreeable, good, well-disposed, charming; she always, however, if opposite or prevailing over Venus in the chart,[15] she produces wrongdoings from women; in day chart, jealousy. **48** The Moon to Mercury: unpleasantness, colds; otherwise a good time; look however if they do not oppose each other or if the Moon is dominated: in that case they bring troubles from documents or money and cause the encounter of contrary opinions.
Transmission of times of 2 stars; distributions from Critodemus.

How the tool works

The aim of the doctrine is to make prognostications about a concrete period in the life of the client. Whereas modern astrologers would generally cast new charts – that is, look at the configurations of the heavens at the desired time – some ancient methods sought to minimized the dependence on astronomical data by making the most of the natal chart.

Different methods for allotting times to the stars were used (see below), which attests to the relative popularity of this kind of doctrine over time, but Critodemus' was probably the most widely used theory, to judge from its attestation in several astrological treatises. Also, applications of this theory to individual natal charts

---

14 Here the astrologer feels the need to remark on the fact that the Sun and the Moon are at best ambiguous (e.g., when combined with the favorable Jupiter and Venus), and clearly negative when distributing to themselves or to each other. An exact parallel to this statement is Firmicus' exposition of the Moon's distributions (VI 39): *semper enim et Solis et Lunae tempora malos habent exitus, praesertim si eos malivolae stellae minaci radiatione respiciant.*

15 This is perhaps equivalent to what Ptolemy calls "to apply" in *Tetrabiblos* I 24, that is, to precede in the zodiac by a small number of degrees. Like Critodemus, Ptolemy uses the concept only in relation to the Moon (cf. *Tetr.* III 13 on the "quality of the soul"). Paulus similarly defines it only for the Moon (*Eisagoge* 17).

have been found in two second-century papyri, which is far from irrelevant, since astrological papyri for the most part just contain a list of planetary positions.[16]

Astrological prognostications were derived from considering the transmitting star and the receiving star: F 7 is the concrete list of effects for each particular transmission. The workings of the mechanism are explained in Firmicus II 26 and Hephaestio II 29, where it becomes clear that two levels of rulership are established, which we may call general and particular:

- The Sun begins the general succession of rulership in day charts (persons born by day) and the Moon in night charts.[17]
- Each star in succession in the chart following the Sun/Moon (in the direction of the signs) is assigned a general rulership of 10 years and 9 months.
- The star holding the general rulership assigns to each of the stars, beginning with itself and following with the others following it in the chart, a number of months, always the same for each star:

Sun	19
Moon	25
Saturn	30
Jupiter	12
Mars	15
Venus	8
Mercury	20

One star remains a distributor or transmitter over the whole succession of the particular rulerships. Thus, if one astrologer consulted the list of effects in F 7, they would stick to the same paragraph (e.g., the transmissions of Jupiter) for the whole period of 10 years and 9 months during which Jupiter holds the general rulership.

The number of months assigned to the planetary subrulers corresponds to the so-called minimum periods, expressed in months. These minimum periods are also used by Firmicus to predict the length of life, along with a set of periods called maximum, and another set called middle (II 25). The minimum periods are

16 For Bodl. MS gr. class. B 24, cf. Zellmann-Rohrer 2023; for PSI inv. 3780+P.Med. inv. 124, see Jones and Perale *forthcoming*.

17 Firmicus' indication that the Sun or the Moon must start the succession (depending on whether it is a day or a night birth), and then the rest of the planets in the order on the chart, is confirmed by Hephaestio (II 29.1 ἀρξάμενοι ἀπὸ πρώτου καὶ αἱρετικοῦ φωτὸς ἐμέρισαν τοῖς ἐφεξῆς ἀστράσι κατὰ τὰ ἐφεξῆς ζῴδια), and by the example set out by Valens in VI 6. A further hint in this direction is that Valens puts the Sun and the Moon first in his own list of the effects of the transmissions (IV 18 and 19). Also, Bodl. MS gr. class. B 24 begins with the Moon, and PSI inv. 3780+P.Med. inv. 124 with the Sun.

derived from planetary cycles in which the planets return to the same position and phase, as used in Babylonian Goal-Year texts. Thus, for Saturn a period of 59 years is attested to have been used – in which the planet completes 2 revolutions and 57 synodic periods, or cycles of solar-related phenomena – out of which the rounded half, 30, is taken as a "minimum period" (in which the planet completes 1 revolution).[18] Similarly, there is a Babylonian Goal-Year period of 71 years for Jupiter, in which the planet completes 6 revolutions, out of which a "minimum period" of 12 years is taken, in which Jupiter completes 1 revolution (71/6≈12).

For the Sun and the Moon, other numbers were used. In the case of the Sun, this was the period of 19 years of the Metonic cycle, which was used in the Babylonian calendar because it contains an exact number of lunations or lunar months (235). The period of 25 years for the Moon has less clear origins, but there is probably some common ground, since 25 years also contain a fairly exact number of lunations (309).[19]

These numbers are used in the distributions not as years but as months, so that each particular ruler governs for a period very roughly approximating a year (between Venus' 8 months and the Moon's 25).[20] In total, they amount to 129 months, or 10 years and 9 months.

Hephaestio in II 29 presents the method including, in addition to the subdivision of the minimum periods, a further division nested in each of the particular minimum periods. Thus, the subrulership of Saturn, taking 30 months, comprises smaller periods in which Saturn subdistributes to all the stars again with the same proportion of the minimum periods. Thus, Saturn takes 210 days (= 30 months × 30 days/month × 30/129), Jupiter 84 days (=30 months × 30 days/month × 12/129), and so on.

Firmicus does not refer to these shorter periods, but the fact that Hephaestio attributes them, together with the other two levels of time-rulers, to "some of the ancient Egyptians" (τινες τῶν ἀρχαίων Αἰγυπτίων), and that Valens also uses them in his own exemplification of the system in VI 6, demonstrates that they were frequently used in the tradition, even if they were probably not yet present in the work of Critodemus.

---

18 See the exposition in *HAMA*, 605–606.

19 This is the reason why this period was frequently used as the layout of astronomical tables: cf. *HAMA*, 816.

20 In the list of effects of the distributions in Valens (IV 17–18/20–24) we find references to the period of rulership at the second level as ἔτος ("year"), for all the stars except Venus, the only one whose minimum period is shorter than 1 year (8 months). While ἐνιαυτός often refers to a longer, undefined cycle, ἔτος is more strictly indicative of the length of one year.

Other methods

The popularity of these methods is also attested from the many variations found in the manuals of Valens and Hephaestio.[21] I summarize the ones that are most clearly presented.

(1) The method bearing the greatest relation to that of the minimum periods is explained in Valens IV 1, and in more detail in IV 30. It is called the method of "fourths" because each planet is assigned the fourth of its minimum period, in years, so that Saturn, whose minimum period is 30 years, gets 7 years and 6 months. For the subdivision within each of the general rulerships, each planet is assigned a number of days in the same proportion of the minimum periods, so that, for example, Saturn within his own rulership gets $7.5 \times 365.25 \times 30/129 \approx 637$ days. Presumably the order of the assignments is the same as in the standard method.

(2) Another method in Valens IV 30 entails multiplying the minimum periods by 4, interpreting the result as days – obtaining, e.g., for Saturn $4 \times 30 = 120$ days – except for those of the Moon and the Sun, which are given 25 days and 6 h, respectively. The aim of this scheme is to reach the 365.25 days of a mean year, which is the total length of these rulerships. The general rulerships thus take an exact year, and the order is probably established in the same manner as in the standard and the preceding method.

(3) Another method in Valens IV 30, in which the general rulerships take an exact year, assigns the subrulerships a number of days corresponding to the total number of the terms in the Egyptian system – 57 days for Saturn, 79 for Jupiter, 65 for Mars, 82 for Venus, 76 for Mercury (cf. Ptol. *Tetr.* I 21.11) – with a grand total of 359, plus the remaining days that complete the year assigned to the Moon and the Sun. Here, the order does not derive from the position of the planets in the chart: the ascending sign with its sign ruler governs the first year of life and the following signs with their rulers are assigned to the following years in succession. Then, within a given year, the subrulerships follow the order of the terms of the ruling sign (e.g., for Libra: Saturn, Mercury, Jupiter, Venus, and Mars).

---

21 Valens alludes to the existence of these numerous variations at IV 11.6: διαβολῆς δὲ γενομένης περὶ τὰς καθολικὰς τῶν διαιρέσεων ἀγωγάς, αἷς τινες μὲν ἐχρήσαντο πρὸς τὰς τῶν ὁρίων ἀκολουθίας, ἕτεροι δὲ πρὸς τὰς μικρὰς περιόδους, οἱ δὲ πρὸς τὰ δωδεκατημόρια, ἃ συνάγεται ἔτη ῑ καὶ μῆνες θ̄, ἄλλοι δὲ πρὸς τὰ ὑψώματα ("Since a quarrel has arisen on the method of the general divisions, which some use following the terms, other following the minimum periods, others the twelve signs which amount to 10 years and 9 months and other the exaltations. . .").

(4) A method dismissed by Hephaestio but which he allegedly explains for the sake of completeness[22] assigns to the seven stars the minimum periods counted as days, beginning with the planet ruling the sign of the year (probably as in method 3) and continuing with the planets ruling the following signs in succession; the assignment is performed twice for each cycle in order to obtain a period close to 1 year. Hephaestio remarks that Saturn, which rules the succeeding signs of Capricorn and Aquarius, should only be counted once, thus resulting in a total of 184 × 2 = 368 days.

(5) Dorotheus IV 1: The first year is governed by the ruler of the ascendant, the second by the lord of the next sign, and so on.[23]

Discussion of another method is left for the next section, since it involves the following fragment of Critodemus.

Difference between ἐπιμερισμοί and παραδόσεις?

In addition to "distributing" (ἐπιμερίζω), Critodemus' F 7 and Valens in his parallel account of the effects of the distributions (IV 17–18/20–24) also use the term "transmitting" (παραδίδωμι), apparently without distinction. For example, in the case of F 7, we find the phrase "Saturn transmitting to himself" (Κρόνος ἑαυτῷ παραδιδοὺς) or "Saturn transmitting to Jupiter" (Κρόνος Διὶ παραδιδοὺς), and similarly in Valens' text. While this seems to imply that both terms are used indistinctly, other passages in Valens suggest that "transmitting" (παραδίδωμι) can also refer to the transmission of the rulership from one star to another at the same level (general rulership to general rulership/particular rulership to particular rulership):

22 Heph. II 36: οὗτοι οἱ μερισμοὶ οὐ πάνυ ἡμῖν συμφωνοῦσιν, ὑπὲρ δὲ τοῦ μὴ λαθεῖν αὐτοὺς ἐξεθέμεθα ("These distributions do not match our own, but we have set them out so as not to forget").

23 Perhaps we should not regard this method as related to that of Critodemus, since it does not present two levels of time-lords. Another system where the order of the terms is used, as in method 3 (but not the total number), is attested for Dorotheus (*apud* Hephaestio II 26.25–34), which also establishes a single succession of chronocrators or time rulers. It takes the rulers of the successive terms beginning with the place of the *aphetes* or starter (cf. the calculations of the length of life, Chapter 6), assigning to them the number of degrees of the terms converted to time-degrees and interpreted as years (i.e., for the first term of Aries [20 t-dg in Alexandria], the 6° of Jupiter are converted to 6 × 20/30 = 4 years). The *aphetes* is selected as the term ruler, or sign ruler, or triangle ruler of the Sun if it aspects the Sun and is not aspected by malefics in an *aphetic* place (usually ascendant, midheaven, and descendant); or else, of the Moon, of the Lot of Fortune, of the previous syzygy, or of the ascendant (in this order).

καθολικὸς οὖν κοσμοκράτωρ Ζεύς, δεύτερος Ἑρμῆς ἀπὸ Διὸς παραλαβών, τρίτος Ἄρης ἀπὸ Ἑρμοῦ . . . (IV 10.17)

The overall ruler is Jupiter, secondly Mercury receiving [the rulership] from Jupiter, third Mars from Mercury

However, the much lengthier list of effects in Hephaestio II 29–36 uses only the verb "distribute" (ἐπιμερίζω). The wording in the list of effects in Firmicus' manual (VI 33–36) is similarly clear (33):

Si Saturnus temporum dominus fuerit effectus, Iovi ex tempore suo XII menses deputat.

If Saturn is made the lord of times, it allots to Jupiter 12 months from his own time.

The wording of the two papyri containing applications of the chronocrators (the time rulers) to particular horoscopes (see above) also emphasizes vertical transmission, from general ruler to subruler, using the verb παραλαμβάνω ("receive").[24] Finally, Valens' own additions to Critodemus' list of effects – those related to the ascendant (IV 19) and of the four lots (25)[25] – use only "transmit" (παραδίδωμι) and "receive" (παραλαμβάνω).[26] Ultimately, we should probably interpret that such ambiguous use of παραδίδωμι and a horizontal interpretation of the effects are idiosyncrasies of Valens.

The term ἐπιμερισμός and its associated verb, found in F 7 and in Hephaestio's account, is quite rare. The noun in the plural was used for the titles of Roman and Byzantine grammatical works analyzing epic texts word by word (i.e., dividing the lines into words),[27] but the verb had already been used by Strabo (XIII 1.10) in reference to the way in which Homer "distributes in succession the places that lie

24 In recurring expressions such as παραλαμβάνει ὁ τοῦ Διὸς ἀστὴρ μῆνες ιβ̅ ἐκ τῶν καθολικῶν χρόνων τῆς Ἀφρωδίτης ("Jupiter receives 12 months from the general times of Venus").

25 The lots were different points on the zodiac with different names, obtained from adding to the ascending point a distance between two stars. For example, the Lot of Fortune for day births was obtained by adding the distance from the Sun to the Moon (in night births, from the Moon to the Sun). On the lots as relevant points of the chart, treated almost as planets, see the deluxe horoscope POxy 4277, cf. Tolsa 2017b.

26 It is possible that the effects of these new concepts (ascendant, exaltations, and lots) were considered together with those of the stars ruling the signs concerned, since in the examples of Valens IV 8 we see that the lengths of the periods are assigned according to the sign rulers. Valens sometimes uses the lots, along with other places different from the standard Sun/Moon, as starting points for the succession of chronological rulers in the examination of different aspects of the clients' life. Cf. Valens IV 11.47–52: from the ascendant when researching bodily and mental activities; from the Sun when investigating rank; from the Moon for diseases; from midheaven for occupations and livelihood; from the Lot of Fortune for good fortune; from the descendant for change and trouble, etc.

27 *Epimerismi Homerici*, Priscian's *Partitiones*, and in Herodian's *Epimerismoi*. Cf. Dyck 1981.

along the coast after the Aesepus river" (ἐπιμερίζει δὲ συνεχῶς τὰ κατὰ τὴν μετὰ τὸν Αἴσηπον). Since Strabo frequently resorted to the commentary on the Homeric *Catalogue of Ships* by the grammarian Apollodorus of Athens for his remarks on Homeric geography, he could have taken the neologism from this work,[28] and it is likely that the term was already used in the technical grammatical sense by Apollodorus himself.[29] The parallel is quite precise, since in the epic lines we have the two levels formed by the hexameters (level 1) and the words in each verse (level 2), as in Critodemus' system of time rulerships.[30]

A possible Orphic background

Among the many works attributed to Orpheus in Antiquity, a hexametric poem with the title *Dodecaeterides* has been transmitted sketchily through various testimonia and fragments, including various versions in prose.[31] The text is organized in 12 paragraphs corresponding to the 12 zodiacal signs:

> (*CCAG* V.1: 241–242 = *OF* 732, part)
> Κριός. Τὸ δὲ ἔτος, ὅπερ αὐτὸς κυριεύσει, ὑπάρχει πολεμικὸν ἀπ' ἀρχῆς, πάμφορον πάνυ, παντοίας τροφὰς ἐπιφέρον, πτώσεις δὲ ἀνθρώπον. καὶ τετρἀποδα θάλψουσι πάντα.
>
> Aries. The year which it governs is warlike at the beginning, very fertile, bringing all kinds of foods. Falling of human beings, and all four-footed animals will suffer heat.

The ruling planets for each sign are mentioned in the version of the *Geoponica* (I 12), as well as the planet Jupiter transiting the 12 signs in succession:

> Ἐπὰν ἐπιστῇ τῷ κριῷ ὁ τοῦ Διὸς ἀστήρ, οἴκῳ ὄντι Ἄρεος
>
> When Jupiter is in Aries, the house of Mars

---

28 Another scholar of the *Iliad* and contemporary of Strabo, Aristonicus of Alexandria, also used the word in one scholion (to *Il.* XXII 120), again in a nontechnical sense, "to distribute the possessions" (ἐπιμεριζόμενοι τὰ κτήματα).

29 The first attested use of the word in this technical sense is Apollonius Dysc. (second century AD), *Synt.* 491.13. Cf. again Dyck 1981.

30 Critodemus could have known Apollodorus for the famous mythological expositions of *On the gods*, which, as we know from the testimony of Philodemus, contained Orphic material such as Orphic cosmogonies, as well as names of texts and authors: cf. Obbink 2011.

31 See an inventory of the *dodecaeterides* in the Greek manuscripts in Cumont and Boll 1905 (=*CCAG* V.1): 171–172. See also a Spanish translation of the testimonia and fragments of one prose version in Martín Hernández 2015, 108–118. She follows Bernabé's edition (in turn relying on Heeg), but Boll's (*CCAG* V.1: 241–242) is preferable, since Bernabé includes without justification alterations inspired by the version of the *Geoponica* (1.12).

However, these are probably later additions because of the late date of the text and the absence of this feature in the other versions, and because it is astronomically incorrect that Jupiter travels for 1 year through each of the signs.[32] In fact, the 12 signs are also absent in some texts, which suggests that they could also have appeared later. Nevertheless, the Jupiter period of 12 years was used in Babylonia for the prediction of meteorological phenomena recurring every 12 years, which suggests that Jupiter was already associated with the *Dodecaeterides* at the beginnings of the tradition, and this association was perhaps mentioned in the original poem.[33]

Censorinus, in his chronological compendium *De die natali*, informs us that one type of Great Year is the so-called *dodecaeteris* or "Chaldean Year" (*anno Chaldaico*), which astrologers (Censorinus 18.6):

> *non ad solis lunaeque cursus sed ad observationes alias habent adcomodatum, quod in eo dicunt tempestates frugumque proventus ac sterilitates, item morbos salubritatesque circumire.*
>
> have not determined according to the Sun and the Moon, but from other observations, because they say that in it tempests and abundance of fruits and sterile periods travel in circle, as well as maladies and healthy times.

Therefore, the Orphic poem was likely known to Varro and could therefore be earlier than Critodemus. We cannot say whether the doctrine was known to Critodemus and influenced his choice of the theory of the distributions for his last book, but it is interesting to note the general affinity between two doctrines that divided time into a double partition (recurring periods of 12 years and subpartition by years in the *Dodeceterides* vs. periods of 10 years 9 months and planetary subrulerships close to one year) and employed the concept of minimum period.

Relationship between the different textual witnesses

(a) The Additamenta (F 7), Valens IV 17–18/20–24, and Hephaestio II 29–36

Between the lists of effects in the Additamenta and in Valens, the similarities in structure, vocabulary, and length are great. The effects are virtually the same, albeit more detailed in Valens' account. Thus, doubts could arise as to whether the remark in the title and at the end of the Additamenta, indicating that the text comes from Critodemus, does in fact imply derivation from a text by Critode-

32 One of the versions can be dated to the Julio-Claudian period, since it contains references to month names dedicated to the family of Augustus: cf. Fujii 2013: 144–150.
33 See Tolsa *forthcoming*.b.

mus independent to that of Valens. However, Valens' list includes sections on the effects when the transmitter/receiver is the ascendant (19) or the four lots (20), not appearing in the Additamenta (nor in any of the other witnesses), which are relevant to Valens' theory, and which are probably his own addition. Secondly, and more conclusively, very particular comparable expressions appear in the Additamenta, Valens, and Hephaestio, suggesting that Valens developed his distributions from the Additamenta (or a similar text) rather than the other way around.[34]

Further evidence to support this hypothesis is that almost every section of the distributions in the Additamenta contains an indication similar to that of Venus, to "remember to compare the birth chart against this" (μεμνῆσθαι μέντοι τοῦ κατὰ γένεσιν σχήματος ἵνα πρὸς τοῦτο συγκρίνῃς, 34). If in Valens we occasionally find remarks on the additional effect of malefics positioned diametrically and in other situations, we never encounter such a general admonition to check the chart. However, in both Firmicus and Hephaestio, an introduction to the general rulership of each star – though very different between the two authors in terms of details – is devoted to just this kind of inspection of the transmitter's situation in the chart.[35] Thus, it again seems more likely that these repeated suggestions were present in Critodemus' text and not added by Valens.

In the case of Hephaestio, it is probable that he found the doctrine in an intermediary source, for various reasons: firstly, even though he is no original writer, his introductions to the general rulerships are quite long and detailed, and the effects of the subrulerships are always divided for day- and night-births, a distinction that is not present in the other sources. Furthermore, Hephaestio refers to "some of the ancient Egyptians" (τινες τῶν ἀρχαίων Αἰγυπτίων) as responsible for introducing the doctrine. It seems unlikely that he is referring to Critodemus. In fact, from his only explicit quotation of Critodemus (F 11), it seems that the compiler only had access to an epitome (that he calls *Pinax*), which probably presented the effects of the distributions as given in the Additamenta (F 7).

---

34 See the comparison of the distribution from the Moon to Mercury, showing a specific parallel with Hephaestio and not with Valens. Additamenta (4.48): **ἕνεκεν γραπτῶν** ἢ χαλκῶν ἐπιφέρουσι ταραχὰς καὶ ἐναντίαις γνώμαις περιπίπτουσιν. Hephaestio (II 36.24): εἰ δὲ ὑπὸ Ἄρεως θεωρηθῇ **ἕνεκεν γραπτῶν** ποιεῖ δίκας. Valens (IV 17.11): **χάριν ἀργυρίων ἢ γραπτῶν** ἢ ψηφικῶν πραγμάτων καὶ ἀγῶνα μέγαν ὑπομενοῦσι.

35 For example, for the distribution of Saturn, see Firmicus VI 33.2 "Look then where in the chart it is situated, and whether it is a day or a night chart . . . and when you have collected this with a diligent spirit . . ." (*respice itaque quatenus sit in genitura positus, et an sit diurna genitura an vero nocturna . . . quae cum omnia diligenti ratione collegeris. . .*).

(b) The witness of the *Epitome Parisina* (*CCAG* VIII 3, 102)

The name of Critodemus appears in another text that attributes to him the authorship of the distributions. It is found in a Parisian manuscript (Par. gr. 2425) containing several summaries of ancient astrological treatises. The author of the text, however, did not have independent access to the work of Critodemus, since the content clearly derives from Valens and the Additamenta.

> ὅσα ἴδαμεν συγκεφαλαίωσις
> α̅. Περὶ παραδόσεως Ἡλίου διέξεισιν, ὅπως παραδίδωσιν τοῖς λοιποῖς πλανωμένοις ἕξ, καὶ τί σημαίνει ἑκάστῳ αὐτῶν παραδιδούς. – β̅. διαλαμβάνει περὶ ἐπιμερισμοῦ· τί σημαίνει Ἥλιος ἑαυτῷ ἐπιμερίσας καὶ τίσι τῶν ἄλλων συσχηματισθείς, ἥ τε Σελήνη ἑαυτῇ ἐπιμερίσασα καὶ τοῖς λοιποῖς ἕξ. – γ̅. τί δηλοῖ Κρόνος ἑαυτῷ ἐπιμερίζων καὶ τί ἑκάστῳ τῶν ἕξ. – δ̅. ὁ τοῦ Διὸς ἑαυτῷ ἐπιμερίζων τί σημαίνει καὶ τί τῶν λοιπῶν πλανωμένων ἑκάστῳ μερίδα διδούς. – ε̅. Ἄρης ἑαυτῷ ἐπιμερίζων τί δηλοῖ καὶ τί τῶν ἄλλων ἑκάστῳ τοὺς ἐπιμερισμοὺς διατιθείς· ὡσαύτως δὲ καὶ Ἀφροδίτη καὶ Ἑρμῆς ἑαυτοῖς τε ἐπιμερίζοντες καὶ τοῖς ἄλλοις πλανωμένοις τί ποτε ἀποσημαίνουσιν. εἶτα διαλαμβάνει περὶ παραδόσεως ἀστέρων τὴν ἀρχὴν ἀπὸ Κρόνου καταβαλλόμενος, τί ποτε ἀποσημαίνει ἑαυτῷ παραδιδοὺς καὶ τί τῶν ἄλλων ἑκάστῳ· εἶτα Διὸς παράδοσις παραπλησίως· καταλήγει Ἡλίῳ τε καὶ Ἀφροδίτῃ καὶ Σελήνῃ. ἐν οἷς καὶ ἡ Κριτοδήμου συγκεκεφαλαίωται σπουδή· γέγονε δὲ ὁ Κριτόδημος οὗτος
>
> Summary of what we saw
>
> (1) He goes through the transmission of the Sun, how he gives to the other six planets, and what it means when he gives to each of them. (2) He distinguishes in the distribution: what it means that the Sun distributes to himself and when configured with one of the others, and the Moon distributing to herself and the other six. (3) What it means when Saturn distributes to himself and to each of the six. (4) What it means when Jupiter distributes to himself and what it means when to the other six. (5) What it means when Mars distributes to himself and what when assigning the distributions to each of the other; in the same way Venus and Mercury, what it means when distributing to themselves and to the other stars. Then again he distinguishes about the transmission of the stars making the start from Saturn, what it means when transmitting to himself and to each of the other; then again the transmission of Jupiter; he continues with Sun and Venus and the Moon. And so the doctrine of Critodemus has been summarized. This Critodemus lived in the time (*sic*)

The codex Par. gr. 2425 contains summaries of several ancient astrological works. The first two are summaries of Ptolemy's *Tetrabiblos* and Paulus' *Introduction*. There follow summaries of the lost astrological treatises by Demetrius, Thrasyllus, Critodemus, Kallikrates, Balbillus, and Antiochus. The text above is the chapter dedicated to Critodemus.

There is a clear distinction to be made in this manuscript between long summaries, likely derived from the original works, and the much shorter ones, namely those of Kallikrates, Demetrius, and Critodemus, which were probably salvaged from chapters surviving separately in the manuscript tradition and in other authors' works. A clue appears at the end of Kallikrates' summary (*CCAG* VIII 3, 103):

Ἐν οἷς καὶ τῆς τοῦ Καλλικράτους συναγωγῆς, ἣν ἐκ τῶν παλαιῶν ἤθροισεν, συγκεφαλαίωσις.

And here is the summary of Kallikrates' doctrine, which has been gathered from the ancient treatises/authors.

In the case of the Critodemus summary, the heading is also quite clear: "summary of what we saw" (ὅσα ἴδαμεν συγκεφαλαίωσις).[36] In fact, the summary only deals with the distributions, and what the copyist saw can be deduced from the awkward exposition, since he lists them twice, first in the order Sun, Moon, Saturn, Jupiter, Mars, Venus, Mercury, and then in the order Saturn, Jupiter, <Mars,> Sun, Venus, <Mercury,> and Moon (in the second list the planets in triangular parentheses were skipped). As it turns out, the latter sequence coincides with the order given in the Additamenta, whereas the initial order corresponds to that of Valens IV 17–18 and 20–24.

F 8. An alternative system of distributions versus the standard: Valens IV 26

Type of fragment: Short chapter in Valens' manual whose heading apparently attributes to Critodemus a doctrine on the distributions beginning with the *monomoiria* (the degree-ruler) of the Moon, following the traditional order of the stars, and lasting for a fixed number of years. However, on close inspection we see that the end of the chapter contains the more typical 10-year-9-month variant, which is probably what is referred to in the attribution in the heading.

<κ̅ς̅.> Τὸ δ̄ περὶ χρόνων διαιρέσεως καὶ τὴν ἑπτάζωνον ἀνωφερῶς· κατὰ Κριτόδημον.

1 Σελήνη ᾱ ἔτος ᾱ· Ἑρμῆς β̄ ἔτη β̄· Ἀφροδίτη γ̄ ἔτη γ̄· Ἥλιος δ̄ <ἔτη δ̄>· Ἄρης ε̄ ἔτη ε̄· Ζεὺς ς̄ ἔτη ς̄· Κρόνος ζ̄ ἔτη ζ̄· γίνονται ἔτη κη̄.
2 μονομοιρία δὲ γίνεται οὕτως· ἐν ᾧ ἂν ἡ Σελήνη ζῳδίῳ ᾖ, αὐτὸς γίνεται πρῶτος ὁ κύριος τοῦ ζῳδίου[a] καὶ λήψεται, εἶτα κατὰ ζώνην οἱ ἄλλοι. **3** οἷον ὑποδείγματος χάριν ἔστω Σελήνην εἶναι Ζυγῷ μοίραις ς̄· πρώτη Ἀφροδίτη λήψεται, δεύτερος Ἑρμῆς, τρίτη Σελήνη, τέταρτος Κρόνος, πέμπτος Ζεύς, ἕκτος Ἄρης· γίνεται Ἄρεως μονομοιρία. **4** Ἄρης οὖν πρῶτος λήψεται, ὁ κύριος τῆς μονομοιρίας τῆς Σελήνης, ἔτη ε̄, εἶτα οἱ ἑξῆς κατὰ γένεσιν μετὰ τὸν Ἄρεα κείμενοι ἀστέρες. **5** μετὰ δὲ τὸ συμπληρωθῆναι ἔτη κ̅η̅ ἄρχου πάλιν ἀπὸ τοῦ μετὰ τὸν Ἄρεα κειμένου ἀστέρος.

36 Though it has been corrected by the editors to Κριτοδήμου συγκεφαλαίωσις.

6 ποίει δὲ καὶ τὰ ῑ ἔτη καὶ μῆνας θ̄, ἡμέρας μὲν ἀπὸ Ἡλίου, νυκτὸς δὲ ἀπὸ Σελήνης. 7 ἐὰν δὲ ἡμέρας ὁ Ἥλιος μὲν κακῶς κεῖται, ἀπὸ Σελήνης ἄρχου· ὁμοίως καὶ νυκτὸς ἀπὸ Ἡλίου. **8** ἐὰν δὲ ὁ Ἥλιος καὶ ἡ Σελήνη ἀκυρολόγητοι γένωνται, ἀπὸ τοῦ οἰκοδεσπότου ἢ ἄλλου τινὸς ἀστέρος καλῶς κειμένου.

a ζῳδίου MSS: χρόνου Cumont, Pingree. In my view, this conjectural change is wrong, since Valens is here describing here the mode in which the system of *monomoiriai* is built, taking as the first single degree the ruler of the sign, as is clear in the example given.

<26.> The fourth <method> for the division of the times according to the spheres upwards. According to Critodemus.

1 Moon 1st, 1 year; Mercury 2nd, 2 years; Venus 3rd, 3 years; Sun 4th, <4 years>; Mars 5th, 5 years; Jupiter 6th, 6 years; Saturn 7th, 7 years. The total is 28 years. **2** The single degree is found as follows:[37] whichever star rules the sign in which the Moon is located, that star will be taken as the ruler of the sign, then the rest in the order of the spheres.[38] **3** For example: the Moon in Libra 6°. Venus will be taken first,[39] Mercury second, the Moon third, Saturn fourth, Jupiter fifth, Mars sixth. **4** Therefore, the single degree is assigned to Mars.[40] Now Mars will be taken first as the ruler of the single-degree of the Moon for 5 years, then the stars coming after Mars in the natal chart in order. **5** After the 28 years are completed, begin again with the star coming after Mars.[41]

---

37 This method uses the so-called single degrees (*monomoiriai*): This was a system of rulerships governing parts of the zodiac (like the sign-rulerships, the terms, the triangles, etc.), in this case assigning the seven stars to each of the 360°. There were several such systems, and the one specifically described here, in which the first degree of the sign is assigned to the sign ruler, coincides with the one attested in the horoscope *GH* L 497. Other systems are attested in Paulus (chapter 32, by triangle, similar to the one given here but using the triangle rulers, so that the three signs in the triangle have the same *monomoiriai*) and in *POxy* 4277, for which see Tolsa 2017b.

38 Since the Moon is used, this is probably an example for a night birth. The procedure begins by identifying the planetary ruler of the *monomoiria* of the degree of the Moon.

39 Venus is the ruler of Libra.

40 Each degree is assigned one star in succession in the ‘natural’ order of the spheres, beginning at Libra 1 with Venus.

41 So, after the whole round taking 28 years, with Mars as a general ruler and beginning the series of subrulerships as in the standard system, but with these different time spans. Then, for the next round we take the next star in the chart as the general ruler.

6 Assign the 10 years 9 months counting from the Sun for day births, from the Moon for night births.[42] 7 For day births, if the Sun is unfavorably situated, start from the Moon; similarly, in night births, from the Sun. **8** If the Sun and the Moon are both deemed to be without authority, start from the houseruler[43] or any other favorably situated star.

What does the heading refer to?

Valens IV 26 presents a method for chronocratorships that differs from the standard method of Critodemus, using the single degrees of the zodiac. The *monomoiriai* are assigned beginning with the sign ruler and following the order of the spheres:

Saturn → Jupiter → Mars → Sun → Venus → Mercury → Moon → Saturn . . .

The distribution begins with the *monomoiria* assigned to the Moon's degree. In Valens' example, Libra 6°, the degree where the Moon is situated in the chart, is given to Mars because Venus is the ruler of the sign (Libra 1 to Venus, Libra 2 to Mercury, etc.). The method assigns a number of years to the planets from 1 = Moon to 7 = Saturn in the order of the spheres, which are the durations of the particular rulerships, while the general periods would last a total of 28 years (=1 + 2 + 3 + 4 + 5 + 6 + 7).

Thus, the assignation of times begins with the planet ruling the *monomoiria* of the Moon (probably that of the Sun for day births). It seems that the general planetary ruler is always identical to the first particular rulership, as in the standard method, and that the order of the assignment (both general and particular) is obtained from the position of the planets on the chart (κατὰ γένεσιν). If, then, for example, the planets appear after Mars in the order Venus, Jupiter, Sun, Saturn, and Moon, the general rulerships would be Mars (28 years), Venus (28 years), etc., while the first particular division would be Mars (5 years), Venus (3 years), and Jupiter (6 years), and the second particular division would be Venus (3 years), Jupiter (6 years), etc.

The heading of the chapter is "The fourth [method] on the division of times according to the 7 spheres upwards; according to Critodemus" (Τὸ δ̅ περὶ χρόνων διαιρέσεως καὶ τὴν ἑπτάζωνον ἀνωφερῶς· κατὰ Κριτόδημον). While this could be understood to attribute the method to Critodemus, it seems more likely that the

42 These periods clearly belong to a different method, namely, the standard method of Critodemus.
43 Probably meaning the ruler of the sign of the Sun for day births/of the Moon for night births.

scribe's last words, which are separated from the first part of the title by a strong pause, only refer to the last paragraph of the chapter, which makes a brief, seemingly disconnected reference to the standard method of Critodemus (mentioning the 10-year-9-month period), along with the indication to start with the Sun or the Moon.[44]

F 9. Transits complementing the distributions, with a double horoscope and Valens' method: Valens V 7.17–36

Type of fragment: A passage on time-rulerships in which Valens states that, according to Critodemus, the effects (of the distributions?) are more forceful if the stars involved return to their positions at birth. Valens develops an alternative and quite convoluted method employing a multiplication table probably taken from the work of Critodemus (cf. F 14), which he exemplifies with two charts: a birth chart and another at a later stage in life for the same individual. These charts probably derive from Critodemus' work, but we can deduce that they were introduced at a later stage from their date in the first century AD. I append here diagrams of the two horoscopes, which do not appear in the manuscript tradition. We can assume that the ancient astrologer used an astrological board for this kind of comparison.[45]

17 βεβαιότερα δὲ τὰ ἀποτελέσματα κριθήσεται εἴς τε τὰ πρακτικὰ καὶ τοὺς κλιμακτηρικοὺς λόγους, ἐπὰν οἱ αὐτοὶ ἀστέρες τὸν αὐτὸν σχηματισμὸν ἐπέχωσιν, ὁποῖον καὶ ἐπὶ γενέσεως· τούτου γὰρ καὶ ὁ θειότατος Κριτόδημος μέμνηται. ὑποδείξομεν δὲ δι' ὀργάνου καὶ δι' ἀγωγῆς ἑξῆς τὸ σχῆμα.

< tabella multiplicationis (10×12) >

19 Ὁ προκείμενος κανών ἐστι συναποκαταστάσεως πρὸς τὰς διαστάσεις τῶν ἀστέρων καὶ συμπλοκάς. **20** οἷον ὑποδείγματος χάριν Ἥλιος Ἄρης Ἑρμῆς ὡροσκόπος Τοξότῃ, Σελήνη Λέοντι, Κρόνος Παρθένῳ, Ζεὺς Σκορπίῳ, Ἀφροδίτη Αἰγόκερῳ. **21** κεκλήρωται ἡ Σελήνη τὴν δυάδα, ἐπεὶ ἀφέστηκεν Κρόνου β̅· ὁμοίως καὶ Ἥλιος καὶ Ἄρης Ἑρμῆς ὡροσκόπος ἐπὶ Ἀφροδίτην· τὴν δὲ τριάδα Κρόνος καὶ Ζεὺς καὶ Ἀφροδίτη, τετράδα δὲ καὶ πεντάδα Κρόνος καὶ Σελήνη ἑξάδα, ἡ δὲ ἑβδομὰς κοινὴ[a] πάντων, ὀγδοάδα δὲ

44 As for the scribe's labeling "the fourth" (τὸ δ̅), see the heading of Valens VII 3: Ἀγωγὴ δευτέρα περὶ χρόνων διαιρέσεως. In book IV, Valens has so far discussed several methods for the divisions of times.

45 Fragments of boards with two rings on which the stones for the planets and other horoscope markings can be placed have been preserved: see Evans 2004, with reproductions.

Ἀφροδίτη, ἐννεάδα δὲ καὶ δεκάδα ἥλιος Ἄρης Ἑρμῆς ὡροσκόπος, καὶ Ζεὺς δὲ τὴν δεκάδα καὶ ἑνδεκάδα, καὶ Ἀφροδίτη τὴν δωδεκάδα.

22 ἄγει δὲ τὸ θέμα ἔτος λα· εὑρίσκονται οἱ χρηματίζοντες ἀστέρες καὶ οἱ κλιμακτηρίζοντες οὕτως. **23** ἀρχὴ δέ ἐστι[b] τῶν προκειμένων κλιμακτήρων ἐπὶ τοῦ τρίτου στίχου τῆς τριάδος· οἱ γὰρ προκείμενοι δύο, ἥ τε μονὰς καὶ ἡ δυάς, ἀχρημάτιστοι διὰ τὸ τὴν μονάδα χρηματίζειν μέχρι ιβ, τὴν δὲ δυάδα μέχρι κδ, τὴν δὲ τριάδα μέχρι καὶ λϛ, καὶ ἑξῆς ὁμοίως.

24 θεωρεῖται δὲ οὕτως· ἐπεὶ τὸ λα ἔτος πίπτει εἰς τὸ[c] ια τῆς τριάδος, κεκλήρωται δὲ Κρόνος καὶ Ζεὺς καὶ Ἀφροδίτη ἐπὶ γενέσεως τὴν τριάδα, ἐπιζήτει τοὺς ἐπὶ καιροῦ παροδεύοντας, μήποτε τὴν ια παραδῶσιν ἑτέρῳ ἢ καὶ ἀλλήλοις. **25** οἷον ἐπὶ τῆς προκειμένης γενέσεως ἐπὶ τοῦ καιροῦ ἦσαν οἱ ἀστέρες Ἥλιος Ζεὺς Ἑρμῆς Διδύμοις, Κρόνος Παρθένῳ, Ἄρης Ἀφροδίτη Ταύρῳ, Σελήνη Ἰχθύσιν, οἱ δὲ κληρωσάμενοι ἀστέρες τὴν ια ἦσαν Ζεὺς καὶ Κρόνος καὶ Ἀφροδίτη. **26** εὑρίσκομεν δὲ Σελήνην ἀποκαθισταμένην τῇ Ἀφροδίτῃ, Δία δὲ οὐδενί. **27** εὐθέως μεταβαίνω ἐπὶ τὸν τέταρτον· εὑρίσκω τὰ λβ ἐν τῇ ὀγδοάδι· οὐδεὶς τῶν κυριευσάντων τῆς τετράδος κλιμακτηρίζει. **28** μεταβαίνω ἐπὶ τὴν πεντάδα· χρηματίζει δὲ τῆς πεντάδος ἡ Σελήνη καὶ Κρόνος καὶ εὑρίσκονται οὗτοι ἀλλήλοις ἀποκαθιστανόμενοι. **29** ἔρχομαι ἐπὶ τὴν ἑξάδα· οὐδεὶς διὰ τῶν ϛ διέστηκεν. **30** μεταβαίνω ἐπὶ τὸν στίχον τῆς ἑβδομάδος· εὑρίσκεται δὲ τῆς πεντάδος ὁ χρόνος ἀγόμενος· κοινὴ[d] δὲ πάντων ἀστέρων ὡς εἴρηται ἡ ἑβδομὰς εὑρέθη. **31** Ἄρης καὶ Ἀφροδίτη Κρόνῳ. **32** ἔρχομαι δὲ ἐπὶ τὸν τῆς ὀγδοάδος· Ἀφροδίτη διὰ δ οὐκ ἀποκαθίσταται οὐδενί. **33** ἑξῆς ἐπὶ τὸν τῆς ἐννεάδος κλιμακτῆρα· κυριεύσουσι δὲ τῆς ἐννεάδος Ἥλιος Ἄρης Ἑρμῆς ὡροσκόπος Ἀφροδίτη· ἔστι δὲ ἐν τούτῳ τῷ στίχῳ τὰ λϛ. **34** ἐπὶ τῆς τετραετηρίδος εὑρέθησαν Ἥλιος Ζεὺς Ἑρμῆς ἀποκαθιστάμενοι Κρόνῳ. **35** πάλιν μεταβαίνω ἐπὶ τὴν δεκάδα· κυριεύσουσι δὲ τῆς δεκάδος Ἥλιος Ἄρης Ἑρμῆς Ζεὺς ὡροσκόπος· ἐν δὲ τῷ στίχῳ τούτῳ δ. διὸ καὶ εὑρίσκονται παραδιδόντες Ἥλιος Ἕρμης Ζεὺς Κρόνῳ.

36 Ἀπαραβάτως οὖν ἐνεργεῖς καὶ χρηματιστικοὶ οἱ χρόνοι οὗτοι γίνονται τῶν διαστάσεων, ὅταν οἱ ἐπὶ γενέσεως κυριεύοντες αὐτῶν ἐν ταῖς ἐπικαίροις παρόδοις ἣν εἶχον ἐπὶ γενέσεως διάστασιν ἔχουσιν.

a κοινή: κενή Pingree: an unnecessary change, as again below (see note *ad loc.* in translation).

b ἐστι: ἐπί MSS

c τό: τήν Kroll, Pingree: an incorrect change, to my mind, since we should understand τὸ σελίδιον.

d κοινή: κενή Pingree

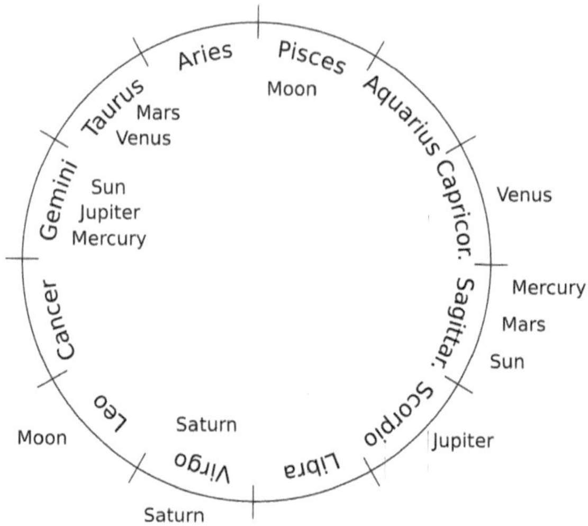

Birth chart outside the circle/death chart inside.

17 Forecasts will be more secure with respect to actions and critical points when the same stars come into the same configuration as at birth, as the most divine Critodemus reminds us. We will show the method through a table and through the following procedure.

<

1	2	3	4	5	6	7	8	9	10	11	12
2	4	6	8	10	12	14	16	18	20	22	24
3	6	9	12	15	18	21	24	27	30	33	36
4	8	12	16	20	24	28	32	36	40	44	48
5	10	15	20	25	30	35	40	45	50	55	60
6	12	18	24	30	36	42	48	54	60	66	72
7	14	21	28	35	42	49	56	63	70	77	84
8	16	24	32	40	48	56	64	72	80	88	96
9	18	27	36	45	54	63	72	81	90	99	108
10	20	30	40	50	60	70	80	90	100	110	120

>

19 The present table is the table of the stars’ mutual return to the same intervals and configurations. **20** For example: Sun, Mars, Mercury, ascendant in Sagittarius; Moon in Leo; Saturn in Virgo; Jupiter in Scorpio; Venus in Capricorn. **21** The Moon is

allotted the number 2, since it is two signs away from Saturn.[46] The same is true of the Sun, Mars, Mercury, and the ascendant with respect to Venus.[47] Saturn, Jupiter, and Venus receive the number 3.[48] Saturn has the 4 and the 5 and the Moon has the 6. Number 7 is common to all.[49] Venus has the number 8. The Sun, Mars, Mercury, and the ascendant have the 9. Jupiter 10 and 11, and Venus the number 12.
22 The chart is in its 31st year: the operative stars and the critical points are found as follows. **23** When calculating the present critical points, begin at the third row, corresponding to the number 3, because the preceding two intervals, number 1 and 2, are inoperative: number 1 is operative to year 12, number 2 to year 24, number 3 to year 36.[50]
24 It is calculated thus: since the 31st year falls in the 11th column of the number 3, and since Saturn, Jupiter, and Venus are allotted the number 3 at the nativity, investigate the stars in transit at the time in question to see if they transmit to another star or to themselves at a distance of 11 signs. **25** Take the preceding nativity: the stars' positions at the time in question were as follows: Sun, Jupiter, Mercury in Gemini; Saturn in Virgo; Mars, Venus in Taurus; Moon in Pisces.[51] Now the stars allotted the 11 were Saturn, Jupiter, and Venus.[52] **26** And we find at the time in question that Venus has returned to a position 11 signs from the Moon,[53] but that no star has returned to a position 11 signs from Jupiter. **27** Immediately I move to the fourth row. I find 32 in the eighth position. None of the ruling stars are critical in the number 4.[54] **28** I move to the number 5: the Moon and Saturn are operative

46 Valens counts inclusively: from Leo to Virgo, counting in this way, there are two signs.
47 From Sagittarius to the next sign, Capricorn.
48 Saturn, in Virgo, has Jupiter in Scorpio (three signs away counting inclusively); Jupiter has Venus in Capricorn, also three signs away. There is no star in Pisces, three signs away (counting in the direction of the signs) from Venus, so Valens considers "away" contrary to the direction of the signs here (which is not generally the case).
49 No stars are at a distance of seven signs (opposition): therefore, this number is not exclusive of any star and is thus "common" (κοινή) to all.
50 The first and second rows of the table are not used because the multiplication table has columns only up to 12, and in this procedure, the "active" stars in a given year (here the 31st) are determined by going through the rows and identifying the corresponding factor in the columns: in the third row, we have 3 × 11 = 33 (taking the first column that surpasses the year number).
51 This is the chart in the 31st year.
52 At birth: The same that are allotted the number 3 (they are the stars bearing sextile aspect).
53 In the chart of the 31st year, Venus is in Taurus and the Moon in sextile in Pisces: The procedure consists in identifying stars bearing the same aspect to another star (not necessarily the same star) at birth and at the year in question.
54 Because in the chart of the 31st year, there are no stars at a distance of eight signs.

in the number 5 and are found to be returning to each other.[55] **29** I move to the number 6: no stars are 6 signs apart. **30** I move to the row of the number 7. The chronocratorship is found to be passing through the number 5.[56] The number 7 is found to be common to all, as mentioned above.[57] **31** Mars and Venus to Saturn.[58] **32** I move to the row of the number 8: Venus does not have a distance of 4 to any star.[59] **33** Then to the critical point of the number 9: the Sun, Mars, Mercury, the ascendant, and Venus rule the number 9;[60] 36 is in this row. **34** At a 4-year-interval the Sun, Jupiter, and Mercury are so distanced to Saturn.[61] **35** Next I move to the number 10: the Sun, Mars, Mercury, Jupiter, and the ascendant rule the tenth;[62] in this row is the number 4; therefore the Sun, Mercury, and Jupiter are found to be transmitting to Saturn.
36 These times will be incontrovertibly active and operative in relation with the distances: when their rulers at the nativity have the distance in their transits at the time in question which they had at the nativity.[63]

Valens' procedure

The initial paragraph, where Critodemus is cited, as well as the example horoscope seem to be taken from Critodemus' treatise, although the horoscope must have been a later addition because of its dating (first century AD). The procedure is Valens' own, but it is based on a multiplication table from Critodemus (ὑποδείξομεν δὲ δι' ὀργάνου καὶ δι' ἀγωγῆς ἑξῆς τὸ σχῆμα, "we now will show the method through using a table and a procedure"), which was probably also an addition, related to

55 Here it seems that Valens changes the criterion, because the Moon and Saturn are five signs apart at birth (not 7) and seven signs apart at the 31st year (5 × 7 = 35).
56 Because 7 × 5 = 35 (7 × 4 = 28, not reaching 31).
57 Cf. above: "Common" here means that at birth there are no stars at a distance of seven signs. But this is irrelevant here because we are seeking stars at a distance of five signs (7 × 5 = 35).
58 Saturn is irrelevant here: It only means that in the 31st year Venus and Mars (which are five signs apart at birth) are also five signs from one star at the 31st year and are therefore also related to this interval (that star now turns out to be the same for the two, namely Saturn).
59 There are no stars with this aspect, either at birth or at the 31st year, neither Venus nor any other.
60 At birth, but this is irrelevant, because we are looking for the four-sign interval.
61 That is, the Sun, Jupiter, and Mercury are allotted the four-sign interval at birth and at the 31st year, and in this second chart the three hold this interval with respect to Saturn. See below for the apparently mistaken designation "4-year interval" for the four-sign interval.
62 At birth. Again, this is irrelevant, but Valens might have changed the criterion as in row 5, above.
63 This is a summary of the procedure (without going into the details of the factorization).

the table of *klimakteres* which Valens copies shortly after the present fragment (cf. F 14). Factorization of the year of life is essential for the method of *klimakteres* ascribed to Critodemus, and the corresponding table extends up to 120; the multiplication table used in this fragment, of 10 rows and 12 columns, probably appeared in Critodemus' work in close association with that table.[64]

Valens' procedure is at first sight difficult to understand. It resembles a method found at the end of book IX using a pair of tables (see Appendix III), in that the characteristics of the tables (the proportion between height and length of the cells in the tables described in book IX, the number of columns here) have a strong bearing on both procedures. Here, Valens inspects the chart of an individual at birth (15 December 37) and later in his 31st year (11 June 68), whose dates coincide with the birth and death of Emperor Nero;[65] astrologers often used the charts of famous individuals from the past, but this is not Valens' modus operandi elsewhere.[66]

Valens uses the table in the following way: He begins by identifying the first row that can reach the number 31, namely the third, since 3 × 11 = 33 (i.e., 3rd row 11th column); next, he looks for any stars that have another star 11 signs away in the birth chart, and which also have a star (not necessarily the same) 11 signs away in the second chart; finally, he repeats the procedure for rows 4 to 10, finding the column in which it reaches 31. For example, in row 9 we find 9 × 4 = 36 (9 × 3 = 27 would be too low), so we look for stars with a 4-sign interval at birth and in the chart of the 31st year (sections 33 and 34):

at birth
the Sun, Mars, Mercury, and the ascendant are in Sagittarius
Saturn is in Virgo, 4 signs before (inclusive counting)

Jupiter is in Scorpio
the Moon is in Leo, 4 signs before

in the recasting in the 31st year
the Sun, Jupiter and Mercury are in Gemini
Saturn is in Virgo, 4 signs after

64 H. Peter was the first to say that this had to be a simple multiplication table, *pace* Pingree in his edition of Valens (Pingree 1986), but he supposes that it extends to 12 rows and does not connect it to the table of *klimakteres*. Cf. Peter 2001, 122.
65 This was discovered by Peter 2001, 122–123. The horoscopes are included in *GH*, 78 (*GH* L 37/68).
66 Cf. the alleged charts of Paris, Demosthenes, Homer, Plato, Pindar, Archilochus, and Archimedes in Firmicus VI 30. All these were strongly idealized charts, but real charts of more recent individuals were surely included, albeit generally anonymously: cf. Hübner 2005 for the possible horoscope of Sulla in Firmicus VI 31.1.

The stars that are common to the two charts are the Sun, Jupiter, Mercury, and Saturn. The implications of this are unclear. On the one hand, Valens seems to be looking for time rulers at the 31st year of life of an individual, but his procedure identifies so many chronocrators for this year of life that the effort seems to have been wasted. Another – perhaps more reasonable – possibility is that he is attempting to find time rulers active in the years of life which are a multiple of each row in question. In other words, the rulers of row 4 would be active in the 4th year, but also in the 8th, the 12th, and so on. The doctrine established in the table of *klimakteres* (Chapter 7) just below in the same book, which he found in Critodemus' work, works in this way, and Valens could be trying to develop a similar method here. The designation "4-year period" (τετραετηρίς) for an interval of 4 in the inspection of row 9, above, would support this hypothesis. How, though, would the astrologer pick up the year of life to do the recasting? If the horoscope was really that of Emperor Nero, the date of the recasting in the example would coincide with the date of death,[67] so the solution would consist in choosing this point for the recasting. If this were the case, the procedure could obviously not be used for a living client.

Critodemus' doctrine and the horoscope example

According to Valens' introduction to this procedure, Critodemus reminds us (μέμνηται) that the effects or predictions (ἀποτελέσματα) will be more certain "when the same stars reach the same configuration" (ἐπὰν οἱ αὐτοὶ ἀστέρες τὸν αὐτὸν σχηματισμὸν ἐπέχωσιν, 17). This refers to the transits, the positions of the stars at some later point in life in comparison with the positions in the birth chart.[68] Since Valens refers unequivocally to time rulers, if this doctrine on transits is from Critodemus it could mean that the effects of the distributions (as described in F 7) are more significant when the stars concerned return to the same position: that is, at the time of the transmission. In Nero's chart, Saturn returns to exactly the same degree in Virgo. If we assume that it was a night birth,[69] the order of the distribution would be, according to the chart: Moon → Saturn → Jupiter → Sun, Mars, Mercury (these three in unknown order) → Venus. Then, using the 10 years and 9 months method, the first three periods would be:

67 Peter 2001, 122.

68 Cf. Dorotheus' doctrine on transits: Dor. IV 1.186–235 (Pingree 1976, 379–383).

69 The Sun is in the same sign as the ascendant, so it is not certain whether it was a day or a night birth.

Moon: First period of 10 years 9 months
Saturn: Second period of 10 years 9 months (10¾–21½ years)
Jupiter: Third period of 10 years 9 months (21½–32¼ years)

The 31st year would therefore correspond to the distribution of Jupiter (general ruler) to Saturn (subruler).[70] Now, if in the 31st year the "receiver" is Saturn, its deadly influence would be enhanced by its returning to the same degree of the zodiac, matching the circumstance of death. This, then, was probably Critodemus' original doctrine, for which Nero's horoscope was an appropriate example.

70 Jupiter would begin distributing to itself at 12 months, then to the Sun . . . and finally to Saturn, 30 months. Saturn would then be the subruler (the "receiver") from 29¾ years to 32¼.

Chapter 5
The "terms" (F 10)

The terms (ὅρια) are a standard doctrine of ancient astrological practice. Most astrological manuals include an exposition of them, and they are used in a number of papyrus horoscopes. Each one of the 12 signs is divided into sections that are assigned to each of the stars in turn. These divisions, the terms, are generally of different sizes for each of the signs, and the order of the planets to which they are assigned is also different. For example, in Aries in the standard system, the first 6° belong to Jupiter; the next 6° (from the 7th to 12th) to Venus; the next 8° to Mercury (from the 13th to the 20th); the next 5° to Mars (from the 21st to the 25th), and the last 5° to Saturn (from the 26th to the 30th):

Aries (30°)				
Jupiter 6° (0–6)	Venus 6° (6–12)	Mercury 8° (12–20)	Mars 5° (20–25)	Saturn 5° (25–30)

However, in Taurus the first division is of 8°, and it is assigned to Venus, and so on: see, for example, the classical exposition in Ptolemy *Tetrabiblos* I 21. Their utility in Greek astrology was to provide a system of regions of the zodiac with which the planets were familiar. Thus, in the same way that Mars was the ruler of the whole sign of Aries, Mars was also the term ruler of the 22nd degree of Aries. This was of course useful for the horoscopes that were calculated not just in signs (giving, for example, only the ascendant in Aries, Saturn in Taurus, and so on) but also in degrees (e.g., Ascendant in Aries 5° and Saturn in Taurus 10°).[1]

Critodemus F 10 provides the extension and the ruler for each term in the 12 signs. His terms coincide with the prevailing system in Hellenistic astrology, the "Egyptian" system described by Ptolemy (*Tetrabiblos* I 21), which is in fact of Babylonian origin.[2] An idiosyncratic feature of Critodemus' exposition is that, for each term, he also gives an astrological interpretation, which in most cases takes the appearance of a character description: for example, the terms of Mars in Aries signify "violent deaths, pirates, robbers". Sometimes, the effects have a chronological dimension, as in the terms of Jupiter in Aries ("progresses from low to better"), but even those cases could be interpreted as life-defining events. The effects are

1 The positions of the planets were much easier to calculate from astronomical tables called sign-entry tables, which gave only the sign where they were positioned: see Jones 1999 I, 42–44. Most of the surviving ancient birth charts only record the signs, which obviously makes it impossible to use the terms in the interpretation.
2 See Jones and Steele 2011.

https://doi.org/10.1515/9783111329147-006

usually defined by the nature of the term ruler, but sometimes the ruler of the sign also seems to have an influence. As a norm, the terms were used just for the familiarity of the planets: giving them a precise interpretation was not a standard practice. Only Valens described their specific effects, probably following Critodemus’ example, even if – in contrast to his exposition of the distributions – Valens did not follow Critodemus’ list closely.

It does not seem a coincidence that we find a long list of the effects of the terms in Critodemus. After all, the terms are a double partition with a similar structure to the distributions we have seen in the previous chapter: here, the general rulership is that of the sign ruler, which, as in the case of the distributions, is of fixed length (30°); the subrulers are now the particular terms.

F 10. *CCAG* VIII 1, 257–261

Type of fragment: A highly standardized and formulaic text presenting the extension and rulers of the terms sign by sign, and explaining one by one their effect on the individual’s life. We might reasonably assume that this is the effect of the term in which the Moon is located at birth, but this is not specified. The explanations are unique, in the sense that there is no other characterization of the effects of the terms in the ancient literature apart from those of Valens (likely influenced by Critodemus) and the Byzantine Kamateros (who follows Critodemus very closely). The text is found in Byzantine astrological manuscripts containing compilations of astrological fragments. Authenticity is assured by specific lexical coincidences with other fragments of Critodemus (e.g., μονή, cf. F 3). Similarly to F 7 on the distributions, the terms are presented in a homogeneous phrasing which is typical of astrological epitomes/paraphrases. We cannot be certain that this is the form in which Valens, Hephaestio, and Firmicus read Critodemus, but the fact that Valens’ account of the distributions closely reflects the form and content of the version in the Additamenta (F 7) suggests that this is the case.

Ἀποτελέσματα ὁρίων Κριτοδήμου.

1 Κριοῦ ἀπὸ πρώτης μοίρας ἕως ἕκτης ὅρια Διός· δίαρμα καὶ προκοπαὶ ἀπὸ ταπεινῶν εἰς μείζονα· τινὲς δὲ καὶ ἐν ὄχλοις εὐδοξοῦσιν ἢ καὶ διπολῖται γίνονται. ἀπὸ μοίρας ζ̅ ἕως ι̅β̅ ὅρια Ἀφροδίτης· ἔρως, φιλία καὶ συμπλοκαὶ πρὸς ὑπερεχούσας γυναῖκας. ἀπὸ μοίρας ι̅γ̅ ἕως κ̅ ὅρια Ἑρμοῦ· εὐτροφία, φιλοπονία, εὐδοξία, προκοπαί. ἀπὸ μοίρας κ̅α̅ ἕως κ̅ε̅ ὅρια Ἄρεως· ἄκρα ἔκπτωσις, ἐξορισμοί[a], κατάκρισις, φυγή, μοιχεῖαι, σίνη καὶ ἄπρακτοι περιπλοκαί. ἀπὸ μοίρας κ̅ς̅ ἕως λ̅ ὅρια Κρόνου· διπρόσωποι, ἐπίτροποι, ἐπὶ κριτηρίων, ἔνιοι δὲ παιδευταί, οἱ δὲ καὶ ἐν ὅροις τάσσονται.

2 Ταύρου ἀπὸ μοίρας α ἕως η ὅρια Ἀφροδίτης· ἔρως, φιλοκαθάριοι, εὔμουσοι, ἐπιχαρεῖς. ἀπὸ μοίρας θ ἕως ιδ ὅρια Ἑρμοῦ· εἰρήνη, ἡγεμονία, ἀρχοντικοὶ ἢ ἐπίτροποι. ἀπὸ μοίρας ιε ἕως κβ ὅρια Διός· καλλονή, εὐδοξία, πρόθεσις, προκοπαὶ ἀνέλπιστοι, οὐ φιλίαι πολλαί. ἀπὸ μοίρας κγ ἕως κζ ὅρια Κρόνου· λύσεις, ἀνέσεις πάντων, συνεχεῖς ἀπολαύσεις, ἐλάττωσις ὑπαρχόντων. ἀπὸ μοίρας κη ἕως λ ὅρια Ἄρεως· βιαιοθάνατοι, λῃσταί, ἅρπαγες, κατάκριτοι, τινὲς καὶ σινούμενοι.
3 Διδύμων ἀπὸ μοίρας πρώτης ἕως ἕκτης ὅρια Ἑρμοῦ· γαλήνη, ἐνστάσεις ἡσύχιοι[b]. ἀπὸ μοίρας ζ ἕως ιβ ὅρια Διός· εὐεργεσία, εὐεργετοῦσι καὶ εὐεργετοῦνται καὶ ἐπὶ προκοπαῖς γίνονται. ἀπὸ μοίρας ιγ ἕως ιζ ὅρια Ἀφροδίτης· ἄνθος, εὐμορφία, εὐτυχία ἐξ ἀνελπίστων. ἀπὸ μοίρας ιη ἕως κδ ὅρια Ἄρεως· δύναμις, ἐξουσία ἡγεμονική, προκοπαί. ἀπὸ μοίρας κε ἕως λ ὅρια Κρόνου· ὑποταγή, ὀρφανία, δεσμοί, στρατιὰ ἄδοξος, καθαίρεσις ἀξίας.
4 Καρκίνου ἀπὸ πρώτης μοίρας ἕως ζ ὅρια Ἄρεως· κινήσεις, ταραχαί, ἔχθραι, ἀντιπολιτεῖαι, κίνδυνοι, ἐξορισμοί. ἀπὸ μοίρας η ἕως ιγ ὅρια Ἀφροδίτης· ἐπιθυμητικοί, φιλοκαθάριοι. ἀπὸ μοίρας ιδ ἕως ιθ ὅρια Ἑρμοῦ· συγκράσεις, φιλίαι ὑπερεχόντων· τινὲς δὲ καὶ φιλόλογοι· ἀπὸ μοίρας κ ἕως κϛ ὅρια Διός· ὅρος ζωῆς, εὐεργετικοί, φιλότροφοι, φιλοκτίσται· τινὲς δὲ καὶ θεῷ τὰ ἑαυτῶν ἀνατιθέντες. ἀπὸ μοίρας κζ ἕως λ ὅρια Κρόνου· ἀνωμαλίαι, ἄστατοι, μὴ μένοντες ἐπὶ τῆς ἀξίας, πολύβουλοι, πολυκίνητοι.
5 Λέοντος ἀπὸ μοίρας πρώτης ἕως ϛ ὅρια Διός· ἀνάβασις, προκοπαὶ ἀπὸ μικρᾶς τύχης. ἀπὸ μοίρας ζ ἕως ια ὅρια Ἀφροδίτης· εὐτυχίαι, προκοπαί, φιλίαι πρὸς ὑπερέχοντας. ἀπὸ μοίρας ιβ ἕως ιη ὅρια Κρόνου, μῆκος, πολυετεῖς, κατεψυγμένοι· τοῖς σώμασι καὶ ταῖς διανοίαις. ἀπὸ μοίρας ιθ ἕως κδ ὅρια Ἑρμοῦ λόγος, τιμαὶ ἕνεκεν παιδείας, ἱερονῖκαι. ἀπὸ μοίρας κε ἕως λ ὅρια Ἄρεως· μόχθος, ἀτυχίαι, κακοπάθειαι· τινὲς δὲ καὶ σωματικῶς κινδυνεύσουσι περὶ τὰ ἄκρα.
6 Παρθένου ἀπὸ πρώτης μοίρας ἕως ζ ὅρια Ἑρμοῦ· ἀκρίβεια, φιλοσοφία, γεωμετρία, ῥήτορες, ἀστρόλογοι, οἱ ἀπὸ παιδείας ζῶντες. ἀπὸ μοίρας η ἕως ιζ ὅρια Ἀφροδίτης· ἀθλιψία, εὐφροσύνη καὶ ἀμεριμνία· ἀπὸ μοίρας ιη ἕως κα ὅρια Διός· ὄνομα ἔνδοξον, προκοπαί, φιλίαι, ὄχλων ἀφηγοῦνται, ἐπίσημοι. ἀπὸ μοίρας κβ ἕως κη ὅρια Ἄρεως· φῶς, λαμπροὶ στρατιωτικοὶ ἢ ἐπὶ κριτηρίων. ἀπὸ μοίρας κθ ἕως λ ὅρια Κρόνου· ὑψοταπείνωμα[c] τῇ ἀξίᾳ, καὶ κινδυνεύσουσιν.
7 Ζυγοῦ ἀπὸ πρώτης μοίρας ἕως ϛ ὅρια Κρόνου· ἀφαίρεσις στεργομένων προσώπων, ἀπώλεια καὶ καθαίρεσις ὑπαρχόντων. ἀπὸ μοίρας ζ ἕως ιδ ὅρια Ἑρμοῦ· ἔργον, πολύτεχνος, πολυάδελφος. ἀπὸ μοίρας ιε ἕως κα ὅρια Διός· μερισμοί, κληρονομίαι καὶ χώρας δόσεις. ἀπὸ μοίρας κβ ἕως κη ὅροι Ἀφροδίτης· τροφή, φιλοκαθάριοι, εὐεργετικοί, μεμουσωμένα καὶ καθάρια τρέφουσιν, φιλοπάλαιστροι καὶ φιλοκύνηγοι.ἀπὸ μοίρας κθ ἕως λ ὅρια Ἄρεως· ἰσχύς, ἡγούμενοι, αἱ στρατιαὶ ἔνδοξοι καὶ ὑπεραύξησις.**8** Σκορπίου ἀπὸ πρώτης μοίρας ἕως ζ ὅρια Ἄρεως· προσθέσεις ἀλλοτρίων, ἁρπαγαί, καὶ ἐκ τούτων διάκρισις· ἀπὸ μοίρας η ἕως ια ὅρια Ἀφροδίτης· τύχη, προκόπτουσι δι’ ἄλλων ἀνελπίστως. ἀπὸ μοίρας ιβ ἕως <ιθ> ὅρια

Ἑρμοῦ· μονή, ἐπίτροποι γίνονται καὶ διοικηταί. ἀπὸ μοίρας κ̅ ἕως κ̅δ̅ ὅρια Διός· αὔξησις, πρόβασις ἐπὶ τὸ μεῖζον, κληρονομίαι κτημάτων κυριωτέρων καὶ προκοπαί. ἀπὸ μοίρας κ̅ε̅ ἕως λ̅ ὅρια Κρόνου· ἀνάγκη, συνοχή, καὶ ὠμοτοκία[d], τινὲς δὲ καὶ τῶν σπορίμων στεροῦνται.

9 Τοξότου ἀπὸ πρώτης μοίρας ἕως ι̅β̅ ὅρια Διός· δυνάμεις, ἡγεμόνας, κυρίους ποιεῖ ζωῆς καὶ θανάτου, ἀκαταπλήκτους, βασιλικούς. ἀπὸ μοίρας ι̅γ̅ ἕως ι̅ζ̅ ὅρια Ἀφροδίτης· ῥύσις, ἀδημονία, τινὲς δὲ κατωφερεῖς. ἀπὸ μοίρας ι̅η̅ ἕως κ̅α̅ ὅρια Ἑρμοῦ· πράξεις ἐπίσημοι, φιλόλογοι ῥᾴθυμοι, εὐφυεῖς. ἀπὸ μοίρας κ̅β̅ ἕως κ̅ϛ̅ ὅρια Κρόνου· εὐψυχία[e]. ἀπὸ μοίρας κ̅ζ̅ ἕως λ̅ ὅρια Ἄρεως· ξενιτεία, μετάπτωσις, ἐξορισμός.

10 Αἰγόκερω ἀπὸ α̅ μοίρας ἕως ζ̅ ὅρια Ἑρμοῦ· γένεσις στομαχική, καθάριοι, ὀλιγότροφοι, ἀπὸ μοίρας η̅ ἕως ι̅δ̅ ὅρια Διός· κράτησις συντηρητική, εὐσχήμονες βίου. ἀπὸ μοίρας ι̅ε̅ ἕως κ̅β̅ ὅρια Ἀφροδίτης· ἡδυπάθεια, κατωφέρεια. ἀπὸ μοίρας κ̅γ̅ ἕως κ̅ϛ̅ ὅρια Κρόνου· ἐπίκλησις, ὑπὸ ἐξουσίας τάσσονται, τινὲς δὲ καὶ γένους μετέχουσι, [τραχεῖς, ἀνυπόστολοι, αὐθεντικοί, δίκας κινοῦντες][f]. ἀπὸ μοίρας κ̅ζ̅ ἕως λ̅ ὅρια Ἄρεως· κρύψεις, φυγαί, δεσμοί, φυλακαί, ἐξορισμοί.

11 Ὑδροχόου ἀπὸ πρώτης μοίρας ἕως ζ̅ ὅρια Ἑρμοῦ· ἐγκράτεια, ἐμφιλόσοφοι τῷ ἤθει. ἀπὸ μοίρας η̅ ἕως ι̅γ̅ ὅρια Ἀφροδίτης· πρόνοια, ἐπίτροποι μεγάλων γυναικῶν. ἀπὸ μοίρας ι̅δ̅ ἕως κ̅ ὅρια Διός. νίκη, ἐπίσημοι, λαμπροὶ ταῖς τύχαις ἐκ κακοπαθειῶν, τινὲς καὶ ἱερονῖκαι. ἀπὸ μοίρας κ̅α̅ ἕως κ̅ε̅ ὅρια Ἄρεως· ἀπαγωγὴ ἀπὸ ἐξουσίας, αἰχμάλωτοι ἢ βιοθάνατοι, τινὲς δὲ καὶ σινοῦνται. ἀπὸ μοίρας κ̅ϛ̅ ἕως λ̅ ὅρια Κρόνου· ἀνατροφή, ἔκπτωσις ἐπὶ νεότητος.

12 Ἰχθύων ἀπὸ α̅ μοίρας ἕως ι̅β̅ ὅρια Ἀφροδίτης· φθόνος. πολλοὺς ἔχθρους ἔχουσιν. ἀπὸ μοίρας ι̅γ̅ ἕως ι̅ϛ̅ ὅρια Διός· προκοπή, λαμπρούς, ἡγεμονικούς. ἀπὸ μοίρας ι̅ζ̅ ἕως ι̅θ̅ ὅρια Ἑρμοῦ· σύνεσις, εὐφυεῖς, ῥᾳθυμοτάτους. ἀπὸ μοίρας κ̅ ἕως κ̅η̅ ὅρια Ἄρεως· πλεονασμός, πλεονεξία, ἅρπαγες, ληστρικοί, φονικοί, δίκας πολλὰς ἔχουσιν. ἀπὸ μοίρας κ̅θ̅ ἕως λ̅ ὅρια Κρόνου· ἐναρέτους, σοφοὺς καὶ ἀρχικούς.

a ἐξορισμοί Cumont: ἐξορκισμοί P, διορισμοί VR
b ἐνστάσεις ἡσύχιοι: ἐνστάσεις ἡσυχίαι P, καὶ στάσις ἥσυχος V, καὶ στάσις ἡσύχιος R
c ὑψοταπείνωμα Cumont: ἡψωταπεινώματι V, ὕψωμα ταπείνωμα P
d καὶ ὠμοτοκία coni. Cumont: κωμοτροφία codd.
e ἐυψυχία P: ἀψυχία V
f [] likely a late ancient or Byzantine accretion. Most of these words and expressions are attested rather late in the history of Greek language, and the sequence comes after the clause τινὲς δὲ καὶ, which frequently closes the list of effects of the term: it is therefore reasonable to think that what comes thereafter is an addition.

1 Aries from the first degree to the sixth, terms of Jupiter; passage[3] and progresses from low to better; some receive fame from the crowd or become double citizens.[4] From the 7th degree to the 12th, terms of Venus: love, friendship and relationships with prominent women. From the 13th degree to the 20th, terms of Mercury: thriving, good work, fame, progresses. From the 21st degree to the 25th, terms of Mars: the extreme degrees[5] signify falling away, exiles, condemnation, escapes, adulteries, injuries and useless complications. From the 26th degree to the 30th, terms of Saturn: ambiguous men, governors, judges, some tutors of children as well, and some are placed in memorials.[6]
2 Taurus from the 1st degree to the 8th, terms of Venus: love, lovers of purity,[7] musical, cheerful. From the 9th degree to the 14th, terms of Mercury: peace, domination, rulers or governors. From the 15th to the 22nd, terms of Jupiter: beauty, fame, prominence, unexpected progresses, not many friendships. From the 23rd degree to the 27th, terms of Saturn: relaxing, loosening of everything, frequent enjoyment, loss of possessions. From the 28th to the 30th, terms of Mars: violent deaths, pirates, robbers, condemned, some plunderers, too.
3 Gemini from the first degree to the sixth, terms of Mercury: calm, tranquil situations. From the 7th degree to the 12th, terms of Jupiter: public donation; they give and receive donations and make progresses. From the 13th degree to the 17th: terms of Venus: flower, good shape, good luck from unexpected things. From the 18th degree to the 24th, terms of Mars: strength, governing capacity, progresses. From the 25th degree to the 30th: terms of Saturn: subordination, orphanhood, chains, inglorious military campaign, demolition of reputation.
4 Cancer from the first degree to the 7th, terms of Mars: movements, troubles, enmities, political opposition,[8] dangers, exiles. From the 8th degree to the 13th, terms of Venus: full of desire, loving purity. From the 14th degree to the 19th, terms of

3 The Greek word is δίαρμα. Kamateros 270 changes it for ἅρμα ("chariot", but also a Pythagorean word for unity in *Theol. arithm.* 6).
4 Implying an extraordinary honor. Elsewhere only attested in the astrological poem of Manetho, also in reference to people born under the influence of Jupiter (V 291).
5 Extremely synthetic expression, ἄκρα ("extremes"). The last degrees tend to be bad. The last terms in each sign generally belong to the malefic, and here by analogy the last degrees in the terms of the malefics are worse still. Cf. below the same case with the terms of Mars in Leo (περὶ τὰ ἄκρα). Cf. Kamateros 281, who changes this for ἀκρασία.
6 A somewhat rare expression (ἐν ὅροις τάσσονται), but found again below in the terms of Saturn in Capricorn: ὑπο ἐξουςίας τάσσονται. Cf. ὄχλοις ("crowds") in Kamateros 288.
7 φιλοκαθάριοι: A very rare word, also appearing in Valens I 1 in the effects of Venus, the ruler of these terms (Valens probably drew from Critodemus for that section).
8 ἀντιπολιτεία: Found rarely before late antiquity, mostly in Polybius (second century BC), and also appearing in Strabo and Plutarch. More frequent as a verb (ἀντιπολιτεύομαι), from Aristotle

Mercury: mixtures, friendship of superiors; some lovers of language. From the 20th degree to the 26th, terms of Jupiter: memorial of life,[9] public donors, fond of nurture, fond of building; some offer their properties to the god. From the 27th degree to the 30th, terms of Saturn: anomalies, inconstant men, not remaining in good reputation, shifty, moving a lot.

5 Leo from the first degree to the 7th, terms of Jupiter: flood,[10] progresses from little luck. From the 7th degree to the 11th, terms of Venus: good luck, progresses, friendship with superiors. From the 12th degree to the 18th, terms of Saturn: length, of many years, chilled; both bodies and souls. From the 19th degree to the 24th, terms of Mercury: discourse, honors from education, champions in the games. From the 25th degree to the 30th, terms of Mars: hardship, bad luck, bad experiences; some physically endangered at the extreme degrees.

6 Virgo from the first degree to the 7th, terms of Mercury: exactness, philosophy, geometry, orators, astrologers, those living from education. From the 8th degree to the 17th, terms of Venus: absence of oppression, happiness and absence of anxiety. From the 18th to the 21st degree, terms of Jupiter: honor of the name, progresses, friendships, they lead the masses, significant. From the 22nd degree to the 28th, terms of Mars: light, brilliant military men or judges. From the 29th degree to the 30th, terms of Saturn: ups and downs in reputation, and they will experience danger.

7 Libra from the first degree to the 6th, terms of Saturn: disappearance of beloved persons, destruction and overthrowing of rulers. From the 7th degree to the 14th, terms of Mercury: work, of many arts, of many brothers. From the 15th degree to the 21st, terms of Jupiter: partitions, inheritances, and land transfers. From the 22nd degree to the 28th, terms of Venus: nurture, fond of purity, public donors, nurturing education in the ways of the Muses and purity, fond of wrestling and hunting. From the 29th degree to the 30th, terms of Mars: strength, leaders, expeditions full of honors, and overgrowth.[11]

8 Scorpio from the first degree to the 7th, terms of Mars: increase of enemies, seizures, and quarrels arising from this. From the 8th degree to the 11th, terms of

---

onward. As Cumont suggests (*CCAG* VIII 1, 258), it probably implies a date in which Greek-speaking states were still independent from Rome (*civitates Graecae florebant*).

9 ὅρος ζωῆς: An unusual expression, but cf. the terms of Saturn in Aries, "placed in memorials" (ἐν ὅροις τάσσονται). Kamateros changes this to ζωῆς δῶρος (596).

10 A probable reference to the Nile, which situates Critodemus or his source in Egypt. This is a rare instance in these significations where the effect is not individual, but "mundane", according to the terminology. Hephaestio (I 23.21), drawing from the ancient Egyptians (Nechepsos-Petosiris), prognosticates that when Jupiter is in Leo one can expect good floods. Leo and Jupiter are also considered a wet sign and a wet planet, respectively, in one astrometeorological fragment by a certain Syrus, as well: cf. Tolsa 2019b.

11 ὑπεραύξησις: Not a widely circulated term, either as a noun or as a verb. It is mainly medical.

Venus: good luck, they progress unexpectedly through others. From the 12th degree to <the 19th>, terms of Mercury: salvation, they become governors and administrators. From the 20th degree to the 24th, terms of Jupiter: growth, advancement to the better, inheritances of significant properties, and progresses. From the 25th degree to the 30th, terms of Saturn: necessity, oppression, premature birth[12]; some are deprived of seeds.

9 Sagittarius from the first degree to the 12th, terms of Jupiter: it makes strength, leaders, rulers of life and death, unbreakable, royal men. From the 13th degree to the 17th, terms of Venus: flow down, distress, some vicious.[13] From the 18th degree to the 21st, terms of Mercury: significant actions, easy-tempered lovers of language, of good nature. From the 22nd degree to the 26th, terms of Saturn: high spirit.[14] From the 27th degree to the 30th, terms of Mars: living abroad, change, exile.

10 Capricorn from the 1st degree to the 7th, terms of Mercury: nativity related to the stomach[15], pure, needing little nourishment. From the 8th degree to the 14th, terms of Jupiter: ability to preserve, elegant in lifestyle. From the 15th degree to the 22nd, terms of Venus: pleasant living, downward tendency. From the 23rd degree to the 26th, terms of Saturn: title, appointed by the authority, some being members of the family, [rough, using no concealment,[16] warranted,[17] setting in motion judicial processes[18]]. From the 27th degree to the 30th, terms of Mars: hiding, escapes, chains, guarding, exiles.

11 Aquarius from the 1st degree to the 7th, terms of Mercury: self-control, philosophical in character[19]. From the 8th degree to the 13th, terms of Venus: providence, protectors of great women. From the 14th degree to the 20th, terms of Jupiter: victory, famous men, brilliant in fortune coming from a bad position, some victorious in sacred games. From the 21st degree to the 25th, terms of Mars: withdrawal from

---

12 This is a conjecture (cf. apparatus, ὠμοτοκία), but a plausible one: Mars is associated with this concept in Valens II 34; see also Hephaestio II 13.11 (Venus together with Saturn and Mars).

13 A somewhat surprising effect given the positive nature of Venus and the ruler of the sign (Jupiter).

14 Perhaps surprising for the terms of Saturn, but Sagittarius is the sign of Jupiter and this is surely beneficial. Cf. Valens I 1 εὐψυχείς for the terms of Mars in Pisces (which is also the sign of Jupiter).

15 In Valens' description of the effects and fields of activity of the planets (I 1), Mercury is related with the belly and the intestines (among other body parts), and Saturn (the ruler of Capricorn) with the internal organs.

16 ἀνυπόστολοι: Word first surviving in Philodemus (*Rh.* I 109) (first century BC).

17 αὐθεντικός: First attested in the first century BC (Cicero *Lit. Att.* 9.14.2).

18 δίκας κινοῦντας: Expression not attested in the ancient corpus, but cf. the Byzantine "interpretation" (ἑρμηνεία) to the Aesopic proverb 64 Perry (δίκας κινοῦντες παῖδες).

19 ἐμφιλόσοφοι: Late concept, surviving from Philo of Alexandria onward, but appearing in the astrological tradition (in Ptolemy's *Tetrabiblos* III 14).

authority, prisoners or dead by violence; some are plunderers. From the 26th to the 30th, terms of Saturn: rearing,[20] banishment in youth.

12 Pisces from the 1st degree to the 12th, terms of Venus: jealousy, they have lots of enemies. From the 13th degree to the 16th, terms of Jupiter: progress; makes brilliant and leading men. From the 17th degree to the 19th, terms of Mercury: union, good nature, easy character. From the 20th degree to the 28th, terms of Mars: excess, arrogance, rapacious, pirates, assassins, they have many judicial processes ongoing. From the 29th degree to the 30th, terms of Saturn: virtuous, wise, and regal.[21]

Different systems

Several systems for the terms were in circulation, some of them also assigning terms to the Sun and the Moon. The name “terms” is quite transparent: it may have originated from the astrologers’ expositions of precisely from what point to what point each section extends. The logic of the system could, however, have presented a problem to Greek astrologers, since there is no simple set of rules from which the assignations and the lengths of the terms can be derived. This is a topic discussed in detail by Ptolemy (*Tetrabiblos* I 21) who, while acknowledging that the Egyptian (standard) system shows some regularities, nevertheless complains about its general irrationality, and goes on to describe a system he allegedly found in an old manuscript, which is presented as an updated version of the Egyptian terms.[22]

The Egyptian terms are those presented by Critodemus. Generally speaking, the first term in each sign is governed by a planet that has some sort of familiarity with the sign in question. The most influential rulership here is that of the triangle to which the sign belongs.[23] Thus, Jupiter is a ruler of the triangle Aries/Leo/Sagittarius, and he is also the ruler of the first term in Aries, Leo, and Sagittarius. Mars is a trigonal ruler of the fourth triangle (Cancer–Scorpio–Pisces), and he is the

---

20 ἀνατροφή: Probably referring to a well-nurtured infancy: cf. the following, negative effect.

21 The terms of Saturn in Sagittarius, the other sign of Jupiter, are also positive in nature.

22 See Tolsa 2018: The finding of a manuscript containing this system claimed by Ptolemy was probably an invented story, and Ptolemy in fact described the system very succinctly without going into its details. As I argue in the cited article, Ptolemy probably did not really provide the table containing the specific terms found in the manuscripts in the *Tetrabiblos*, which would be a later addition.

23 The signs of the zodiac are distributed in four equilateral triangles, Aries-Leo-Sagittarius, Taurus-Virgo-Capricorn, etc. (cf. Ptolemy *Tetrabiblos* I 19 and Appendix II). The triangles are also called triplicities. Both this system of terms and the division of the zodiac in triangles were imported from Babylon: see Jones and Steele 2011 for the terms, and Koch-Westenholz 1995, 167–168, for the triangles.

first term ruler in Cancer and Scorpio. The system is not entirely consistent even here: only 8 of the 12 signs follow this rule. In Virgo and Pisces, the first term is ruled by a planet whose exaltation is in the sign (Mercury and Venus, respectively). Two other signs – Leo and Capricorn, whose first terms are assigned to Jupiter and Venus, respectively – have no obvious rule for establishing the first term ruler. As for the lengths of the terms, there seems to be even less rationale. Ptolemy proposes a system of terms with the following ordering (I 21.22–23):

> ἐπὶ μὲν γὰρ τῆς τάξεως τῆς καθ' ἕκαστον δωδεκατημόριον παραλαμβάνεται τά τε ὑψώματα καὶ τὰ τρίγωνα καὶ οἱ οἶκοι. καθόλου μὲν γὰρ ὁ μὲν δύο τούτων ἔχων ἀστὴρ οἰκοδεσποτείας ἐν τῷ αὐτῷ ζῳδίῳ προτάσσεται κἂν κακοποιὸς ᾖ. ὅπου δὲ τοῦτο οὐ συμβαίνει, οἱ μὲν κακοποιοὶ πάντοτε ἔσχατοι τάσσονται, πρῶτοι δὲ οἱ τοῦ ὑψώματος κύριοι, εἶτα οἱ τοῦ τριγώνου, εἶτα οἱ τοῦ οἴκου ἀκολούθως τῇ ἐφεξῆς τάξει τῶν ζῳδίων, πάλιν τῶν ἐξῆς ἀνὰ δύο ἐχόντων οἰκοδεσποτείας προτασσομένου τοῦ μίαν ἔχοντος ἐν τῷ αὐτῷ ζῳδίῳ.

> For their arrangement within each sign, the exaltations, triangles, and houses are taken into consideration. For, generally speaking, the star that has two rulerships of this sort in the same sign is placed first, even though it may be maleficent. But wherever this condition does not exist, the maleficent planets are always put last, and the lords of the exaltation first, the lords of the triangles next, and then those of the house, following the order of the signs. And again in order, those that have two lordships each are preferred to the one which has but one in the same sign.[24]

For example, in Aries, no planet has two rulerships, so the malefics are assigned the last terms. Then, for the first term we look for a planet with an exaltation in Aries: only the Sun is exalted there, but since the luminaries are not used in this system, we must now look for a planet with a trigonal rulership: the Sun and Jupiter are rulers of the triangle of Aries, so Jupiter is taken as the first term ruler. So far this coincides with the Egyptian system. Then, for the second term ruler in Aries, Ptolemy suggests choosing a ruler using the same rules but for the next sign ("following the order of the signs"), Taurus. This is repeated until the sequence of five terms has been completed. However, the application of these rules inevitably leads to several *aporiai*, making it impossible to build a system completely consistent with Ptolemy's indications.[25]

---

24 Trans. Robbins.

25 For example, Ptolemy assigns 7° to the benefics (Jupiter and Venus), 5° to the malefics (Saturn and Mars), and 6° to Mercury, but an extra degree is added when the planet in question holds two rulerships, which should be taken from a malefic (Ptolemy does not say from which of the two). See the efforts to deduce a completely clear rationale for the system in Bezza and Fumigalli (undated) and Houlding 2010; in Tolsa 2018 I propose that the anonymous commentary on the *Tetrabiblos* represents an early effort of this kind, from which the table in the manuscripts ultimately derives.

Other astrologers proposed far easier systems. For example, Ptolemy records a "Chaldean" system in which Aries, Leo, and Sagittarius (the signs of the first triangle) are assigned first the planetary ruler of the first triangle (Jupiter), then the planetary ruler of the second triangle (Venus), then the rulers of the third triangle (Saturn and Mercury), and finally the ruler of the fourth triangle (Mars). The same sequence is followed in the signs of the second triangle, now beginning with the ruler of the second triangle and ending with that of the first; and so on for the remaining triangles. In this system, the lengths of the terms are also simple: in all signs, the first term is given 8°, the second 7°, the third 6°, the fourth 5°, and the fifth 4° (always completing 30°). In Chapter 8 we will see that Critodemus' treatise, in the version read by Valens, contained an alternative system similar to these "Chaldean" terms, which was added to the table for calculating the length of life (F 16), itself a later addition.

Significations of the terms: how they work

As we have seen, in addition to providing the terms themselves – that is, from which degree to which degree each term extends in each sign – Critodemus gives several significations in each case, consisting mainly of a general description of the person born under the influence of the term in question.

We are not told how the astrologer determines what terms are operative in the native person. In the chapter of Ptolemy's *Tetrabiblos* dedicated to the general prediction of the kind of life (III 14), we are told that Mercury rules the intellectual part of the soul, while the Moon rules the irrational and sensory. Ptolemy, then, draws the individuals' characteristics from the signs containing Mercury and the Moon in the chart, and indeed at this point he provides a cursory description of the signs' effects. The effects of Critodemus' terms could also function in this way so that the terms where the Moon and Mercury were positioned at birth would be the ones to look at.

The terms were also used in many other ways. One of them was to provide the maximum length of life allotted by each of the planets, which was calculated as the sum of all the terms (in degrees) assigned to each planet through all the signs: thus, Saturn allots 57 years of life (Firmicus II 25). We have also seen a system of distributions in which the durations of the chronocratorships were calculated as these very sums (interpreted as days), and in which the order of the sequences was determined by the order of the terms in the sign ruling the year. But the two more typical uses of the terms were similar to the case of other rulerships: (1) to provide another level of dignity to the planets so that a planet's influence was enhanced by being in his own terms; and (2) to provide an interpretation of a certain position in the chart by inspecting its term ruler. For example, in the next chapter we will see that Critodemus employed the term ruler of the descendant to inquire whether the native would die of a violent death (F 13); similarly, in F 20 the inspection of the death chart involves the term ruler of the places in opposition to the planets.

Parallels

Valens is the only other ancient author apart from Critodemus to provide such a list (I 3) and, although his significations do not reveal much influence of his predecessor, it is reasonable to suppose that the choice to include them arose from seeing them in Critodemus' manual. Occasionally, as in the term of Venus in Capricorn, we find a meaningful coincidence:

Val. I 3.48
αἱ δὲ ἑξῆς η̅ Ἀφροδίτης ἀσώτων λάγνων καὶ **κατωφερῶν** ἀκρίτων ἐπιψόγων, εὐμεταβόλων περὶ τὰ τέλη, οὐκ εὐθανατούντων οὐδὲ περὶ τοὺς γάμους εὐσταθῶν.

The next 8° belong to Venus: profligate, lecherous, **downward-trending**, thoughtless, censured, having their ends very much in doubt, not dying well, nor steady in marriage.

Critodemus F 10.10
ἀπὸ μοίρας ι̅ε̅ ἕως κ̅β̅ ὅρια Ἀφροδίτης· ἡδυπάθεια, **κατωφέρεια.**

From the 15th degree to the 22nd, terms of Venus: pleasant living, **downward trend**.

The influence of the terms of Venus is generally positive, even in signs governed by malefics like Aries and Scorpio (ruled by Mars) and Aquarius (ruled by Saturn), so it is significant that here in the sign of Capricorn (also governed by Saturn), the terms assigned to Venus are pernicious. The concept of "downward tendency" (κατωφερῶν, κατωφέρεια), synonymous with a hedonistic and thoughtless way of life, is used nowhere else in Valens, and only once in Critodemus (in the term of Venus in Taurus).

Much clearer than the case of Valens is the astrological poem of the Byzantine official John Kamateros, drinking companion of Manuel I Komnenos (twelfth century), to whom the poem is dedicated, which presents an evident rephrasing of Critodemus' list of terms and effects:

Kamateros, *In astr.* 269–272
ἀπὸ μὲν πρώτης τοῦ Κριοῦ μέχρι τε καὶ τῆς ἕκτης
ὁ Ζεὺς ἐπέχει καὶ κρατεῖ, καὶ φέρουσι τὴν κλῆσιν
ἅρμα τε καὶ ἀκρίβεια τοῦ γεννωμένου ὧδε,
ἐκ ταπεινώσεως ὑψοῖ τύχης τε καὶ πρὸς δόξαν·
μεγίστων ὄχλων ἀστικῶν γίνεται ἀρχηγέτης.

From the first of Aries to the sixth
Jupiter dominates and rules, and these [degrees] have the name
raising-up and efficiency of the one being born so,
he is elevated in luck from low condition to good fame;
he becomes leader of great masses of citizens.

Critodemus F 10 (part)
Κριοῦ ἀπὸ πρώτης μοίρας ἕως ἕκτης ὅρια Διός· δίαρμα καὶ προκοπαὶ ἀπὸ ταπεινῶν εἰς μείζονα· τινὲς δὲ καὶ ἐν ὄχλοις εὐδοξοῦσιν ἢ καὶ διπολῖται γίνονται.

Aries from the first degree to the sixth, terms of Jupiter; raising-up and progresses from low to better; some receive fame from the crowd or become double citizens.

Kamateros introduces a distinction between what he calls the "name" or "appellation" (κλῆσις) and the explanation of the term, reflecting to a certain degree a division already found in Critodemus' sequences of effects. Indeed, Critodemus generally lists abstract substantives (here δίαρμα, προκοπαί) first, followed by more specific descriptions of personal characters or occupations. One could theoretically deduce that Kamateros was reading the original version in which Critodemus referred to the abstract substantives as κλῆσις, like Kamateros, but the wording throughout is so close to F 10 that we should probably assume the label is entirely due to Kamateros. In fact, it is likely that Kamateros did not even have the whole epitome of Critodemus, but only the same material that has survived to the present day.[26]

General patterns of the significations

The traditional features of the planets that rule the terms generally explain the effects listed in F 10. Jupiter, Mars, and Mercury seem to maintain a small and stable set of characteristics:

Jupiter: Fame, public career and honors, unexpected donations and inheritances, social progress

Mars: Military men, pirates, hardship, bad luck, enemies, few children, exile and escape, violence

Mercury: Activities and professions related to knowledge

26 Cf. the case of the *Epitome Parisina* (Chapter 4), in which it is clear that the author attempts to reconstruct Critodemus' work from Valens. Kamateros only acknowledges following Ptolemy for the triangles (2306), Hephaestio for the *katarchai* ("initiatives", i.e., horoscopes for initiating an activity), and Rhetorius for the regions associated with the signs (1287), but it is clear that he was not entirely precise nor exhaustive: for example, the latter topic is in fact taken from the initial chapters of Hephaestio. In Chapter 2, we have seen that Kamateros imitated the beginning of Critodemus, but, like us, he may have read it in Valens.

Mars is somewhat more variable than the other two, his more positive characteristics emerging when his terms are in signs ruled by certain planets. Thus, the terms of Mars in Gemini, which is ruled by Mercury, indicate "strength, governing capacity, progresses" (δύναμις, ἐξουσία ἡγεμονική, προκοπαί), and similarly those in Virgo, ruled by Mercury, indicate "light, brilliant military men or judges" (φῶς, λαμπροὶ στρατιωτικοὶ ἢ ἐπὶ κριτηρίων). The terms of Mars in Libra, governed by Venus, read "strength, leaders, expeditions full of honors, and overgrowth" (ἰσχύς, ἡγούμενοι, αἱ στρατιαὶ ἔνδοξοι καὶ ὑπεραύξησις).[27]

Saturn has a generally pernicious influence, though it is much more malleable by the influence of the signs than in Valens' system (I 3), where he is mostly just cold and dry, indicating peril and barrenness. In Critodemus' list, Saturn can be barren (Scorpio), but he can also make ambiguous men and governors (Aries). In the terms of the signs governed by Venus, Taurus, and Libra, erotic characteristics are evident, but in very different senses:

Terms of Saturn in Taurus: Solutions, enjoyment
Terms of Saturn in Libra: Disappearance of beloved persons

In the two signs ruled by Jupiter, a positive planet associated with kingship, the terms of Saturn are positive:

Terms of Jupiter in Sagittarius: Good soul
Terms of Jupiter in Pisces: Regal, wise

The terms ruled by Venus are quite varied. The significations most closely related to Venus' intrinsic characteristics (love, purity) are found in the signs ruled by herself, Libra and Taurus. Curiously, the more negative influences are in the signs ruled by Jupiter – Pisces and Sagittarius – as if the positive effects of the two cancel each other out:

Terms of Venus in Sagittarius: Distress, some vicious
Terms of Venus in Pisces: Jealousy, enemies

Clues on Critodemus' sources

Occasionally we find significations that might reveal the nature of Critodemus' sources. In Gemini, the first term (assigned to Mercury) is partly described as "calm

27 But in Taurus, also ruled by Venus, the term of Mars indicates "violent deaths, pirates, robbers, condemned, some destroyers, too" (βιαιοθάνατοι, λῃσταί, ἅρπαγες, κατάκριτοι, τινὲς καὶ σινούμενοι).

of the sea" (γαλήνη). This meteorological meaning was transferred to a personality type, which is probably what is intended here.[28] Similarly, the first term of Leo (ruled by Jupiter) indicates "flood" (ἀνάβασις), which is probably meant as growth in general, since effects related to progress and upward tendency are elsewhere ascribed to Jupiter in Critodemus' terms. However, its more common context is water and rivers. In Hephaestio, ἀνάβασις refers primarily to the Nile flood, as in I 23 on the yearly influences of the planetary positions at the rising of Sirius. One of the meanings attached to the first term of Libra (of Saturn) in F 10 is of a more sociohistorical nature: "destruction and overthrowing of rulers" (ἀπώλεια καὶ καθαίρεσις ὑπαρχόντων). Although an individual interpretation is still possible – the client could be one of those rulers – the most natural interpretation seems to be that this is meant for the whole country. Thus, it seems that Critodemus gathered the meanings of the terms from sources containing individual, collective, and meteorological significations,[29] with an evident preference for interpretations related to individuals. In particular, the flood reference points to an Egyptian source.

28 Valens similarly describes the degrees of this term as "producing good weather" (εὔδιοι), a probable adaptation of Critodemus' γαλήνη unconscious of the personal meaning of the word.

29 For example, Hephaestio's chapter I 20 summarizes the influences of each of the planets, divided in effects for humans (individual and collective), animals, and agriculture. Cf. also Julianus of Laodicea's excerpt listing planetary influences, *CCAG* I 134–137.

Chapter 6
On the time and kind of death (F 11–13)

In Chapter 4 we have seen that one method to deal with the different periods of life was to divide the time of life into intervals, each governed by one star. A related preoccupation of ancient astrology, which occupied an important place in the manuals, was that of determining the time of death.[1] The intense dedication of ancient astrology to matters of death may be striking to us, but this is surely in part because of contemporary attitudes toward death.

F 11 deals with charts of short-lived individuals (ἄχρονοι, ὀλιγοχρόνιοι), also described as "non-nurtured" (ἄτροφοι), a common topic in ancient astrology no doubt reflecting the high infant mortality in the Graeco-Roman world (common to all premodern societies).[2] This kind of chart is often treated by ancient astrologers (e.g., Ptolemy *Tetrabiblos* III 10) as a special, simpler case of the investigation of the length of life. According to Ptolemy, in these charts, the luminaries (Sun/Moon) are badly afflicted by malefics. A possible reason for this is that the luminaries often represent the *aphetes* ("starter") in calculations of the length of life, which measure in a specific way an interval starting from this point (cf. Ptolemy *Tetrabiblos* III 11). According to the chapter heading, Valens records a succinct account of Critodemus' version of this doctrine (called the "method of *aphesis*"), which I include here as F 12. It is noteworthy that Critodemus seems to prioritize the Moon both in his discussion of short-lived infants (F 11) and in his version of the method of *aphesis* (F 12).[3] The Moon had a special place in the Orphic tradition, since Orpheus is said to have stopped the Moon with his music, and Musaeus is said to be the son of the Moon.[4] F 13 deals with charts indicating violent death (βιαιοθάνατοι), which Critodemus identifies by looking at the lord of the terms of the descendant.

As for the transformations of Critodemus' work, F 11 gives us important insights, since Hephaestio, our source for this fragment, implies that he is quoting Critodemus literally, and yet the passage has the unmistakable form of a highly

1 Ptolemy affirms in his chapter devoted to the calculation of the length of life (III 11.1), perhaps referring to Nechepsos and Petosiris: "according to the ancients, it is ridiculous to attach all predictions to the person who by the constitution of the years of life will never reach the years of those predictions" (κατὰ τὸν ἀρχαῖον γελοῖόν ἐστι τὰ καθ' ἕκαστα τῶν ἀποτελουμένων ἐφαρμόζειν τῷ μηδ' ὅλως ἐκ τῆς τῶν βιωσίμων ἐτῶν ὑποστάσεως ἐπὶ τοὺς ἀποτελεστικοὺς αὐτῶν χρόνους ἥξοντι).

2 See the classical study Bagnall and Frier 1994, based on the census returns of Roman Egypt.

3 See also the doctrine of the critical years in the next chapter, calculated from the distance to the Moon.

4 Molina Moreno 2009, 42 (for Orpheus' lyre); Bernabé and Casadesús 2009, 42 (Orpheus' lyre); Martínez Nieto 2009, 565 (for Musaeus and Epimenides).

https://doi.org/10.1515/9783111329147-007

standardized and monotonous paraphrasis. I will also address the fact that Hephaestio calls Critodemus' work *pinax*, which I interpret as a description of the style of the text rather than a new title.

F 11. Charts of non-nurtured children: Hephaestio II 10.41–46

Type of fragment: A literal quotation by Hephaestio from a passage of Critodemus on characteristic birth charts of non-nurtured babies. The text is repetitive and standardized, which is to be connected with Hephaestio's label *pinax* ("a register").

41 <Κριτόδημος δὲ>[a] ἐν τῷ αὑτοῦ πίνακι[b] ταῦτά φησιν. **42** οἱ φωστῆρες ἐν ταῖς ἐσχάταις μοίραις τῶν ζῳδίων ὄντες ἀχρόνους ποιοῦσιν· Σελήνη ἐν τῷ ἐκλειπτικῷ ὑπὸ Ἡλίου καὶ Ἄρεως ὁραθεῖσα ἀχρόνους ποιεῖ· Σελήνη ἐν τῷ βορείῳ μειουμένη ὑπὸ Κρόνου ὁραθεῖσα μάλιστα κατὰ τετράγωνον ἢ διάμετρον ἢ ἐν τῷ νοτίῳ μειουμένη ὁμοίως ὑπὸ Ἄρεως ὁραθεῖσα ἀχρόνους ποιεῖ· Σελήνη δύνουσα πληρουμένη τοῦ Ἄρεως ὡροσκοποῦντος, μειουμένη δὲ Κρόνου, ἀχρόνους ποιεῖ. **43** ἐὰν δὲ οἱ κακοποιοὶ τὰ κέντρα ἔχωσιν ἐν ἀλλοτρίοις τόποις, οἱ δὲ φωστῆρες ἐν ἀργῷ τόπῳ, ἄχρηστα καὶ κυνόβρωτα ποιεῖ· ἐὰν οἱ κακοποιοὶ ἀνατολὰς ἔχωσιν, οἱ δὲ ἀγαθοποιοὶ κρύψεις, ἄχρονα ποιοῦσι τὰ τικτόμενα. **44** ἐν ταῖς ἐκλείψεσιν Ἡλίου καὶ Σελήνης ἐὰν ὁ κύριος τοῦ ὡροσκόπου κρύψιν ἔχῃ καὶ μὴ σχηματίζῃ ἀχρόνους <ποιεῖ>. **45** ἐὰν δὲ τοῦ ζῳδίου τὰ ἔσχατα μέρη ὡροσκοπῇ, τὰς δὲ προτέρας μοίρας κατέχωσιν οἱ κακοποιοί, ἄχρονα γεννᾶται, ὡ<ς ἐπίπαν δὲ καὶ οἱ>[a] τικτόμενοι ἐν ταῖς ἐκλείψεσιν ἄχρονοι ἢ τυ<φλοὶ τίκτονται>[a]. **46** καὶ ἐὰν Σελήνη ἀστέρι κακοποιῷ στηρίζ<οντι συνάψῃ ὀλιγοχρόνια τὰ τικτόμενα ποιεῖ.>[a]

a *lacunae* reconstructed from the epitome IV; the first instance also in the margin of the only manuscript covering this section.
b Πίνακι (capitalized) in Pingree's edition, denoting the title of a work. See discussion for my interpretation.

41 Critodemus in his summary[5] says the following: **42** The luminaries in the last degrees of the signs make short-lived [infants].[6] The Moon on the ecliptic, aspected by the Sun or Mars, makes short-lived.[7] The Moon waning in the northern turn-

5 See discussion below.
6 Perhaps related to the fact that the last degrees of the signs usually fall in terms of the malefics (cf. Chapter 5).
7 The Moon does not travel exactly on the ecliptic, but on the slightly tilted circle crossing it at two points called nodes, which in turn move along the ecliptic at the approximate rate of 1 turn every 18 years. Thus, the Moon is on the ecliptic when she is on one of the nodes, which is when there is

ing-point[8], aspected by Saturn especially in square or diameter, or equally waning in the southern turning-point and aspected by Mars, makes short-lived.[9] The full Moon in the descendant with Mars in the ascendant, or waning with Saturn [in the ascendant] makes short-lived. **43** If the malefics hold the centers[10] and are in the places of other stars,[11] and the luminaries are in the inactive place,[12] this makes useless persons devoured by dogs. If the malefics hold the ascendant, and the benefics are invisible, they make the new-born short-lived. **44** In the eclipses of Sun and Moon, if the lord of the ascendant is invisible and not aspected, it makes short-lived. **45** If the extreme sections of the sign are in the ascendant,[13] and the malefics hold the first sections, short-lived are born, as well as generally those born in eclipses are born short-lived or blind. **46** And if the Moon is touching a stationed star,[14] she makes the new-born short-lived.

Critodemus' *pinax*

Is πίναξ the title of another work by Critodemus? Valens only mentions the title Ὥρασις, and we have no other information about other titles or works. The term πίναξ refers primarily to a tablet, which can lead to several derivative meanings,

---

a possible eclipse, and eclipses are generally interpreted as bad signs. Hephaestio reports a similar procedure from Dorotheus in predictions about the parents (II 4.20–23): things are not good when the luminaries are in the last degrees of the signs, or when the Moon is on the ecliptic (ἐν τοῖς ἐκλειπτικοῖς). "To be aspected" is equivalent to "to be in aspect with", meaning to hold a square, triangular, or hexagonal relation with another planet or luminary.

8 The text says ἐν τῷ βορείῳ. The four points, the maximum northern, southern, and the nodes, are bad in Hephaestio's chapter on injuries (II 13.15): "especially when the Moon is on the nodes or on the turning-points, that is, in the southern extreme or in the northern one" (μάλιστα τῆς Σελήνης οὔσης ἐπὶ συνδέσμων ἢ ἐπὶ καμπίων, τουτέστιν ἐν τῷ νοτίῳ πέρατι ἢ <τῷ> βορείῳ).

9 The waning phase of the Moon is generally worse than the waxing Moon, and here its decreasing visibility might be suggestive of shorter life span. Cf. Appendix II for the detrimental effect of the square aspect and opposition.

10 τὰ κέντρα ἔχωσιν: This just means to be in the centers or cardines (cf. Appendix II). See Heph. II 19.25, quoting Dorotheus: κἂν κέντρον ἔχῃ πτολιπόρθιος Ἄρης, which Hephaestio interprets as ἐπίκεντροι (II 19.21). The centers enhance the effects of the planets they contain.

11 That is, in signs that are ruled by other stars, not themselves. If they were in their own signs, their positive effects would be enhanced.

12 This is the eighth domicile, also called "death" (Valens IV 12.1), which is crucial in Dorotheus' doctrine on the kind of death, on which see below, F 13.

13 Meaning the last degrees of the sign: cf. the first prediction.

14 A planet that is precisely changing from direct motion (eastward) to retrograde (westward) or vice versa.

such as a writing tablet or a painted one. Interestingly, another ancient Greek astrological manual apparently had the same title, to judge from the heading in the *Epitome Parisina* (*CCAG* VIII 3, 99):

> Συγκεφαλαίωσις τοῦ πρὸς Ἱεροκλέα Θρασύλλου πίνακος
>
> Summary of the *pinax* of Thrasyllus, [dedicated] to Hierocles

A natural assumption is that here πίναξ refers to the astrological board displaying the circle of the zodiac and some of its supplementary divisions such as the terms or the decans, upon which gems representing the planets were positioned to recreate the natal chart (cf. introduction); this is undoubtedly the meaning of the word in the *Alexander Romance*, in which the pharaoh-turned-astrologer Nectanebo employs a zodiacal board, named πίναξ in the text, to prophesize for the Macedonian Queen Olympias.[15]

However, the other principal meaning of πίναξ was "summary" or "index", a transference from the tablets in libraries containing the title and contents of the works, as in the title of Callimachus' *Pinakes*. Many ancient works are preceded in the manuscripts by indexes introduced with the term πίνακες. Closer inspection of the *Epitome Parisina* shows that this meaning is probably the one intended both for Thrasyllus and for Critodemus in this fragment: we should look not only at the titles of the excerpts – which were probably introduced later – but also the endings. The title of the summary of Thrasyllus in fact corrects the name of the astrologer and of his dedicatee, which in the closing sentence of the text are given as Thrasyboulos and Herakles (!).[16] Despite these errors, in his closing statements the epitomizer is quite precise about his own procedure. At the end of Balbillus' summary, the scribe is keen to signal that he has produced the summary with the treatise in hand.[17] This was not always the case, as we know from the previous chapter. Furthermore, he always indicates in one way or another that what he has written is a summary: for example, he writes "has been summarized" (συγκεφαλαίωται) in the case of Cri-

15 Pseudo-Callisth. I 4: "Just after saying that he produced a regal and precious board" (Ἅμα δὲ τῷ ταῦτα εἰπεῖν προενεγκάμενος πίνακα πολυτίμητον βασιλικόν). A probable, but less clear instance is Plutarch's reference to an astrologer friend of Varro and his περὶ τὸν πίνακα μέθοδος (Plut. *Romul.* 12.3). A similar concept is found in the geographical πίνακες (cf., e.g., Strabo II 2.10, Ptolemy *Geogr.* passim), which can be simply translated as maps.

16 Sometimes the title provides information that is not present in the text: in the title of the summary of a certain Callicrates we find the name of the dedicatee (Timogenes) and the title of the work (ἐπιγράφεται ὁ λόγος Καλλικράτους Θησαυρός, "the work is titled 'Thesaurus'").

17 *CCAG* VIII 3.104: "These are the chapters of Balbillus, which we have seen ourselves, from his astrological writings" (ταῦτα καὶ τὰ τοῦ Βαρβίλλου κεφάλαια ὧν ἡμεῖς εἴδομεν αὐτοῦ ἀστρολογούντων γραμμάτων).

todemus, "summary" (συγκεφαλαίωσις) for Callicrates, and "headings" (κεφάλαια) for Balbillus. Now let us compare the closing statements in Balbillus' and Thrasyllus' epitomes:

> ταῦτα καὶ τὰ τοῦ Βαρβίλλου κεφάλαια ὧν ἡμεῖς εἴδομεν αὐτοῦ ἀστρολογούντων γραμμάτων (*CCAG* VIII 3, 104)
>
> These are the headings of Balbillus' [work], which we have seen ourselves, from his astrological writings
>
> ταῦτα μὲν καὶ ὁ Θρασυβούλου πρὸς Ἡρακλέα πίναξ (*CCAG* VIII 3, 101)
>
> This is the *pinax* of Thrasyboulos' [work dedicated] to Herakles

Seeing that Thrasyllus' text is not otherwise characterized as a summary, it seems at least plausible that πίναξ could have been used here in the sense of "index". Later, the author of the title may have interpreted the term incorrectly as the title of the astrological work, perhaps because of the "board" meaning and/or because the sequence of the wording is ambiguous (ὁ Θρασυβούλου πρὸς Ἡρακλέα πίναξ can mean "the *Pinax* of Thrasyboulos dedicated to Herakles").

Further support for this argument comes from Hephaestio. This compiler was clearly not interested in the titles of the astrological works from which he drew; he gives only the name of the author, with one possible exception, if we disregard the case of Critodemus: the *Synagogai* ("Compilations") that Pingree takes for the title of a treatise by Protagoras of Nicaea (III 47.52).[18] This is, however, more probably a generic name for a manual or treatise. Hephaestio also refers to his own work by this name,[19] which, incidentally, the summarizer of the *Epitome Parisina* also used for Callicrates' manual, the real title of which was Θησαυρός ("Treasure").[20] It would be strange for Hephaestio to have used a true title in only case, that of Critodemus. It is far more likely that by ἐν τῷ αὐτοῦ πίνακι he meant "in his summary", especially if we consider the form of the text that he quotes. This was probably the same long summary from which the fragments on the distributions (Chapter 4) and the terms (Chapter 5) were excerpted.

---

18 On this author, tentatively dated to the third or fourth century, see Pingree 1987, 438–439.

19 At the beginning of his third book: "from my collection of what has been said by the ancients" (ἐκ συναγωγῆς ἡμῖν τῶν παρὰ τοῖς ἀρχαίοις εἰρημένων, III 1.1).

20 *CCAG* VIII 3.103: "and here is the summary of the collection of Kallikrates, which he has gathered from the ancients" (ἐν οἷς καὶ τῆς τοῦ Καλλικράτους συναγωγῆς, ἣν ἐκ τῶν παλαιῶν ἤθροισεν, συγκεφαλαίωσις).

Context in Hephaestio: other astrological theories on short-lived children

Chapter II 10 of Hephaestio, where the passage is quoted, is titled Περὶ ἀτρόφων, "On unnurtured children". His text makes it clear that ἄτροφος and ἄχρονος are synonymous as far as astrology is concerned (ἄτροφον ἢ ἄχρονον, par. 7). This seems to have been a typical chapter of ancient astrological treatises, since Ptolemy devotes a long section to this topic (III 10), also called Περὶ ἀτρόφων, as does Firmicus VII 2 (mixing exposed and non-nurtured infants). Hephaestio fills most of the chapter with his paraphrasis of Ptolemy's theory, as he does in the preceding and the following chapters (on monsters/on the length of life). Ptolemy's theory for the non-reared child is as follows (III 10.2):

> ἁπλῶς γάρ, ἐάν τε κεκεντρωμένον ᾖ τὸ ἕτερον τῶν φώτων, καὶ τῶν κακοποιῶν ὁ ἕτερος συνῇ ἢ διαμηκίζῃ, ταῦτα δὲ μοιρικῶς καὶ κατ' ἰσοσκελίαν μηδενὸς μὲν ἀγαθοποιοῦ συσχηματιζομένου, τοῦ δὲ οἰκοδεσπότου τῶν φώτων ἐν τοῖς τῶν κακοποιῶν τόποις κατειλημμένου, τὸ γεννώμενον οὐ τραφήσεται, παρ' αὐτὰ δὲ ἕξει τὸ τέλος τῆς ζωῆς.
>
> In short: if one of the luminaries is angular and one of the malefics is in conjunction with it, or in opposition, both in degrees and with equality of distance,[21] while no benefic bears any aspect, and if the houseruler of the luminaries[22] is found in the places of the malefics, the child that is born will not be reared, but will at once experience the end of life.

There is a vague resemblance with the conditions of Critodemus. In both theories, for a chart to be distinctive of non-reared children, the luminaries should be badly afflicted by malefics, but whereas for Ptolemy it is the luminaries that must be in the angles, and Critodemus reserves this condition for the malefics. Critodemus, however, gives many more possibilities, focused instead on the Moon and often with an emphasis on the extreme and the first degrees of the signs: thus, the luminaries are afflicted when they are positioned in the last degrees, possibly because these tend to contain the terms of the malefics (cf. Chapter 5). The first degrees of the sign are important, so if they are held by malefics the prediction is negative.

The opinion of Apollinarius and the Egyptians was, to judge from what Hephaestio records (par. 14–16 in the same chapter), similar to Ptolemy's and Critodemus' in that the interacting forces are those of the luminaries and the malefics: if malefics are in the place called "Life" (i.e., the ascendant) and they succeed the

21 I take this second statement to be synonymous with μοιρικῶς, "to the degree" (it would then be a *hendiadys*). The text that follows the quotation also suggests this interpretation: "If this comes about without the equality of distance, but the rays of the maleficent planets succeed closely upon the places of the luminaries. . ." (ἐὰν δὲ μὴ κατ' ἰσοσκελίαν μὲν τοῦτο συμβαίνῃ, ἀλλ' ἐγγὺς ἐπαναφέρωνται τοῖς τῶν φώτων τόποις αἱ τῶν κακοποιῶν βολαί. . .).

22 The ruler of the sign where the Sun and Moon are situated.

luminaries, implying that the Sun or the Moon is in the place known as Bad Daimon, then the child will be ὀλιγοχρόνιος.

Dorotheus' theory, also recorded by Hephaestio just before that of Critodemus (par. 37–40), was completely different. For day births, it is necessary to count the degrees from Saturn to Mars and then transpose these degrees counting from the ascendant,[23] and if the lord of the sign where the count falls is Mars and Mars is in one angle, then the child is ἄχρονος (for night births, the procedure is the same, substituting Mars for Saturn). A very similar technique is recorded in a section on chronic illnesses in the Arabic version of Dorotheus (IV 1.75).

In light of the disparity of theories on the subject, it is significant that Critodemus' theory bears a close affinity with some predictions of Firmicus (VII 2). Like Critodemus, Firmicus does not propose a single way of determining short life for an individual, but a series of possible configurations. The clearest parallel is perhaps the following (VII 2.11, cf. F 11.42):

> *Cum Sol et Luna in quocumque signo XXX signi possederint partem, is qui natus fuerit nulla habebit vitae tempora. Si vero Solem et Lu nam in XXX parte constitutos malivolae stellae aut <de> quadrato aut <de> diametro respexerint, is qui natus fuerit expositus a canibus laceratus extinguitur.*
>
> When the Sun and the Moon hold the 30th degree of a sign, the native will have no time of life. If malefic planets are in square aspect to or in opposition to the Sun and Moon in the 30th degree, the native will be torn by dogs and perish.

Also very similar to one of the prognostications in Critodemus, according to which the Moon in the descendant with a malefic in the ascendant makes short-lived individuals (42), is the following prognostication in Firmicus' chapter (VII 2.16):

> *Si Luna in occasu fuerit inventa, et Mars aut cum ipsa in isdem partibus fuerit inventus, aut in horoscopo partiliter constitutus, et sit diurna genitura, is qui natus fuerit exponetur.*
>
> If the Moon is on the descendant and Mars is with her in the same degree or in the ascendant in a diurnal chart, the native will be exposed.

Thus, some of the predictions in Firmicus may directly come from Critodemus, who, as we know, was a direct influence on Firmicus in other doctrines. Furthermore, the position of Firmicus' chapter at the beginning of book VII, which he opens with his version of Critodemus' oath (Chapter 3), supports this possibility. This being said, Firmicus incorporated numerous predictions either of his own or from other

23 This kind of calculated point is called a "lot" (κλῆρος): The best known is the Lot of Fortune, determined by adding to the ascendant point the distance in longitude from the Sun to the Moon (from the Moon to the Sun for night births).

sources, involving (a) consideration of the houses, which play a limited role for Critodemus here,[24] and (b) the Lot of Fortune, which is completely absent from F 11. He also ignored configurations that Critodemus deemed important, containing elements such as the latitude and the phase of the Moon.

F 12. Critodemus' method of *aphesis*: Valens III 5.18–20

Type of fragment: A doctrine of *aphesis* (also called prorogation) for determining the length of life using a table of rising times, ascribed to Critodemus in the chapter heading. Such attributions in Valens' manuscripts are generally reliable, and nothing indicates otherwise in this case. The text in the fragment identifies the most appropriate starting points (*aphetai*) and ending points (*anairetai*) for the prorogation according to Critodemus; instead of counting the degrees from the *aphetes* to the *anairetes* on the zodiac, astrologers looked up in the table of rising times the difference in time-degrees between the two, following the procedure that will be described below. The final part of the text seems to be Valens' own contribution, adding more possible points and conditions to Critodemus' theory.

Ἄλλως περὶ ἐχθρῶν τόπων καὶ ἀφέσεων ἐκ τῶν Κριτοδήμου. ἀπὸ Σελήνης καὶ ὡροσκόπου[a]

18 ἀλλ' ὁπόταν ἡ Σελήνη ἀφέτης εὑρεθῇ, παραφυλάττεσθαι χρὴ τὰς κολλήσεις[b] καὶ τὰς ἑξαγώνους πλευρὰς καὶ τετραγώνους καὶ διαμέτρους τὰς πρὸς τὸν ὡροσκόπον κατὰ ἀναφοράν, αὗται γὰρ ἐνεργητικαὶ κριθήσονται, καὶ μάλιστα ἐν τοῖς ἰσανατόλοις ἢ ἰσαναφόροις ἢ ἰσοδυναμοῦσιν ἢ τοῖς ἀκούουσιν ἢ βλέπουσι ζῳδίοις ἢ ταῖς ἀντισκίοις μοίραις. **19** ὁμοίως δὲ κἂν ὁ ὡροσκόπος ἀφέτης εὑρεθῇ, τὰς πρὸς τὴν Σελήνην διαστάσεις ὡσαύτως σκοπεῖν κατὰ ἀναφοράν. **20** ἔδοξε δέ μοι ἐκ πείρας θανατικὰς μοίρας κρίνειν καὶ δυναστικὰς τὰς μεσουρανούσας τοῦ τε ὡροσκόπου καὶ τῆς Σελήνης πρὸς ἀλλήλους καὶ τὰς τούτων διαμέτρους· αὗται γὰρ κεντρωθεῖσαι οὐ τὴν τυχοῦσαν κέκτηνται δύναμιν.

a Pingree puts the heading in square brackets, wrongly to my mind.
b κολλήσεις coni. Riley, κωλύσεις MSS. Indeed, the reading in the manuscripts ("hindrances") does not make sense in the context. For "contact" (κόλλησις) in astrological context, cf. e.g. Valens II 38.12.

24 He mentions the inactive place, the angles, and the ascendant, but no house by number. Cf. Appendix II for the houses.

Otherwise: on hostile places and *apheseis*, from the [doctrines] of Critodemus. From the Moon and the ascendant.
18 Whenever the Moon is found to be the *aphetes*, one must pay attention to the contacts, the hexagonal sides, the squares, and the oppositions to the ascendant according to the rising times.[25] These will be judged to be effective, and especially in signs of the same ascensional or rising times, signs of the same power, the listening or beholding signs,[26] or the degrees of the antiscia.[27] **19** Likewise, if the ascendant is found to be the *aphetes*, look at its distances with respect to the Moon according to the rising times, too. **20** From my experience, I would consider the deadly and the powerful degrees those at midheaven, those at the ascendant and the Moon against each other, and those in opposition to these points. If they are at a center, they have an extraordinary influence.

Context in Valens

This fragment is found in Valens' text just before another passage related to Critodemus, whose title reveals its relation to the first table of book VIII (Περὶ ἐχθρῶν ἀστέρων καὶ κλιμακτηρικῶν τόπων περὶ τὸ α' ὄργανον Κριτοδήμου), and which uses the terms written therein for one horoscope that exemplifies a method for finding the time of death (see Chapter 8). Pingree prints the title of our fragment in square brackets, to indicate that it should be suppressed, adducing that it refers to the next chapter. I, however, see no grounds for hypothesizing a scribal confusion. Generally, the titles of Valens' manuscripts completing the information of the chapters are reliable.[28] Furthermore, the text clearly reveals that the doctrine is adopted from some source, since it first proposes the Moon and the ascendant with their related *anairetai*, and thereafter the midheaven and other points as Valens' own contribution (20).

---

25 That is, when the Moon is *aphetes*, consider these points as a possible *anairetes*: the ascendant, its hexagonal sides, squares, or oppositions. "According to the rising times" just means to calculate the length of life in the usual way (see below).
26 All these concepts are synonymous: cf. Ptol. I 15. The pairs of signs of equal rising times are found drawing parallel lines joining the signs in the following way: Aries and Pisces, Taurus and Aquarius, Gemini and Capricorn, etc.
27 Such pairs are also organized in parallel lines, but the joining lines are perpendicular to those of the beholding signs: thus instead of Aries–Pisces, Taurus–Aquarius, and so on, the antiscia are Gemini–Cancer, Taurus–Leo, etc. The idea is that if the candidates to *aphetes* and *anairetes* are of the same rising times or according to the rays of antiscia, then they are more probably the ones we are looking for.
28 Names of astrologers as sources occur in the titles of II 29 (Hermippos), II 32 (Timaios), and IV 27 (Seuthos and Hermeias).

The method of *aphesis* and the anaphoric table (or table of rising times)

Most of Valens' book III is dedicated to the method of the vital sector (also called *aphesis*, or prorogation), a widely used calculation of the length of life consisting in adding the rising times between two points on the zodiac: the *aphetes* ("starter") and the *anairetes* ("destroyer"), which was found up to 90° ahead along the zodiac. Pliny calls the method *tetartemorion*, "fourth part", from the maximum distance in degrees between both points (a quarter of the zodiac), when recording the alleged opinions of Epigenes and Berossus on the greatest length of life:

> *Epigenes CXII annos inpleri negavit posse, Berosus excedi CXVI. durat et ea ratio, quam Petosiris ac Necepsos tradidere (tetartemorion appella<n>t a trium signorum portione), qua posse in Italiae tractu CXXIIII annos vitae contingere apparet.*
>
> Epigenes says it [human life] cannot exceed 112 years; Berossus, that it exceeds 116. The system that Petosiris and Nechepsos transmitted to us (they call it "tetartemorion", from the portion of three signs), according to which in Italy it may be extended to 124 years.

The method behind these numbers became the most popular doctrine for calculating the length of life. Most variations of the method consist of the different doctrines for finding these points. The time it would take for this sector to rise was calculated in time-degrees using the table of ascensions, and the result was interpreted as years of life (see Appendix II for the table of rising times and its use). The underlying idea is that the years of life correspond to the time-degrees that the destroying point takes to travel to (and impact) the point where the *aphetes* is, using the clockwise, daily rotation of the heavens and maintaining a fixed starting point.[29] If one was very lucky, his or her vital sector corresponded to the 90° portion of the ecliptic rising in the maximum time, that is, the two slowest rising signs, Virgo and Libra, plus a half sign extending beyond them before and after. In Alexandria, both Virgo and Libra took 40 time-degrees, while Leo and Sagittarius took 36. So, the maximum length of life in Alexandria could be calculated as:

$$18 + 40 + 40 + 18 = 116 \text{ years of life}$$

An analogue calculation gives the numbers for the maximum length of life in Babylon and Italy recorded by Pliny.[30]

---

29 When the starter is not the ascendant, the precise astronomical calculation (i.e., how long it takes for the destroyer to hit the aphetic point) is much more complicated (cf. Ptolemy III 11), but we should probably assume that the average astrologer carried out the calculation in the same way.
30 The passage is discussed in *HAMA*, 721.

Critodemus' method compared with others

Much of Valens' book III is devoted to explaining and exemplifying variants of the *tetartemorion* method for calculating the length of life. In III 3, Valens considers the Sun, the Moon, or the ascendant as possible *aphetai*. For the luminaries to be the *aphetes*, they must be in *aphetic* places (meaning "appropriate for the *aphetes*") which, judging from his examples, seem to be the angles above the horizon.[31] One variant consists in observing if what Valens calls the "controller" (ἐπικρατήτωρ), defined as the ruler of the terms of the Sun or the Moon (III 1.3–4), bears an aspect toward the *aphetes*, or is well positioned (e.g., in one angle): in such a case, the *aphetes* will give the full number of years (III 3.1–6). The presence or the aspects of malefics affecting degrees in the vital sector either shorten or terminate life at that point. Ptolemy (III 11.9) considers more possibilities for the *aphetai*: in addition to the Sun and the Moon, he also includes the ascendant, the Lot of Fortune, and the rulers of those parts of the zodiac, establishing a clear hierarchy for the choice point, which must always be above the horizon. The ruler of the terms of the Sun or the Moon is the first option for the *aphetes* in Dorotheus, who requires that it aspect the Sun/Moon.[32]

Critodemus follows the general conventions of the method of *aphesis*. From F 12, it appears that he only considered the Moon and the ascendant (and their aspects) as candidates for *aphetic* and *anairetic* places. If the Moon is the *aphetes*, the possible life sectors reach up to the ascendant or its aspects (never surpassing 90°). Sextile, square, and opposition are mentioned, but not triangle. It seems striking at first sight not to consider the triangle aspect of the ascendant as a possible *anairetic* point, which would be a natural option when the Moon is close to the descendant: see illustration below, taking into account that the destroyer should be above the horizon and more advanced along the signs – that is, more to the left – than the starter. The problem is solved considering that, for cases like this one in which the starter would be found in the occidental quadrant, Ptolemy prescribes a slightly different method, *horimaia* (III 11.9), which consists in calculating the life sector in the sense contrary to the signs (i.e., clockwise), from the *aphetes* to the descendant, which symbolizes death (the descendant is now fixed and the starter impacts it with the daily motion of the heavens). That Critodemus had this method

31 Cf., e.g., Ptolemy III 11. What Valens calls the ruler of the chart in book III, playing a similar role as the *aphetes*, must also be above the horizon: cf. III 5, first example.

32 *apud* Hephaestio II 26. Dorotheus, like Ptolemy, establishes a hierarchy of choices: If the condition is not met, one has to try with the ruler of the triangle. If not, try in the same order with the other luminary, with the Lot of Fortune, or with the previous syzygy. If not, then it will be the ascendant.

in mind is suggested by the fact that the descendant – the point in opposition to the ascendant – is included in the fragment as a possible destroyer.

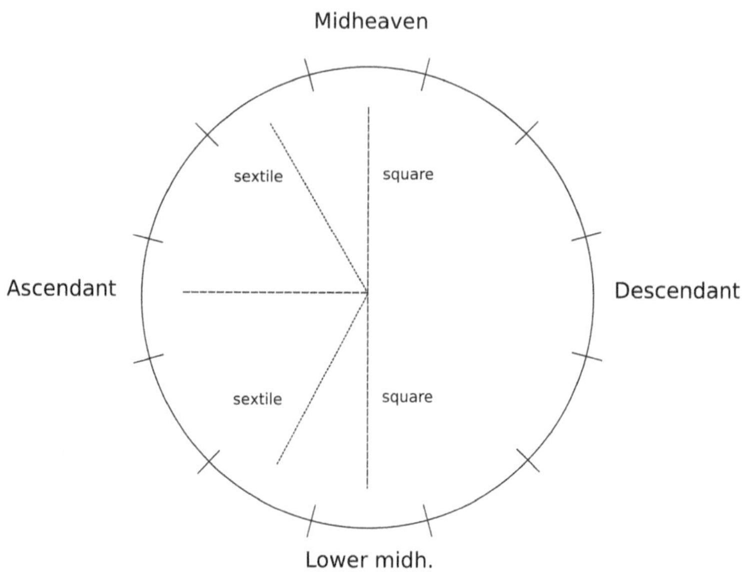

Conversely, if the ascendant is the *aphetes*, the places corresponding to the Moon and its aspects should be considered as hostile and potential destroyers.

F 13. Charts of violent deaths: Rhetorius 77 (*CCAG* VIII 4, 199)

Type of fragment: Characteristics of the birth charts of individuals who experience violent deaths, listed in a highly schematic, formulaic style.

Ἄλλως κατὰ Κριτόδημον[a]. ὁ ὁριοκράτωρ τοῦ δύνοντος[b] κακοποιὸς ὢν καὶ ὕπαυγος ἐν ταῖς τοῦ Ἡλίου μοίραις τυχὼν ἐκ δόλου καὶ ἐξ ἐνέδρας τὸν θάνατον σημαίνει, ἔξαυγος δὲ ὢν βιαιοθανάτους φανερῶς ποιεῖ. ὁ ὁριοκράτωρ τοῦ δύνοντος[c] κακοποιὸς ὢν καὶ στηρίζων[d] ἢ ἀναποδίζων ἀπὸ σίνους καὶ φαρμάκων ἢ τραυμάτων ἀναιρεῖ. ὁ ὁριοκράτωρ τοῦ δύνοντος κακοποιὸς ὢν καὶ ἐν ἀνθρωποειδέσιν ζῳδίοις ὑπὸ κακοποιῶν θεωρούμενος ἐξ ἀνθρώπων τὸν θάνατον ποιεῖ, ἐν δὲ παρύγροις ζῳδίοις ἀπὸ ὑγρῶν, ἐν δὲ θηριώδεσιν ἀπὸ θηρίων ἢ ὕψους.

a Ἄλλως κατὰ Κριτόδημον om. **P**. But this manuscript systematically omits references to authors: cf. e.g. the references to Valens in p. 188 lines 12 and 15.

b δύνοντος κέντρου **P**; ζ̄ **A**. The more recent manuscript **P** (=Par. gr. 2425, 15th c.) seems to have corrected the abbreviated expression "7th" (=7th place) here and below in the passage, with good judgment: the wordings are equivalent if we are dealing with the whole sign occupying the 7th place, but not here where the text is concerned with the terms in the point of the descendant. Most of the text in **A** (=Par. gr. 2506, 13th c.) represents an older and less complete account: (a) **A** does not have the "Lot of Death" (p. 199, lines 11–15); (b) some (but not all) of the instances on the ruler of the 8th place (p. 200, lines 2–6 and 14–25) do not appear in **A**, whereas **P** has all of them, probably gathered from a previous chapter (p. 162). It seems that **A** had also collected some instances on the 8th place from that chapter, but **P** completed the job.
c δύνοντος **P**; ζ̄ **A**
d στηρίζων – δύνοντος om. **P** (leap from same to same)

Otherwise according to Critodemus: the ruler of the terms of the descendant being a malefic and under the rays – situated in the degrees of the Sun[33] – means death by trickery and ambush, and not under the rays makes violent deaths and dying in public.[34] The ruler of the terms of the descendant being a malefic and in station or retrograde destroys from injury or drugs or blows. The ruler of the terms of the descendant being a malefic in the humanlike signs, and aspected by malefics, makes deaths by humans; in the watery signs, from water; in the animal-like signs, from beasts or heights.

Textual analysis

A sixth-century astrological compilation under the name of Rhetorius has been fragmentarily preserved in an epitome preserved in a dozen manuscripts.[35] F 13 belongs to a large section of the epitome dealing with astrological doctrines by topic. The present chapter is called "general configurations for violent deaths"

33 ὕπαυγος ("under the rays") is a technical term indicating close proximity with the Sun, which hides the planet with its potent rays: cf., e.g., Hephaestio II 19.8: "I think one should not take the planet in conjunction with the Sun, since it will be under the rays" (ἡγοῦμαι ἐνταῦθα μὴ δεῖν τὸν συνάπτοντα τῷ Ἡλίῳ λαμβάνειν, ὕπαυγος γὰρ ἔσται).
34 The descendant is here indicative of death, probably through a simple association of ideas relating to "setting", and also by opposition with the label "Life" given to the place of the ascendant. In the method for the length of life called *horimaia* (cf. above, discussion of F 12), the descendant represents the endpoint of life.
35 For a list of the manuscripts and editions, see László 2020.

(καθολικὰ σχήματα βιαιοθανάτων). The next chapter, numbered 78 in the manuscripts, explores "general configurations for banished persons" (καθολικὰ σχήματα ἐκπτώτων, *CCAG* VIII 4, 202), and the preceding ones also present a similar approach, listing possible configurations for representative personal outcomes: 65 is on mentally disturbed persons (περὶ μαινομένων καὶ ἐπιληπτικῶν, p. 192), while 66–76 deal with configurations of licentious persons grouped according to the technical nature of the predictions.

The text of the compilation as we have it shows little interest in preserving literal quotations of its sources. Occasionally, as in this case, names of the ancient authors responsible for the contents appear in some of the manuscripts.[36] The part attributed to Critodemus is not very different in style to F 11, above. It is likely that Rhetorius had access to a similar paraphrasis of Critodemus' work, but this cannot be deduced from the text, because, unlike in the case of Hephaestio, most chapters in the Rhetorius compilation show this kind of formulaic style that is typical of epitomes. Nevertheless, a new hypothesis situating Rhetorius well before the Byzantine compendium, in the fourth century AD, would seem to favor the possibility that Hephaestio, Rhetorius, and Firmicus had roughly the same version of Critodemus.[37]

Determining the end of the fragment is a more difficult task than in most cases. A cut is found much later in the same chapter (p. 201, line 10), where the text refers to other "paragraphs" (κεφάλαια) following other principles, presumably in the original sixth-century compendium (remember, this is an epitome):

> ἔστι δὲ καὶ ἄλλα κεφάλαια βιαιοθανασίαν ποιοῦντα ἀλλ' οὐχ οὕτω μονοσχηματοῦντα ἀλλ' ἀπὸ πολλῶν ἀστέρων συσχηματιζόμενα.
>
> There are other chapters also dealing with violent death, but they involve several stars, rather than approaching it from a simple configuration.

The compiler does not consider it necessary to write down all these complex configurations: he just mentions one of them involving the Moon in the midheaven in opposition to Mars and the Sun, which is similar in character and content to the first prediction in Firmicus' long chapter VII 23, from which we can deduce that Rhetorius and Firmicus had access to a similar source.

---

36 In the most extensive epitome (in Par. gr. 2425), the sources are suppressed (cf. apparatus in the Greek text).

37 And probably Valens, too (cf. ch. 4). For the new dating of Rhetorius, cf. László 2020: His main arguments are very specific doctrines common to both Rhetorius and Paulus of Alexandria, writing in 378 (the effects of a planet's exact conjunction with the Sun and the so-called Hermetic lots). Previously, Pingree 1977 had dated Rhetorius to the early seventh century AD.

Before that point, the text lists predictions that basically involve one star or point in relation to another: at the very beginning of the chapter (lines 1–14), before the mention of Critodemus, the compiler dealt with configurations involving two lots defined in situ (Κλῆρος ἀναιρέτου, "Lot of the destroyer"; Κλῆρος θανάτου, "Lot of death"), otherwise only found in the sixth-century astrologer Heliodorus, who notes that they are not found in previous compendia.[38] This section could thus have been newly incorporated by the sixth-century compiler. Then, after the brief section on the lord of the terms of the descendant by Critodemus (our F 13), we come to extensive treatment of the lord of the eighth place (=house), which coincides almost exactly with a section of the Arabic Dorotheus (IV 1 par. 143–157).[39]

The distinction between the material from Critodemus and the rest of the text has implications for the dating of Critodemus. Since what comes after the section on the term ruler of the descendant seems to have been taken from Dorotheus, Pingree's argument that θηριομαχοῦντας ("fighting with beasts") in this section (p. 200, line 13) would imply an imperial Roman era date[40] should be rejected and would apply instead to Dorotheus, who indeed wrote in the late first century AD.

Another question is whether Critodemus' treatise contained a section specifically on violent death or whether, in fact, Rhetorius selected predictions from the more general topic of the kind of death. Ptolemy and Hephaestio, for example, wrote sections dealing only generally with the kind of death, and among the surviving ancient authors, only Firmicus writes a chapter specifically on violent death (VII 23). The Arabic Dorotheus is a case in between, since the text (IV 1) announces the topic of the kind of death, but most configurations turn out to be of violent death. However, the eighth place does not specifically indicate violent death, only death in general.[41] Could it be that the received text of Dorotheus was an expansion on a short remark regarding a similar doctrine in the original? Indeed, Hephaestio

---

38 Heliod. *Comm. Paul.* p. 55 κλῆρος ἀναιρέτου ἀπὸ τοῦ οἰκοδεσπότου τοῦ ὡροσκόπου ἐπὶ Σελήνην, καὶ ἐπὶ νυκτὸς ὁμοίως, κλῆρος θανάτου ἀπὸ Σελήνης ἐπὶ τὸν ὄγδοον τόπον καὶ τὰ ἴσα ἀπὸ ὡροσκόπου, καὶ ἐπὶ νυκτὸς ὁμοίως. Heliodorus specifies that these lots (and others) are not found among the seven given by Hermes Trismegistus in the *Panaretes*.

39 Only one prediction, about the Moon, is between the predictions on the lord of the terms of the descendant ascribed to Critodemus and the discussion of the lord of the eighth: ἡ Σελήνη συνεχομένη ὑπὸ Κρόνου καὶ Ἄρεος ἐν ἑνὶ ζῳδίῳ ἐπίκεντρος ἢ ἐπαναφερομένη βιαιοθανάτους ποιεῖ. It probably does not derive from Critodemus either, since it seems to complete the series of stars that are considered in the next section as rulers of the eighth domicile, in which the Moon is missing: in the given order, we find Saturn, the Sun, Mars, Venus, Mercury, and Jupiter.

40 Cf. Pingree 1978, 425.

41 This is stated in the introduction to the section on the eighth domicile in the Rhetorius epitome (chapter 57, pp. 126–174), which contains the predictions on the ruler of the eighth house found in this chapter. It is noteworthy that in there we find the prediction of a good death – Jupiter and

has preserved Dorotheus' indication that the lord of the eighth sign from the Moon or from Saturn is used by some astrologers to predict a good or bad death.[42] This very chapter of the Arabic Dorotheus (IV 1, par. 15) quotes Valens (second century AD), and the former book contains one horoscope from the fourth century AD: it is therefore possible that the long passage on the eighth domicile came from a fourth-century paraphrase. As such, it is reasonable to think that by this time, authors had become specifically interested in the features of charts indicative of violent deaths, and that they selected predictions related to this topic from more general theories in older authors. The doctrine of Critodemus as we have it in F 13 seems to support this hypothesis. His predictions on the kind of death (of any kind) would have consisted in inspecting the identity and position of the lord of the descendant's terms: if it was a malefic, as in the predictions that are shown here, it indicated a violent death.

Different theories

Doctrines on violent death were varied. Here are the predictions found at the beginning of the chapter, before those of Critodemus:

> Κλῆρος ἀναιρέτου ἀπὸ τοῦ οἰκοδεσπότου τοῦ ὡροσκόπου ἐπὶ Σελήνην καὶ τὰ ἴσα ἀπὸ ὡροσκόπου, νυκτὸς δὲ ἐναλλάξ· ἡ Σελήνη ὁρῶσα τὸν κλῆρον τοῦ ἀναιρέτου βιαιοθανάτους ποιεῖ, χεῖρον δὲ ἐὰν ἡ Σελήνη εὑρεθῇ ἐν τοῖς μελεοκοπουμένοις ζῳδίοις· ὁ κύριος τῆς προγενομένης συνόδου ἢ πανσελήνου ἀπόστροφος ὢν τοῦ ἰδίου οἴκουκαὶ ὑπὸ κακοποιῶν θεωρούμενος· ἡ συνοδοπανσέληνος ἡ προγενομένη ὑπὸ κακοποιῶν θεωρουμένη μόνων· ἐὰν ὁ κύριος τοῦ ὡροσκόπου ἢ τοῦ κλήρου τῆς τύχης ὑπὸ κακοποιῶν θεωρούμενος εὑρεθῇ· ἐὰν ὁ πρῶτος τριγωνικὸς δεσπότης τοῦ ὑπογείου κέντρου κακῶς πέσῃ καὶ ὑπὸ κακοποιῶν θεωρηθῇ δίχα Διὸς καὶ Ἀφροδίτης. Κλῆρος θανάτου ἀπὸ Σελήνης ἐπὶ τὸν ὄγδοον τόπον τοῦ ὡροσκόπου ἐπὶ ἡμέρας καὶ νυκτός· καὶ ὅπου ἂν ἐκπέσῃ ὁ κλῆρος, σκόπει τοὺς ἐπιθεωροῦντας· ἐὰν γὰρ κακοποιὸς μόνος ἐπιθεωρήσῃ τὸν κλῆρον, βιαιοθανάτους ποιεῖ.
>
> The Lot of the Destroyer: from the ruler of the ascendant to the Moon and the same distance counted from the ascendant; if the Moon aspects the lot, it makes violent deaths, and worse if the Moon is found in the mutilated signs.[43] The ruler of the preceding conjunction[44] or full

Venus together in the eighth indicate a good death (εὐθανασία) – that is obviously not copied in the chapter on violent death.

42 Heph. II 25.15–16: τῶν ἄπο δὴ φράσσαντο τέλος αἰσχρόν τε καὶ ἐσθλόν.

43 The mutilated signs are the animal signs that lack a part of the body: Taurus (only the upper half appears), Cancer (which is said to be blind), Scorpio (with its claws in Libra), and Sagittarius (appearing in profile). Cf. Bouché-Leclercq 1899, 151.

44 Of the Sun and the Moon, that is, a new Moon.

Moon being outside from its own sign, aspected by malefics; the preceding syzygy[45] aspected by malefics only; if the lord of the ascendant or the Lot of Fortune is aspected by malefics; if the first trigonal ruler[46] of the lower midheaven is badly situated and aspected by malefics, separately from Jupiter and Venus. The Lot of death: from the Moon to the 8th domicile, from the ascendant both by day and night; and where the Lot falls, look at the aspecting stars: if only one malefic aspects the Lot, it makes violent deaths.

According to this doctrine, malefics or the Moon aspecting these three lots (destroyer, fortune, death) or other points (syzygy and syzygy ruler, ascendant ruler, first trigonal ruler of lower midheaven) signify violent death. Firmicus Maternus (VII 23) similarly attributes a prominent role to the malefics in this matter, collecting a considerable number of predictions (about four pages in the edition) which are fairly difficult to systematize: they mostly feature the malefics, the Moon, and the Sun in prominent places such as the angles, aspecting each other, and in certain domiciles.

In contrast with these rather unstructured doctrines, the theories of Critodemus and in the Arabic Dorotheus are clear and concise. For Critodemus, it was the identity of the ruler of the terms of the descendant that indicated whether death was violent or not: if it was a malefic, this naturally implied a violent death. In particular, depending on the kind of sign where this ruler was situated, the type of violent death would have different causes:

Kind of sign	**Kind of violent death**
Human-like (Gemini, Virgo, Sagittarius, Aquarius)	By humans
Watery (Cancer, Aquarius, Pisces)	Water
Beasts (Aries, Taurus, Leo, etc.)	Beasts or heights

However, certain circumstances of the malefic term ruler also determine the kind of death for Critodemus:

45 Conjunction or opposition of the Moon with the Sun (i.e., new Moon or full Moon). The usual Greek term is συζυγία. Here we find συνοδοπανσέληνος, whose first appearance is in the fourth-century work of Pappus (*Comm. Alm.* pp. 211, 18).

46 Cf. Appendix II for the triangle rulers: the first is understood as the ruler by day if it is a day birth, and the second is the ruler by night; vice versa for night births. For the trigonal ruler of the lower midheaven indicating the kind of death, cf. Dorotheus Ar. I 24.16.

Circumstance	Prediction
Under the rays of the Sun	Death by ambush
Not under the rays of the Sun	Death in public[47]
Stationed or in retrogradation	By blows or drugs

As we have seen, the Arabic Dorotheus picks the ruler of the eighth place. Any such ruler under certain circumstances might indicate a violent death. For example:

> Κρόνος οἰκοδεσποτῶν τὸν ὄγδοον τόπον καὶ μαρτυρῶν αὐτόν δίχα τῶν ἀγαθοποιῶν ἐξ ὑγρῶν ἀναιρεῖ.
>
> Saturn ruling the eighth place and aspecting this very place separately from benefics[48] kills from water.

Here, the nature of the ruling planet plays the most prominent role in determining the kind of violent death. Just as Saturn's natural humidity determines a watery death, Venus ruling the eighth and afflicted by certain configurations kills by the influence of women or wine, Jupiter kills by the violence of kings or other rulers, and Mars by the force of beasts or enemies.

As for the meaning of the eighth domicile, the introduction to the section on this domicile in the Rhetorius epitome is sufficiently clear (ch. 57, *CCAG* VIII 4, 161):

> ἔστι δὲ τὸ ζῴδιον ἀπόστροφον τοῦ ὡροσκόπου. διὰ τοῦτο καὶ περὶ θανάτου λόγον σημαίνει τὸν ἀπόστροφον τῆς ζωῆς.
>
> It is the sign turned away from the ascendant. Because of this, it governs the cause of death, what is turned away from life.

Note that the eighth sign from the ascendant is not opposed to it (that would be the seventh, the descendant); it is the next one, which Ptolemy calls a "disjunct" sign (ἀσύνδετος, I 17) because it bears no aspect to it. The fact that Critodemus focuses instead on the descendant is consistent with his label for the eighth place in F 11, which is not "Death" as in Valens IV 12.1, but "the Inactive Place" (ἀργὸς τόπος, also recorded by Valens ibid.). There were good reasons to relate the descendant to death: most evident of all is that the descendant is in opposition to the ascendant, the traditional place of life. In the same way that the ascendant – the zodiacal point on the horizon from which the stars emerge at the time of birth – is interpreted as the first and most important point of the chart, the descendant could be easily associated with the place of death because of its position on the horizon where the

47 The proximity of the Sun hides the planet; that is how the analogy works, so when under the rays, death comes from a hidden place, while otherwise it is public.
48 That is, without the positive influence of the presence of benefics in the eighth place.

stars set and disappear. This connotation is found in the method for calculating the length of life called *horimaia*: see the discussion on F 12, above.

Finally, let us see how Ptolemy addresses this topic. In the *Tetrabiblos*, investigation of the kind of death – including whether it will be violent – and calculation of the length of life are closely linked, since he proposes to inspect the planets that are found in the destructive place, considering the contributions of the planets in aspect as well as the nature of the sign and the terms of the destructive place (IV 9). Unlike the Arabic Dorotheus or Critodemus, Ptolemy does not resort to a "second level" of planetary positions: he does not inspect the place of the ruler of the sign or of the terms. One could say that, in combining the same doctrine calculation of the length of life and investigation of the kind of death, he presents a more "physical" and "direct" theory, which is entirely consistent with his self-declared aim to salvage the most plausible and empirically reasonable parts of astrology (I 1–3).[49]

---

49 This is not always so: in III 14 on the quality of the soul, where Ptolemy determines general individual types by looking at the planets dominating what he defines as the places of the soul (the positions of Mercury and the Moon), the kind of domination is sign rulership, and the position of the sign rulers (of the sign of Mercury and of the sign of the Moon) is considered.

Chapter 7
The *klimakteres* (F 14–15)

The long fragment F 14 can be divided in two parts. The first, probably deriving from the original Critodemus, is an explanation of a method for critical years or *klimakteres* according to the planetary distances from the Moon. According to Valens, Critodemus combined this doctrine with a series of planetary periods – not coinciding with the minimum periods of the distributions – which define critical years through their multiples (e.g. period 3 defines the following points: 3rd year, 6th year, 9th year, etc.). The second part is formed by a large table setting up the *klimakteres* established through these periods up to the 120th year. The format of a multiplication table not given by Valens but announced in his text (measuring 10×12, cf. F 9) suggests that it was first employed for factorizing the years of life in order to apply the table of *klimakteres*, since the procedure involves finding two multiplicative factors that give the year of life concerned (e.g. 9×4=36th year), and because the dimensions of the multiplication table give a maximum of 120, precisely the oldest age covered in the table of critical years.

The presence of tables in Critodemus' work is consistent with Valens' ambiguous criticisms examined above, but at the same time it is hardly compatible with the original form of Critodemus' manual, if it was written in iambic trimeter (Chapter 2). Technical manuals were written in verse partly for the sake of memorization, and large tables were not easily memorized. Furthermore, the table of *klimakteres* contains numerous specific medical terms, which are completely absent from the rest of the fragments and from the astrological genre as a whole. A reference to the astrological *klimakteres* in Pliny's *Natural History*, probably deriving from Varro's reading of Critodemus (see Chapter 1), explicitly mentions the medical origin of the doctrine. Varro may have been reading a version of Critodemus that already contained the addition of the large table of *klimakteres*, but it is more likely that he only had Critodemus' explanation in metrical form.

A second fragment (F 15) from a chapter attributed to a certain Qīṭrnūs in the Arabic version of Dorotheus features a doctrine on illness mentioning *klimakteres*, which prescribes a similar combination of multiplicative cycles. It is unclear whether this is meant to be Critodemus, but it is certainly plausible given the similarity of the name and doctrine.

https://doi.org/10.1515/9783111329147-008

F 14. Valens V 7.18 + 8.1–100

Type of fragment: Valens explains the doctrine of the *klimakteres* through a horoscope example, the diagram of which (not present in the text) I provide in the translation.

< tabella multiplicationis (10×12) >

(. . .)

<η̅.> Ἄλλως περὶ κλιμακτήρων, καθὼς Κριτόδημος ἀπὸ Σελήνης τὴν ἄφεσιν ποιεῖται.

1 Οἷον Ἥλιος Ὑδροχόῳ, Σελήνη Λέοντι, Κρόνος Καρκίνῳ, Ζεὺς Διδύμοις, Ἄρης Σκορπίῳ, Ἀφροδίτη Κριῷ, Ἑρμῆς Ἰχθύσι. **2** τὰ μὲν οὖν ι̅β̅ ἔτη πρὸς τὴν τοῦ Ἄρεος ἀποδιάστασιν διὰ τῶν δ̅· ἐστὶ δὲ ἁπλοῦς ὁ κλιμακτήρ. **3** τετράκις γὰρ τὰ δ̅ γίνεται ι̅ϛ̅ · οἱ γὰρ τετράγωνοι ἁπλοῖ, οἱ δὲ ἑτερομήκεις σύνθετοι. **4** ὁ δὲ ι̅η̅ πρὸς Ἀφροδίτην σύνθετος ἐκ τοῦ [τὰ] δὶς θ̅. **5** τὰ μὲν οὖν δύο ἐν Παρθένῳ, ἐν Παρθένῳ δὲ οὐδείς· ἐν δὲ τῷ θ̅ ἐστὶν Ἀφροδίτη. **6** εἰ δὲ ἦν τις καὶ ἐν Παρθένῳ, συναπεκαθίστατο ἂν διὰ τὸ καὶ †ἑνός ι†.
7 ἐπεὶ δὲ καὶ τρὶς ϛ̅ ὁμοίως ι̅η̅ καὶ ἑξάκις <γ̅>[a]· ἀλλ' οὔτε Ζυγῷ τῷ τρίτῳ ἔστι τις ἀστὴρ οὔτε ἐν Αἰγόκερῳ τῷ ἕκτῳ. **8** πάλιν τὰ κ̅ σύνθετος· τετράκις γὰρ ε̅ γίνεται κ̅ καὶ τὰ πεντάκις δ̅ ὁμοίως. **9** Ἄρης ὁ ἐν Σκορπίῳ χρηματίζει, ἐν τῷ ε̅ Τοξότῃ οὐδείς. **10** τὰ δὲ κ̅α̅ ἔχει τὸ τρὶς ζ̅· Ζυγῷ οὐδείς, Ὑδροχόῳ Ἥλιος· πρὸς Ἥλιον οὖν. **11** τὰ δὲ κ̅δ̅ ἔχει τετράκις ϛ̅· ἐν Σκορπίῳ πάλιν Ἄρης, Αἰγόκερῳ δὲ οὐδείς. **12** τὰ δὲ κ̅ε̅ ἔχει τετράγωνον ἁπλοῦν ἀριθμόν· ἀλλ' οὐδεὶς ἐν Τοξότῃ. **13** τὰ δὲ κ̅ζ̅ συνέστηκεν ἐκ τῆς τρὶς θ̅· διὰ τριῶν ἐστιν οὐδείς, μόνη δὲ ἡ Ἀφροδίτη διὰ θ̅. **14** ὁ κ̅η̅ διὰ δ̅ πρὸς Ἄρεα καὶ διὰ ζ̅ πρὸς Ἥλιον. **15** ὁ μ̅ ἔχει μὲν δ̅ καὶ ι̅ καὶ ε̅ καὶ η̅, συμφωνεῖ δὲ πρὸς Ἄρεα διὰ τῶν δ̅, πρὸς Ἑρμῆν διὰ τῶν η̅. **16** ὁ μ̅δ̅ πρὸς Ἄρεα καὶ Δία. **17** συναποκατασταθήσονται δὲ καὶ πλείονες ἔσθ' ὅτε· οἷον ἐπὶ τοῦ μ̅ ὥσπερ ἐν τῷ δ̅ ἐστὶν Ἄρης καὶ ἐν τῷ η̅ Ἑρμῆς· εἰ ἦν τις καὶ ἐν τῷ ι̅ καὶ ἐν τῷ ε̅, συναπεκαθίστατο ἂν αὐτοῖς.
18 Λέγει δὲ τὰ συμβαίνοντα μᾶλλον εὐτονώτερα καὶ ἐμφανέστερα γίνεσθαι, ἐὰν ἴδιος ᾖ ὁ τῶν ἐτῶν ἀριθμὸς τοῦ συμφωνοῦντος πρὸς τὴν ἀποδιάστασιν ἀστέρος, οἷον οὕτως· τὰ μὲν γ̅ εἶναι Κρόνου, τὰ δὲ ε̅ Ἀφροδίτης, τὰ δὲ ζ̅ Ἄρεως, τὰ δὲ η̅ Ἑρμοῦ, τὰ δὲ θ̅ Διός, τὰ ι̅γ̅ Σελήνης, τὰ ι̅η̅ Ἡλίου.
19 ἐὰν οὖν ἅμα συνεμπίπτῃ οὗτός τε ὁ ἀριθμὸς καὶ ἡ διάστασις εἰς τὸν αὐτὸν ἀστέρα, γίνεται χρηματίζων ἐν χρηματιστικῷ· ἐὰν δὲ ὁ ἐνιαυτὸς μὴ συνεμπίπτῃ εἴς τινα τοῦ διαστήματος ἐπ' αὐτὸν φέροντος, ἀχρημάτιστος ἐν ἀχρηματίστῳ. **20** καὶ ἐὰν μέν τις διάστασις χρηματίζουσα ἐν δὲ τοῖς μεταξὺ ἔτεσι μὴ εὑρεθῇ, τῇ πρώτῃ χρηστέον, ἕως ἑτέρα εὑρεθῇ. **21** οἷον ἐπὶ τοῦ προκειμένου θέματος ὁ κ̅η̅ πρὸς Ἄρεα Ἥλιον· ὁ κ̅θ̅ οὐκ ἔχει διάστασιν· ὁ λ̅ ἔχει μὲν γ̅ καὶ ε̅ καὶ ϛ̅ καὶ ι̅ καὶ † α̅ καὶ ταῦτα τὰ ζῴδια· ὁ δὲ λ̅α̅ πάλιν οὐχ ἁρμόζει τινὶ διαστήματι. **22** ταῦτα τὰ ἔτη ἐπικρατήσει ὅ τε Ἄρης καὶ Ἥλιος οἱ ἐν τῷ κ̅η̅ ἔτει χρηματίσαντες ἕως λ̅β̅, ὧν συναποκαθιστῶσι Ἄρης διὰ δ̅ καὶ Ἑρμῆς διὰ η̅.

23 Ὑποτάξομεν δὲ καὶ τὰς διαφορὰς τῶν κλιμακτήρων κατὰ τὴν τῶν ἀστέρων χρονογραφίαν καὶ τὴν πρὸς ἀλλήλους ἀποκατάστασιν.

α̅ ἀσθενήσει καὶ ἐπίφοβος ἔσται.
β̅ κινδυνεύσει δι' ὑγρῶν <ἢ>[b] σπασμῶν.
γ̅ Κρόνου κλιμακτὴρ. ἐπισφαλής.
ε̅ Φωσφόρου, ἀσθενήσει.
ς̅ Κρόνου δεύτερος.
ζ̅ Ἄρεως πρῶτος ἐπικίνδυνος, πυρετοῖς αἵμασι τραύμασι πτώμασιν ἑλκώσεσι περιτρέπων ἢ σιδήρου τομαῖς.
η̅ Ἑρμοῦ πρῶτος ἀσύνθετος.
θ̅ Διὸς πρῶτος, Κρόνου τρίτος ἐπικίνδυνος· ἀσθενήσει ἢ ῥιγοπυρέτοις ὀχληθήσεται καὶ τῶν ἐντὸς ἢ κοιλίας πόνοις.
ι̅ Ἀφροδίτης δεύτερος· ἀσθενήσει ἐκ πληθώρας.
ι̅β̅ κλιμακτὴρ Κρόνου τέταρτος· ἀπροσδοκήτως ἢ δι' ὑγρῶν.
ι̅γ̅ Σελήνης πρῶτος· πυρετὸς δύσκολος ἐπιγενήσεται ἢ κατάπτωσις καὶ τῶν ἐντὸς ἢ θώρακος πόνοι.
ι̅δ̅ Ἄρεως δεύτερος· ἐπικίνδυνος δύσκολος.
ι̅ε̅ Κρόνου πέμπτος, Ἀφροδίτης τρίτος, ἀνετικός.[c]
ι̅ς̅ Ἑρμοῦ δεύτερος κλιμακτήρ, σύνθετος διὰ χολερᾶς ἢ ἀρτηρίας καὶ δυσαναληψίας.
ι̅η̅ Διὸς δεύτερος, Κρόνου τρίτος, Ἡλίου πρῶτος, χαλεπὸς λίαν.
κ̅ Ἀφροδίτης τέταρτος, ἀκίνδυνος κατὰ τὸ πλεῖστον· νόσοι δὲ ἐκ πληθώρας ἢ κόπου παρακολουθοῦσιν.
κ̅α̅ Ἄρεως τρίτος, Κρόνου ἕβδομος· δύσκολος καὶ ἐπικίνδυνος.
κ̅δ̅ Κρόνου ὄγδοος Ἑρμοῦ τρίτος, δύσκολος διὰ μελαγχολίας καὶ ὑγρῶν.
κ̅ε̅ Ἀφροδίτης πέμπτος, σύνθετος.
<κ̅ς̅ Σελήνης β̅. ἐπικύνδυνος.>
κ̅ζ̅ Διὸς τρίτος, Κρόνου ἔνατος, μέσος.
κ̅η̅ Ἄρεως τέταρτος κλιμακτήρ, ἐπισφαλής.
λ̅ Κρόνου δέκατος, Ἀφροδίτης ἕκτος, ἀκίνδυνος κατὰ τὸ πλεῖστον.
λ̅β̅ Ἑρμοῦ τέταρτος, σκυλτικός.
λ̅γ̅ Κρόνου ἑνδέκατος, δύσκολος.
λ̅ε̅ Ἄρεως πέμπτος, Ἀφροδίτης ἕβδομος, ἐπικίνδυνος καὶ εὐεπιβούλευτος.
λ̅ς̅ Διὸς τέταρτος, Κρόνου δωδέκατος, χαλεπὸς καὶ ἐπικίνδυνος.
μ̅ Ἀφροδίτης ὄγδοος, Ἑρμοῦ πέμπτος, οὐ χαλεπός.
μ̅β̅ Ἄρεως ἕκτος, Κρόνου ι̅δ̅, χαλεπὸς καὶ ἐπικίνδυνος.
μ̅ε̅ Διὸς πέμπτος Ἀφροδίτης ἔνατος, Κρόνου ι̅ε̅ · οὗτος ὁ κλιμακτὴρ καλεῖται Στίλβων καὶ προσέχειν δεῖ, μή πως περὶ τοὺς πόδας γένηται πάθος κατὰ

τοῦτον τὸν χρόνον τοῦ Ἑρμοῦ χρηματίζοντος ἐν τῇ γενέσει· κινδύνους γὰρ ἐπιφέρει ἄρθρων καὶ ἀσθενείας καὶ βιωτικὰ συμπτώματα καὶ ἀηδίας.

μη Ἑρμοῦ ἕκτος, Κρόνου ις, χαλεπὸς λίαν καὶ ἐπικίνδυνος.

μθ Ἄρεως ἕβδομος, ἐπικίνδυνος, ἢ αἰφνίδιος διὰ πυρετῶν ἢ αἱμαγμῶν καὶ βιαίας αἰτίας.

ν Ἀφροδίτης δέκατος, ἐπικίνδυνος.

να Κρόνου ιζ, νόσους βλάβας ἀτυχίας ἐπιφέρει.

νβ Σελήνης τέταρτος, οὐ καλός.

νδ Κρόνου ιη, Διὸς ἕκτος, Ἡλίου τρίτος, χαλεπὸς καὶ κινδυνώδης.

νε Ἀφροδίτης ἑνδέκατος, οὐ κακός.

νς Ἄρεως ὄγδοος, Ἑρμοῦ ἕβδομος, λυπηρὸς σκληρός.

νζ Κρόνου ιθ, χαλεπώτατος.

ξ Κρόνου κ, Ἀφροδίτης δωδέκατος, ἐπισφαλής.

ξγ Κρόνου κα, Διὸς ἕβδομος, Ἄρεως ἔνατος, ἀνδροκλάστης[d], χαλεπὸς καὶ θανατηφόρος.

<ξδ Ἑρμοῦ η. οὐ λίαν κακός.

ξε Σελήνης ε, Ἀφροδίτης ιγ. κοινός.

ξς Κρόνου κβ.

ξθ Κρόνου κγ, χαλεπός.

ο Ἄρεως ι, Ἀφροδίτης ιδ, δύσκολος καὶ χαλεπός.>

οβ Κρόνου κδ, Διὸς ὄγδοος, Ἑρμοῦ ἔνατος, χαλεπὸς καὶ θανατηφόρος.

οε Κρόνου κε, Ἀφροδίτης ιε, ἐπικίνδυνος.

οζ Ἄρεως ἑνδέκατος, δύσκολος καὶ θανατηφόρος.

οη Κρόνου κς, Σελήνης ἕκτος, χαλεπός.

π Ἀφροδίτης ις, Ἑρμοῦ δέκατος, συγκρατικός.

πα Κρόνου κζ, Διὸς ἔνατος, ἐπικίνδυνος.

πδ Κρόνου κη, Ἄρεως δωδέκατος, δύσκολος καὶ κακοποιός.

πε Ἀφροδίτης ιζ, κοινός.

πζ Κρόνου κθ, ἐπικίνδυνος.

ϟ Κρόνου λ, Ἀφροδίτης ιη, Διὸς δέκατος, <Ἡλίου ε>, χαλεπός.

ϟε Ἀφροδίτης ιθ, οὐ καλός.

ϟς Κρόνου λβ, <Ἑρμοῦ ιβ>, δύσκολος.

ϟη Ἄρεως ιδ, χαλεπός.

ϟθ Κρόνου λγ, <Διὸς ια>, μέσος.

ρ Ἀφροδίτης κ, οὐ κακός.

ρβ Κρόνου λδ, χαλεπός.

<ρδ Σελήνης η, Ἑρμοῦ ιγ, οὐ λίαν κακός.>

ρε Κρόνου λε, Ἀφροδίτης κα Ἄρεως ιε, δύσκολος.

ρη Κρόνου λς, <Διὸς ιβ, Ἡλίου ς>, θανατηφόρος.

ρι Ἀφροδίτης κβ, οὐ κακός.

ρια Κρόνου λζ, ἐπισφαλής.
ριβ Ἄρεως ις Ἑρμοῦ ιδ, δύσκολος καὶ δεινός.
ριδ Κρόνου λη, ἐπικίνδυνος.
ριε Ἀφροδίτης κγ, κοινός.
ριζ Κρόνου λθ, Σελήνης ἔνατος, Διὸς ιγ, ἐπικίνδυνος.
ριθ Ἄρεως ιζ, ἐπισφαλής.
ρκ Κρόνου μ, Ἀφροδίτης κδ, <Ἑρμοῦ ιε>, θανατηφόρος.

a καὶ καὶ (sic.) ἑξάκις ιππ **V**, καὶ ἑξάκις **S**, καὶ ταῦτα ἐξετάζομεν Pingree.
b The text in angular parentheses is reconstructed by Pingree from a Latin version of the table in the Liber Hermetis, which he prints in Appendix XX 15 of his edition of Valens.
c ἀνετικός **V**: ἀναιρετικός **S**. The reading in V, the older manuscript from which S is copied, is preferable because it is correct and consistent with the medical vocabulary of the table.
d *androclastes* App., ἀνδροσκάστης codd., *androclas* Firmicus IV 20.3.

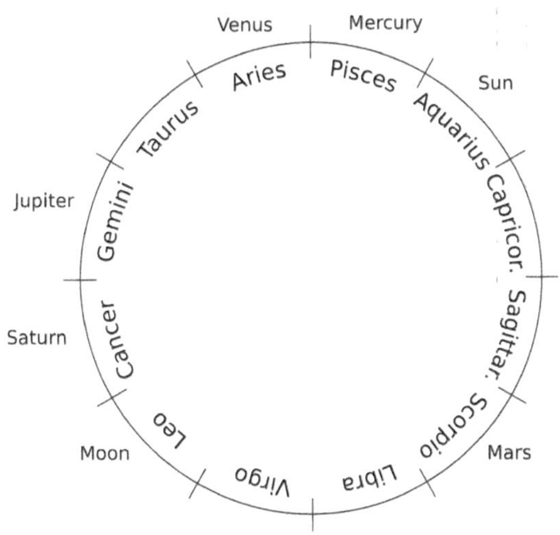

< multiplication table (10×12), see reconstruction in F 9 >

(. . .)

8. Another method for the critical years: Critodemus begins with the Moon.

1 An example: Sun in Aquarius, Moon in Leo, Saturn in Cancer, Jupiter in Gemini, Mars in Scorpio, Venus in Aries, Mercury in Pisces. **2** The 12 years, because of Mars'

distance of 4 signs.[1] The next *klimakter* is *simple.* **3** Because 4 times 4 is 16. Squares are *simple*, rectangles are *compounded.*[2] **4** The 18 for Venus is *compounded*, from 2 times 9. **5** The 2 falls in Virgo, but no star is in Virgo; Venus is 9 signs from the Moon. **6** If a star was in Virgo, it would return because of the . . .[3]
7 Since 3 times 6 equals 18, so does 6 times <3>. However, no star is in Libra, the third sign, or in Capricorn, the sixth. **8** Then 20 is *compounded*, because 4 times 5 is 20, the same as 5 times 4. **9** Mars, which is in Scorpio,[4] is operative; no star is in Sagittarius, the fifth. **10** The 21 has the triple 7: no star is in Libra, but the Sun is in Aquarius: the *klimakter* is for the Sun, then. **11** The 24 has 4 times 6: Mars is in Scorpio, again; no one is in Capricorn. **12** The 25 has a square, a *simple* number,[5] but there is no one in Sagittarius. **13** The 27 is formed by the 3 and the 9: no star is at 3 signs, only Venus at 9. **14** The 28, from 4, to Mars; and from 7, to the Sun. **15** The 40 has 4 and 10, and 5 and 8: it combines with Mars, due to the 4, and with Mercury, due to the 8. **16** The 44, to Mars and Jupiter.[6] **17** Sometimes several stars return to the same point, as in the case of 40, since Mars is in the fourth sign and Mercury in the eighth; if some star had been in the tenth and the fifth signs, they would also have returned together there.
18 He says that the effects will be stronger and more evident if the number of the years is equal to the distance of the concerned star, as follows:[7] Saturn 3, Venus 5, Mars 7, Mercury 8, Jupiter 9, Moon 13, Sun 18.

---

1 See the chart: Mars is 4 signs ahead of the Moon, counting inclusively. The idea is that, because 12 is a multiple of 4, it is a *klimakter* of Mars (as would be 8, 16, etc.).

2 Valens apparently attempts to correct the table's surprising adjectives ἀσύνθετος ("incomposite") for the *klimakter* of the year 8 (first of Mercury), and σύνθετος ("composite") for the *klimakteres* 16 (second of Mercury) and 25 (fifth of Venus). Valens' use is inspired by the normal sense of the terms, as in Nicomachus *Introd. arithm.* I 11, i.e. ἀσύνθετος for prime and σύνθετος for non-prime odd numbers; Valens would accept multiples of one sole factor as incomposite. In *klimakter* 16 we read "composite through cholera, bronchitis. . .", as if these apparently mathematical concepts had been understood as indicating a mixture of effects / a simple effect. Cf. *klimakter* 80 (Venus and Mercury), which is said to be "mixing" (συγκρατικός). My guess is that the original meaning of σύνθετος in the table was "mixed", indicating a mixture of effects, and that ἀσύνθετος was then added for the 8th *klimakter*, thinking mathematically in relation with the multiplication series of Mercury (8, 16, 24, . . .), as if 8 was a prime number.

3 Venus is at a distance of 9 signs from the Moon, counting inclusively. There is no star 2 signs ahead of the Moon (i.e. in the next sign, Virgo); if there were, 18 would also be a *klimakter* of that star. The text is corrupt here, but the meaning is clear ("to return" means to have another critical point or *klimakter*).

4 The fourth sign from the Moon. Sagittarius is the 5th.

5 Here, Valens again corrects the table, which marks 25 as "compounded" (cf. note above).

6 Because $44=11\times4$; Jupiter is at a distance of 11 signs, Mars at 4.

7 That is, if, when analyzing a concrete year, an operative interval (=number of signs from the Moon) coincides with the applicable number for the star as in the table that follows.

19 If the number and the distance to the same star coincide, the star will be operative in the operative period. If the year does not coincide with that of one interval leading to that star,[8] it will be inoperative in an inoperative period. **20** And if one interval is found to be operative, but another interval is not found for the subsequent years, the initial interval should be used until another is found. **21** For example, using the present star chart: 28 is assigned to Mars and the Sun.[9] 29 has no intervals. 30 has 3, 5, 6, and 10: these signs too are empty. 31 again is not associated with any interval. **22** Therefore, Mars and the Sun, operative in the 28th year, control the succeeding years until the 32nd year, when Mars (because of the 4) and Mercury (because of the 8) have returns.[10]

23 We will note down the differences of the *klimakteres* according to the chronology of the stars and their mutual returns.

1. It will weaken and frighten.
2. It will endanger by humidity <or> convulsions.
3. *Klimakter* of Saturn: precarious.
5. Of Phosphorus:[11] it will weaken.
6. The 2nd of Saturn.
7. The first of Mars: dangerous, upsetting with fevers, bleedings,[12] wounds, falls, ulceration, or sword cuts.
8. The first of Mercury: unmixed.[13]
9. The first of Jupiter, the 3rd of Saturn: dangerous; it will weaken and be troubled with shivering fevers and troubles of the insides or the abdomen.

---

8 That is, the interval in signs from the Moon to that star.

9 Cf. previous analysis, and the following note.

10 Mercury is 8 signs ahead of the Moon in the chart, and at the same time its period is also 8. However, Mars has a period in the table not of 4 years but of 7, so it should in theory not be operative, since it must coincide. The same is true of the Sun in the 28th year, since it has a period of 18 in the table. Valens, then, is being inconsistent in the application of the second part of the doctrine. Furthermore, the table must have been created independently, since an interval of 18 signs does not make much sense.

11 That is, Venus. A series of non-theophoric names for the planets – Phaethon (Jupiter), Phainon (Saturn), Stilbon (Merucy), Phosphorus (Venus), Pyroeis (Mars) – began to be used sometime in or before the 1st century BC: Geminos (90–35 BC) 26, Cic. *Nat. deor.* II 53. Dorotheus repeatedly uses all these names except Phosphorus.

12 The text literally says "bloods" (αἵμασιν), but it should probably be understood as a process having to do with blood, e.g. as in *klimakter* 49: αἱμαγμῶν (bleedings). Cf. *klimakter* 16.

13 Probably filled later than *klimakter* 16 and interpreting the term σύνθετος mathematically (cf. note above); so *klimakter* 8 would be conceptualized as the 'prime number' initiating the series.

10. The 2nd of Venus: it will weaken from excessive blood.
12. The fourth *klimakter* of Saturn. Unexpectedly or because of humidity.
13. The 1st of the Moon: a difficult fever or seizure will occur, and troubles of the insides or chest.
14. The 2nd of Mars: dangerous, difficult.
15. The 5th of Saturn, the 3rd of Venus: relaxing.
16. The 2nd *klimakter* of Mercury: mixed, because of cholera, bronchitis,[14] and difficult convalescence.
18. The 2nd of Jupiter, the 6th of Saturn, the 1st of the Sun; very bad.
20. The 4th of Venus: generally safe; the diseases come from fullness or fatigue.
21. The 3rd of Mars, the 7th of Saturn: difficult and dangerous.
24. The 8th of Saturn, the 3rd of Mercury: difficult because of black bile and humidity.
25. The 5th of Venus: mixed.

<26. The 2nd of the Moon: dangerous.>

27. The 3rd of Jupiter, the 9th of Saturn: medium.
28. The 4th *klimakter* of Mars: precarious.
30. The 10th of Saturn, the 6th of Venus: mostly safe.
32. The 4th of Mercury: exhausting.
33. The 11th of Saturn: difficult.
35. The 5th of Mars, the 7th of Venus: dangerous and easily treacherous.
36. The 4th of Jupiter, the 12th of Saturn:[15] grievous and dangerous.
39. The 3rd of the Moon, the 13th of Saturn: precarious and dangerous.
40. The 8th of Venus, the 5th of Mercury: not bad.
42. The 6th of Mars, the 14th of Saturn: bad and dangerous.
45. The 5th of Jupiter, the 9th of Venus, the 15th of Saturn: this critical point is called Stilbon; it is necessary to be careful, lest some infirmity of the foot occur at this time, while Mercury is operative in the nativity, because Mercury brings dangers of the joints, weakness, life-threatening affections and displeasure.[16]

---

14 Again, the name of a body part stands for a process, perhaps a disease related to the trachea or the bronchial tubes (cf. note to *klimakter* 7).

15 It is also the 2nd of the Sun.

16 The poetic name Stilbon, denoting Mercury, is unusual here, because 45 is not even a multiple of 8. Cf. Phosphorus applied correctly to the fifth *klimakter*, of Venus. The next *klimakter* (48) is of Mercury, but also of Saturn, so the name Stilbon would scarcely be appropriate there, either. My guess is that perhaps at some point in the transmission this was the second column, and this entry originally referred to the first *klimakter* of Mercury (8), similarly as in the case of Venus.

48. The 6th of Mercury, the 16th of Saturn: very bad, and dangerous.
49. The 7th of Mars: dangerous or unforeseen from fevers, or bleedings and violent causes.
50. The 10th of Venus: dangerous.
51. The 17th of Saturn: it brings diseases, harm, and misfortune.
52. The 4th of the Moon: not good.
54. The 18th of Saturn, the 6th of Jupiter, the 3rd of the Sun: bad and full of danger.
55. The 11th of Venus: not bad.
56. The 8th of Mars, the 7th of Mercury: distressing and harsh.
57. The 19th of Saturn: the worst.
60. The 20th of Saturn, the 12th of Venus: precarious.
63. The 21st of Saturn, the 7th of Jupiter, the 9th of Mars: man-killer, bad and deathly.
<64. The 8th of Mercury: not very bad.
65. The 5th of the Moon, the 13th of Venus: ordinary.
66. The 22nd of Saturn.
69. The 23rd of Saturn: bad.
70. The 10th of Mars, the 14th of Venus: difficult and bad.>
72. The 24th of Saturn, the 8th of Jupiter, the 9th of Mercury:[17] bad and deadly.
75. The 25th of Saturn, the 15th of Venus: dangerous.
77. The 11th of Mars; difficult and deathly.
78. The 26th of Saturn, the 6th of the Moon: bad.
80. The 16th of Venus, the 10th of Mercury: a mixture.
81. The 27th of Saturn, the 9th of Jupiter: dangerous.
84. The 28th of Saturn, the 12th of Mars: difficult and malefic.
85. The 17th point of Venus: ordinary.
87. The 29th of Saturn: dangerous.[18]
90. The 30th of Saturn, the 18th of Venus, the 10th of Jupiter, <the 5th of the Sun>: bad.
91. The 13th of Mars, the 7th of the Moon: difficult.
93. The 31st of Saturn: bad.
95. The 19th of Venus: not good.
96. The 32nd point of Saturn, <the 12th of Mercury>: difficult.
98. The 14th of Mars: bad.
99. The 33rd of Saturn, <the 11th of Jupiter>: medium.

---

17 It is also the 4th of the Sun.
18 88, the 11th of Mercury, is missing.

100. The 20th of Venus: not bad.
102. The 34th of Saturn: bad.
<104. The 8th of the Moon, the 13th of Mercury: not very bad.>
105. The 35th of Saturn, the 21st of Venus, the 15th of Mars: difficult.
108. The 36th of Saturn, <the 12th of Jupiter, the 6th of the Sun>: deathly.
110. The 22nd of Venus: not bad.
111. The 37th of Saturn: precarious.
112. The 16th of Mars, the 14th of Mercury: difficult and dire.
114. The 38th of Saturn: dangerous.
115. The 23rd point of Venus: ordinary.
117. The 39th of Saturn, the 9th of the Moon, the 13th of Jupiter: dangerous.
119. The 17th of Mars: precarious.
120. The 40th of Saturn, the 24th of Venus, <the 15th of Mercury>: deathly.

The periods

In order to discover which *klimakteres* are active in a certain year, we factorize the number of the year in two multiplicative factors. The multiplication table of 10×12 was presumably used for this purpose: these factors would then be at most 10 and 12, respectively. If *x* is one of the factors, the astrologer checks whether there are any stars *x* signs ahead of the Moon and then consults the table of *klimakteres* to see whether the same star has a critical point at that year, in which case it is accepted as an active *klimakter*. Obviously, the Moon had no *klimakteres* according to this method, since it would be absurd to compute the distance to itself, although it definitely had its climacteric periods in the table.

The periods of the table do not coincide with the so-called minimum periods, but in some cases they seem to be derived from them:

	Min. periods	*Klimakteres*
Sun	19	18
Moon	25	13
Saturn	30	3
Jupiter	12	9
Mars	15	7
Venus	8	5
Mercury	20	8

If the 19 years of the minimum period for the Sun are the years of the Metonic cycle (in which an integer number of lunar cycles is contained), its climacteric period probably reflects the Saros cycle, in which the ecclipse pattern repeats itself (every 18 years). The period of 13 for the Moon was perhaps derived from the typical value for the daily advancement of the Moon in degrees, but it is also probably relevant that it is half the minimum period of 25 years.

The other climacteric numbers are less clear at first sight, but it is noteworthy that the sum of them all is 63, a number the table identifies as a "man-killer" (ἀνδροκλάστης). This number is also the sum of the terms ruled by Mars, so it is probable that the association of this number with death is prior to and indeed determined some aspects of the theory of the *klimakteres*.

I will propose a simple speculative reconstruction of the rationale of these numbers. Greek astrological theories that were not inherited from Mesopotamia, such as the scheme of planetary domiciles, tended to simplicity, so I believe this would be a desirable condition. In his exposition of the doctrine of the *klimakteres* (IV 20), Firmicus mentions only this number and its associated cycles of 7 and 9. This is clearly part of the doctrine of our table of *klimakteres*, in which 7 is assigned to Mars and 9 to Jupiter. Jupiter is perhaps surprising: maybe the combination of the factors 7 and 9 was interpreted as the clash between the most powerful of the malefics (Mars) and the most powerful of the benefics (Jupiter). It is also possible that the number 7 was assigned to Mars because it is approximately half the minimum period (15), analogous with the relationship between the 13-year climacteric period of the Moon and its minimum period of 25 years. Then, the remaining 16 years (=63–18–13–9–7) can only be split as follows:

$$16 = 8 + 5 + 3$$

if we want to use periods that are greater than 2 and different from each other and from the ones already used.[19] The periods of 5 and 8 were assigned respectively to Venus and Mercury probably again because they are approximately half their minimum periods of 8 and 20, and Saturn was given the remaining period of 3 years.

The medical language of the table

Numerous medical terms indicating parts of the body as well as diseases that may befall the native can be clearly observed in the first part of the table.[20] I list here the most obvious examples:

---

19 This is the only possible combination using periods greater than 2. Three combinations would be possible with a period of 2: 16=2+6+8=2+4+10=2+3+11, but a period of 2 would probably be too small, giving too many *klimakteres* for a single planet.

20 In any case, the entries are all extremely brief from the 49th year onwards, comprising just one or two words.

Klim.	*Planet(s)*	*Effects*
5	Venus	He/she will be weakened (ἀσθενήσει)
7	Mars	Fevers, bleeding, wounds, falls, ulceration
9	Jupiter	Troubles in the bowels
10	Venus	He/she will be ill because of excessive blood (πληθώρα)
13	Moon	Fever or seizures will occur; troubles of the insides or chest
15	Saturn/Venus	Relaxing (ἀνετικός)
16	Mercury	Vomiting, bronchitis, difficult convalescence (δυσαναληψία)
24	Saturn/Merc.	Black bile (μελαγχολία) and moist syndromes
45	Mercury	Some infirmity of the foot may occur, because Mercury brings dangers to the joints, sickness, life-threatening and disgusting syndromes.
49	Mars	Sudden dangers through fevers, bleeding, and violent occurrences

Two of the medical terms in the table are so specific that they do not appear outside the work of medical authors (ἀνετικός, "relaxing" *klimakter* 15; δυσαναληψία, "difficult recovery" *klimakter* 16). The combination of medical and astrological tenets is also observable in two specific doctrines contained in the table: first, several of these *klimakteres* (9, 13, 16, 24, and 45) refer to parts of the human body, and, in three cases (13, 16, 45), the association with the particular planet has parallels in the other ancient *melothesiai*.[21]

The idea of establishing equally distributed cyclical crises in the development of diseases is as old as the Hippocratic *corpus* and likely derives from the observation of the recurring onsets of fevers in different kinds of malaria: the tertian fevers (cycles of two days) and quartan fevers (cycles of three days). So, according to Hippocrates (*Progn.* 20.1–16), these periods end on days 4, 7, 11, 14, 17, and 20, forming overlapping periods of 7 days, even if the agreement was not perfect:[22]

21 *Melothesiai* were correspondences between a system of astrological entities such as the planets or the zodiacal signs, and parts of the human body, employed to link afflictions of these parts of the body to astrological influences. So, for example, in Manilius (II 453–465) Aries, the first sign, is associated with the head; Taurus with the neck, and so on down to Pisces, which is associated with the feet. In the first chapter of Valens' *Anthologies* (I 1), some of these correspondences seem to have been reinterpreted as planetary influences, so that Mercury, the ruler of Gemini and Virgo, governs the joints and the belly, the parts assigned by Manilius to these two signs. The chest mentioned in *klimakter* 13 in Critodemus' table coincides with the influences of the Moon in Valens I 1 (breasts), and with Manilius II 453–465 Cancer (sign of Moon). The ἀρτηρία (artery, trachea or bronchial tubes) in *klimakter* 16 is found in the effects of Mercury in Valens I 1. Regarding *klimakter* 45, Valens I 1 mentions shoulders, joints, and belly as belonging to Mercury, and in Manilius II 453–465 Gemini (sign of Mercury) is associated with the arms and Virgo (the other sign of Mercury) with the belly.

22 *Alex. sum.* Arabic 37, 48. The series was then lengthened with 24, 27, 31, 34, 40, 60, 80, and 120: *Alex. sum.* Arabic 20 (Bos and Tzvi-Langermann 2015, 72). So, there was another cycle of 20, giving 20, 40, 60, 80, and 120, and also 4+20=24 and 7+20=27. Cf. Galen K 879.3–879.7 (*CG* 286–287).

0		7		14		21	
0	4		11		18		

A link with astronomy was also apparent in the medical field, since these days were understood to follow the lunar cycle, with 7 and 14 corresponding to the half crescent and the full Moon, respectively.[23]

Pliny's witness

In *Natural History* VII 48, a passage already visited above (Chapter 1) for its connections with Berossus, and which very likely reproduces material from Cicero's polymath friend M. Terentius Varro, Pliny deals with the determination of the length of life from an astrological perspective. As we have seen, he explains the fundamentals of the *tetartemorion* method, which he attributes to Nechepsos and Petosiris, and gives the supposed maximum length of life according to Berossus (obtained through this method for the clima of Babylon), Epigenes (Alexandria), and for a latitude in Italy. Then he says:

> *<A>esc<u>lapi rursus secta, quae stata vitae spatia a stellis accipi dicit, quantum plurimum tribuat, incertum est. rara autem esse dicunt longiora quidem tempora, quoniam momentis horarum insignibus lunae dierum, ut VII atque XV, quae nocte ac die observantur, ingens turba nascatur scansili annorum lege occidua, quam climacteras appella<n>t, non fere ita genitis LIIII annum excedentibus.*
>
> The school of Asclepius came thereafter, which states that the allotted duration of life is regulated by the stars, but that the question of the greatest extent of the period is quite uncertain. These say that long life is uncommon, because a very great number of persons are born at critical moments in specific instances of the lunar days – for example, in the seventh and the fifteenth, both by night and day – who are affected by the ascending scale of the years which is termed the "climacteric", hardly ever, when born under these circumstances, attaining the fifty-fourth year.

Critodemus' theory of *klimakteres* was quite unique, leaving a perceptible impact only in Valens and Firmicus (IV 20), who, as we know, were direct readers of his

23 Bos and Tzvi-Langermann 2015, 23. Sometimes the same numbers were interpreted to refer to years of much longer chronic-disease cycles: Galen (K817ff.) reports Hippocrates' views on longer periods of 7 months, 7 years, 14 years, and 21 years, but this is not found in Hippocrates (Bos and Tzi-Langermann 2015, 18). *Alex. sum.* 68 relates the numbers to star cycles, so that 14 days (diameter Sun-Moon) correspond to 14 years (roughly half the orbit of Saturn).

work.[24] It is unlikely, then, that Varro was thinking of a different doctrine, considering that Varro was also familiar with Critodemus' manual (Chapter 1). Furthermore, Pliny's data perfectly matches the theory. The lunar days mentioned in the passage – the 7th and the 15th – roughly correspond to the half crescent and the diameter (=full Moon), and if we imagine a person born in either of the two situations, counting the signs from the Moon to the Sun not inclusively (unlike Valens), we get 9 and 6 signs, respectively. Since one factorization of the year 54 is 6×9,[25] in either configuration – half crescent or full Moon – that year is governed by the Sun according to the distance from the Moon. If we now examine the 54th year according to the table, 18 is a factor and therefore the Sun governs the *klimakter* here too – the true condition of a *klimakter* according to our fragment – which gives the fatal description "grievous and full of danger".

The horoscope example, Valens' witness, and Critodemus' original theory

A date fitting the stellar configuration which exemplifies the method in the fragment has been found in AD 92, but the apparently artificial nature of its positions raises doubts about its authenticity as a real horoscope: Saturn, Jupiter, Venus, and the Sun are all in signs diametrically opposed to their own domiciles; Mars is in his own domicile, and the Moon is in the sign of the Sun.[26] The chart could, then, have been invented to simulate the horoscope of a very unfortunate person. Valens otherwise uses real horoscopes, so this one was perhaps taken from Critodemus' work. What was its purpose here and why did Valens reuse it? Importantly, it is a full-Moon horoscope, so we are in one of the two cases signaled by Pliny/Varro above. It could have been a good exemplification of the theory, in order to show, for example, that the year 54 in which the Sun would have been a true *klimakter* indicated a very dangerous point in life.

---

24 Hephaestio's *klimakteres*, presented in his description of the signs in the first book (I 1) and adopted by Kamateros (222ff., 423ff., 700ff., etc.), are associated with the decans. In Paulus (*Isag.* 34) they are again a different thing, and neither Manilius nor Dorotheus speak of *klimakteres*. Proving Varro's knowledge of the astrological *klimakteres* of Critodemus' kind, we have an explicit witness in Gellius (*Attic Nights* III 10.9), referring to the disturbing character of the periods of 7: *pericula quoque uitae fortunarumque hominum, quae 'climacteras' Chaldaei appellant, grauissimos quosque fieri affirmat septenarios.*

25 It is in fact the only possible factorization with two factors using the multiplication table of 10×12.

26 *GH* L 92; cf. Peter 2001, 156 for this insight. Peter also notes that, unlike in Valens' normal practice, the ascendant is not given.

Valens would have slightly altered the method in using inclusive counting (whether consciously or not we do not know), but the true *klimakteres* of the Sun at half crescent and full Moon only work with non-inclusive counting. Perhaps it is no coincidence that Valens includes a second horoscope after his exposition of the table (8.101–104, not printed above), dated to AD 104, which presents the lunar phase of half crescent. There, Valens attempts an obscure procedure,[27] perhaps consciously modifying the theory.

As for Critodemus' original doctrine, one could think that it only contained the part on the distance of the Moon, attributing an important role to this celestial object as in F 11 and 12 in the preceding chapter: each star except the Moon would have defined a series of *klimakteres* based on the distance (counted non-inclusively) from the Moon: thus, for example, in a full-Moon chart, the Sun would have critical points at the 6th year, the 12th year, the 18th year, etc. However, it would seem odd to interpret the part on the chart-independent *klimakteres* as an addition incorporated earlier than Varro. I think it is more likely that Critodemus combined an established system of planetary *klimakteres*, independent of the birth chart, with an idiosyncratic theory based on the distances of the Moon.[28] The nature of the next fragment, if it is from Critodemus, would reinforce this argument.

F 15 (possible, but not certain). Dorotheus V 41.36–68

Type of fragment: The most relevant source for this fragment is the 11th-century astrological miscellany **L**=Laur. Plut. 28.34, although it is also copied in **C**=Marc. gr. 334 (14th c.), and there is a modified version in **I**=Laur. Plut. 28.13 by Isaac Argyros. In the three manuscripts it appears after two excerpts, and the three together appear translated into Arabic as chapter V 41 of Dorotheus' astrological manual. The Arabic chapter presents the text without divisions, and the title attributes the whole content to a certain Qīṭrnūs, who may be Critodemus. The Greek version cites Dorotheus, but this is probably because it was excerpted from the manual of Dorotheus, into which the material from Critodemus would have been inserted. It was edited first by A. Olivieri (*CCAG* I, 125) and then by D. Pingree in his edition of Dorotheus (V 41 Greek version, 425–426), with no changes. I use the section numbers used by Pingree for the Greek text, differing from those of the Arabic version.

27 *GH* L 104.7.

28 Valens III 8 attributes theories on 7-day and 9-day periods to Nechepsos and Petosiris, which suggests that this second part is earlier than Critodemus. See also F 15.

6 φησὶ δὲ ὁ Δωρόθεος καὶ ταῦτα τὰ σχήματα ἐπικίνδυνα εἶναι, οἷον τά τε ἑβδομαδικὰ καὶ ἐννεαδικὰ τῶν κλιμακτήρων. **7** εἰ γὰρ αἱ συναγόμεναι, φησίν, ἡμέραι ἀπὸ τῆς γενέσεως αὐτῆς τοῦ νοσοῦντος ἕως τῆς κατακλίσεως καὶ μεριζόμεναι παρὰ τὸν ζ̄ καταλήξωσιν εἰς τὸν ζ̄ ἢ μεριζόμεναι παρὰ τὸν θ̄ καταλήξωσιν εἰς τὸν θ̄, κλιμακτὴρ ἔσται τῇ γενέσει ἐκείνῃ· ᾗ ἐὰν κατάρξηται νοσεῖν, ἐπικινδύνως νοσήσει. **8** σκοπεῖν δή φησιν ὁ αὐτὸς καὶ τοῦτο. **9** λαβὼν ἀπὸ τοῦ κατὰ γένεσιν Ἡλίου ἐπὶ τὴν κατὰ γένεσιν Σελήνην τὸν τῶν ζῳδίων αὐτῶν τοῦ διαστήματος ἀριθμόν, συναριθμῶν καὶ αὐτὰ τὰ ζῴδια ἐν οἷς εἰσιν ἐξ ἀρχῆς οἱ δύο φωστῆρες – ἔχων οὖν, ὡς εἴρηται, τὴν τοιαύτην τῶν ζῳδίων ποσότητα, λάβε καὶ τὰς ἀπὸ τῆς ἀρχῆς τῆς γενέσεως ἕως τῆς κατακλίσεως ἡμέρας, καὶ ταύτας μέρισον παρὰ τὴν ποσότητα τῶν ζῳδίων, ὡς εἴρηται· καὶ ἐὰν ἀπαρτίζῃ ὁ ἀριθμὸς οὗτος τῷ πλήθει τῶν ἡμερῶν, ὁ κατακλιθεὶς ἐπικινδύνως νοσήσει.

10 καὶ ἄλλως δέ φησι· σκόπει τὴν τῆς κατακλίσεως Σελήνην. **11** ἐὰν γὰρ τύχῃ ἢ ἐν τῷ δ̄ τόπῳ τῆς γενέσεως ἢ ἐν τῷ ς̄ ἢ ἐν τῷ η̄, ἢ ἔνθα ἦν ἡ Σελήνη αὐτὴ κατὰ πῆξιν, κινδυνώδη τὴν νόσον καὶ ταῦτα ὑποφαίνει. **12** σκοπεῖν δὲ δεῖ, φησί, τούς τε β̄ Ἡλίους, τόν τε κατὰ πῆξιν καὶ τὸν κατὰ πάροδον, καὶ ἔτι τὴν παροδικὴν Σελήνην τῆς κατακλίσεως. **13** καὶ ἐὰν αὐτὴ ἡ Σελήνη πρὶν ἑαυτὴν τετραγωνίσαι συνάψῃ τινὶ τῶν ἡλίων, ἐλαφροτέραν τὴν νόσον δηλοῖ· ἐὰν δὲ πρὸ τοῦ συνάψαι τινὶ τῶν ἡλίων ἑαυτὴν τετραγωνίσῃ, φαῦλον τὸ σημεῖον.

6 And Dorotheus says that these configurations are dangerous: the seven-day and the nine-day periods of the *klimakteres*. **7** He says: if the division by 7 of the days from the very birth of the sick person until the laying-down is even, or if the division by 9 is even, there will be a *klimakter* for this birth, for which reason if the person has begun to be sick, the sickness will be dangerous. **8** He says one must also look this up: **9** Taking the number formed by the interval of zodiacal signs from the Sun at birth to the Moon at birth, reckoning the very signs in which are the luminaries at the beginning,[29] and having, as he says, this quantity of signs, take the days from the beginning of birth until the laying-down, and divide them up by that quantity of signs, as he says: and if this number divides exactly the quantity of days, the person lying in bed will be dangerously sick.

10 And he also says: look at the Moon of the lying-down. **11** If she is either in the 4th place of the natal chart or in the 6th or in the 8th, or where the Moon herself was according to the calculation, this will signal that the illness is dangerous. **12** One

29 The beginning of the illness is usually termed κατάκλισις, so this probably means looking at the signs of the Sun and the Moon in the natal chart (cf. "the beginning of birth", next line). The Arabic version is also somewhat indefinite (53): Pingree adds "now" in his translation, favoring the illness chart.

must look, he says, at the 2 Suns, the one according to calculation[30] and the one in transit, and the Moon in transit at the time of the lying-down: **13** and if the Moon reaches any of the Suns before she squares herself, it will show that the illness is lighter; if, however, she squares herself before reaching one of the two Suns, it is a bad sign.

What the doctrines say

The doctrines in this fragment can be divided in three parts. The first part (6–9, corresponding to Ar.V 41.47–56) shows a certain affinity with Critodemus' doctrine on the *klimakteres*, which lends support to the hypothesis that the Qīṭrnūs to whom chapter V 41 is attributed is in fact Critodemus. Firstly, one should consider the remainder of the division by seven or nine[31] of the sum of days from birth to the beginning of the illness.[32] If the division is exact, then the time corresponding to the beginning of the illness represents a *klimakter*[33] and the illness is serious. Also, take the number of signs from the Sun to the Moon in the natal chart, perhaps adding the number of signs from the Sun to the Moon in the chart of the beginning of the illness (the text is unclear at this point). The next step is to divide the number of days from birth to the beginning of the illness by this number: if the division is exact, the patient will not recover. Again, as in F 14, we have two doctrines using *klimakteres* derived from multiplicative factors which seem to reinforce one another to produce true *klimakteres*. If in F 14 the distance in signs from the Moon defined the *klimakteres* of one planet throughout life, and then a certain fixed period associated with that planet defined another set of *klimakteres*, here one similarly defines *klimakteres* based on the distance in signs between the Sun and the Moon (the Moon again!) and fixed *klimakteres* (of 7 and 9 days). The structure is remarkably similar, in that in both cases one system of *klimakteres* is defined from the chart and the other from fixed numbers, including 7 and 9.

---

30 That is, the Sun at birth.

31 The Arabic text adds that this is according to whether the native had a gestation of seven or nine months. It was a common tenet in Greek medicine and astrology that the period of gestation was either seven or nine months, and that foetuses of eight months were not viable: cf. Hippocr. *Carn.* 19; *Septim.* 1–2, *Oct.* 10; Soranus *Gyn.* 2.11; Varro *apud* Censorinus *Die nat.* 9, on Pythagoras; Aristotle *GA* 772b states that they sometimes live, but that this is not common.

32 The Arabic text helps the reader with this calculation, noting that it suffices to add 1 day per year in the case of division by 7 (because the remainder of 365 by 7 is 1); and 5 in the case of division by 7 (paragraphs 50–52), but it fails to account for the leap years.

33 The technical term appears only in the Greek version.

The second part (10–11, Ar.V 41.57–65) is straightforward: if the Moon at the beginning of the illness is in the fourth, sixth, or eighth place of the natal chart, or in the sign in which the Moon herself was at birth, then the illness will be severe, with a better or worse outcome depending on whether it is in aspect with benefics or malefics. This theory is analogous and simpler to compute than the one in the first part of the chapter (not reproduced above), in which it is prescribed that several charts be cast along the course of the Moon starting at the beginning of the illness: one when she reaches the left square (after seven days), one for the left triangle (nine days), and so on (V 41.14–18, similar to V 31). In the third part of the fragment (12–13, Ar.V 41.66–68), we are told to mark the position of the Sun in the natal chart and in the illness chart, and then look if from the position of Moon at the beginning of the illness to her left quartile we find either of the two Suns: if so, this is a good sign; otherwise, it will be a bad sign.

That this is the oldest core of V 41

We must now address the fact that the Greek text of F 15 attributes the doctrines to Dorotheus. First, it must be taken into account that book V of the Arabic Dorotheus, though it undoubtedly contains authentic material of Dorotheus, is certainly an amplification thereof. This is suggested by several circumstances, including the sheer length of this book compared to the rest – typically, the lengths of books in a work reflect the extension of the original rolls and are roughly similar to one another – and its lengthy introduction, which repeats virtually all the information of the introduction to the first book. Specifically regarding the present chapter, whenever there is a discrepancy between the Greek and the Arabic, it is always the Arabic that presents more material.[34] These amplifications are probably the combined result of a late ancient Greek redaction (4th century, the date of a horoscope included in III 1) and of a lost intermediary Middle Persian translation.[35]

34 Cf. the passages V 41.42–46, 47, 50, 62–65.

35 Cottrell and Ross 2019, 95–96, argue against Pingree's Middle Persian hypothesis, adducing that the approximately fifteen Middle-Persian words of the Arabic Dorotheus are concentrated in book V. However, al-Nadīm's *Fihrist* clearly lists Dorotheus' translator 'Umar ibn al-Farrukhān as a translator from Middle Persian, and Dimitrov 2020 shows that his translation of Ptolemy's *Tetrabiblos* is probably from that language. Farrukhān's *Tetrabiblos* apparently does not contain any Middle Persian words, and so one could hypothesize a different translator for book V of Dorotheus, one that used more freely Middle Persian terms for techincal astrological concepts. See further on this Tolsa *forthcoming*.c.

However, chapter V 41 *prima facie* does not seem to be an addition, since it has a Greek counterpart. The Greek manuscripts divide the chapter in two parts, the first of which is titled περὶ κατακλίσεων καὶ νόσων ("on lying-downs and illnesses"), and the second ἄλλη σκέψις ("another examination"). The bulk of the first part, corresponding to the Arabic V 41.10–24, in which the taking of successive charts is indicated for particular positions of the Moon in her advancement since the beginning of the illness, is very similar to the doctrine in V 31. Since Hephaestio, who seems always to be looking at Dorotheus' hexameters, refers to this doctrine (III 31), we can be certain that this theory was present in the original Dorotheus. F 15 belongs to the second part (ἄλλη σκέψις) and is preceded by an allegorical interpretation of the illness chart in which the four cardines are identified with four elements of sickness. This latter doctrine is transmitted anonymously, unlike F 15, which is ascribed to Dorotheus. Furthermore, the text that follows, which is not printed by Pingree, collects doctrines on illness by other authors including Julian of Laodicea. Thus, I propose the following scenario for the textual history of this chapter, in chronological order:

(1) Doctrines from Qīṭrnūs (representing a Greek author) corresponding to F 15 are inserted as a chapter in the late ancient Greek redaction of Dorotheus, hence the heading referring to Qīṭrnūs.
(2) In the astrological miscellany represented by **L**,[36] two successive chapters on illness gather doctrines from (1) Dorotheus V 31, and (2) Dorotheus F 15 (=Qīṭrnūs) and other sources. In particular, the second chapter records the source of the theories, after the unattributed allegory functioning as a preface.[37]
(3) Presumably a translator from Greek into Middle Persian (6th or 7th century) uses the Greek astrological miscellany to amplify V 41, leaving the order of the doctrines as it appears in the Greek manuscript. This explains the fact that the chapter is ascribed to Qīṭrnūs whereas the first doctrines are actually derived from Dorotheus V 31.

Qīṭrnūs al-Sadwālī

The other difficulty is the interpretation of the *nisba* (relation-name) that accompanies the name Qīṭrnūs in the Arabic manuscript, for which no interpretation has hitherto been proposed. Since the term *al-Sadwālī* does not seem to correspond to

36 Caballero-Sánchez 2013, 86 argues that the original *Syntagma Laurentianum* goes back at least to the 9th century.
37 The Greek chapter continues with other doctrines attributed to authors such as Nechepsos and Petosiris.

any ancient location, the link which it expresses may have another meaning. One possibility is that the letter *dāl* is a misreading of a *rā'*. Both letters are indeed very similar, and even if in the surviving manuscript (Yeni Cami 748, f. 66r) the *dāl* is clear in this instance, elsewhere in the same manuscript both letters can be easily confused. If such confusion had taken place, *sarwālī* could indicate a relation to *sarw*, "cypress", an alternative form of which is *sarwal*.[38] Now, Qīṭrnūs – the possible corruption of Κριτόδημος – is most naturally understood as an Arabization of κέδρινος, a name related to κέδρος ("cedar"), and thus the *nisba* would represent a sort of gloss: "related to the cypress".[39]

38 Cf. the glossary of Andalusian terms for trees in García and Hernández Bermejo (no date).
39 It could already have been a gloss in the Pahlavi version, since Arabic *sarw* is a Persian loan word. The interpretation would be supported by Ibn Nawbakht's list of Greek astrologers which includes a certain "Qīdrūs the Greek, from the city of Athens remembered for its science": Flügel 1872, 239. Nawbakht would have 'simplified' the name to κέδρος. Given the fictitious nature of Ibn Nawbakht's account (cf. above), I think one should not give too much weight to the assertion that the author was from Athens, even if Dorotheus is surprisingly well-assigned to Syria despite his wrong identification as Egyptian in the Arabic transmission (title in book I; V I 1).

Chapter 8
Tables for calculating the length of life (F 16–20)

The fragments collected in this chapter are all related to the two large tables used for the calculation of the length of life in Valens book VIII. The book begins with the explanation of their construction and use (F 17–18), next presents another method consisting of a variation of the standard *tetartemorion* – "the proven and amazing method of the three signs" (δοκιμαστὴ καὶ θαυμαστὴ τριζῳδία, VIII 5.24), illustrated in a fairly high number of real horoscopes of his own archive – and finally returns to the tables to give some examples of their use (F 19–20). At the end of the book, the tables are reproduced in full (F 16). There are strong reasons to believe that both tables belonged to Critodemus' work, although they were presumably incorporated at some later stage, like the table of *klimakteres* (Chapter 7). Valens devotes considerable space here to the development of methods for correctly establishing the degree of the ascendant, which is fundamental to these procedures on the length of life, but it is probable that such correction methods were not present in the work of Critodemus.

There are values in the tables calculated for the latitude of Alexandria, but, as we will see, this seems to have been Valens' own contribution.

F 20 contains a double horoscope that exemplifies a method described by Valens to determine the time of death by recasting the chart, using a unique system of astrological terms attached to the first table (F 16). It implies birth and death dates in AD 2 and 36, respectively, and was probably taken from Critodemus, where it probably served to illustrate a doctrine on transits as the double horoscope of F 9 (Chapter 4).

Note on the reproduction and the formatting of the tables

I present the tables first, although they are placed at the end of book VIII in Valens' manual. For reasons of space, I do not provide the original Greek version. The method and the intended numbers are clear enough, and they have been verified by Pingree. As in his edition, I retain the numbers given or implied (correcting evident typographical errors) by the manuscript (the tables survive just in **S**), even in the cases in which they disagree slightly with the calculations. While in some cases transmission may be the cause of mistakes, in others (especially when more than one alphanumeric symbol is incorrect) it is clear that the calculation was wrong.

https://doi.org/10.1515/9783111329147-009

Table 1: Chart (κανόνιον ᾱ· πλινθίον).

	Libra					Scorpio					Sagittarius				
Degree		**Numbers**	**Years**	**Months**	**Days**		**Numbers**	**Years**	**Months**	**Days**		**Numbers**	**Years**	**Months**	**Days**
1	Sun	2	6	1	15	Mar	14	39	0		Jup	26	67	6	
2		4	12	3			16	44	5			28	72	6	
3		6	18	3			18	49	0			30	77	6	
4		8	24	4			20	55	2			2	5	2	
5		10	30	3	15		22	60	6			4	10	3	15
6	Mar	12	36	7		Jup	24	65	0		Sat	6	15	5	
7		26	78	0			8	21	10			20	51	2	
8		28	84	0			10	27	2			22	56	2	
9		30	89	11			12	32	7			24	61	3	15
10		2	6	0			14	38	0			26	66	3	
11	Jup	4	11	9		Sat	16	43	3		Mer	28	71	4	
12		6	17	1			18	48	7			30	76	6	
13		20	59	11			2	5	4			14	35	7	15
14		22	64	5	15		4	10	9			16	40	7	15
15		24	70	6			6	16	1			18	45	8	
16	Sat	26	76	8		Mer	8	21	5		Ven	20	50	9	
17		28	81	3			10	26	9			22	55	9	
18		30	87	4			12	32	0			24	60	10	
19		14	40	11			26	69	2			8	20	10	
20		16	46	9			28	74	4			10	25	4	
21	Mer	18	51	5		Ven	30	79	6		Sun	12	30	5	
22		20	57	11			2	5	3			14	35	5	
23		22	62	5	15		4	10	6			16	40	5	10
24		24	68	8			6	15	9			18	45	5	15
25		8	22	9			20	52	6			2	5	3	
26	Ven	10	28	5	15	Sun	22	57	0		Mar	4	10	1	
27		12	33	10			24	62	0			6	15	1	
28		14	39	5			26	67	11			8	20	3	
29		16	44	10	15		28	72	11			10	25	2	
30		18	50	4			30	77	7			12	30	2	

Capricorn					Aquarius					Pisces						
	Numbers	Years	Months	Days		Numbers	Years	Months	Days		Numbers	Years	Months	Days		Degrees
Sat	8	20	6		Mer	20	50	10		Ven	2	5	4			1
	10	25	1			22	55	11			4	10	9			2
	12	30	2			24	61	0			6	16	2			3
	14	35	1			26	66	2			8	21	7			4
	16	40	1			28	71	4			10	27	0			5
Mer	18	45	0		Ven	30	76	6		Sun	12	32	6	10		6
	2	5	0			14	35	8			26	70	7			7
	4	10	1			16	40	10			28	75	7			8
	6	15	1			18	46	1			30	80	2			9
	8	20	1			20	51	4			2	5	5	15		10
Ven	10	25	1		Sun	22	56	7		Mar	4	11	0			11
	12	30	1			24	61	10			6	16	6	15		12
	26	65	2			8	19	3			20	55	4			13
	28	70	3			10	25	3			22	61	2			14
	30	75	4			12	31	1			24	66	10			15
Sun	2	5	0		Mar	14	36	6		Jup	26	72	8			16
	4	10	1			16	41	9			28	78	2			17
	6	15	1			18	47	0			30	94	6			18
	20	50	11			2	5	3			14	39	8			19
	22	55	5			4	10	6			16	45	4			20
Mar	24	60	6		Jup	6	15	9		Sat	18	51	2			21
	26	65	8			8	21	0			20	57	0			22
	28	70	5			10	26	4			22	62	11			23
	30	75	5			12	31	9			24	68	11			24
	14	35	9			26	69	0			8	23	11			25
Jup	16	40	6		Sat	28	74	4		Mer	10	28	11			26
	18	45	8			30	79	10			12	34	5			27
	20	50	8			2	5	4			14	40	5			28
	22	55	9			4	10	8			16	46	8			29
	24	60	11			6	16	6			18	52	7			30

(continued)

Table 1 (continued)

		Aries					Taurus					Gemini			
Degrees		Numbers	Years	Months	Days		Numbers	Years	Months	Days		Numbers	Years	Months	Days
1	Sun	14	41	1		Ven	26	83	6		Mer	8	27	3	
2		16	47	1			28	90	2			10	34	1	
3		18	53	2			30	97	0			12	41	0	
4		20	59	3			2	6	6			14	47	11	
5		22	65	5			4	13	3			16	54	10	
6	Ven	24	71	7		Mer	6	19	7		Sat	18	61	9	15
7		8	24	0			20	65	0			2	6	10	15
8		10	30	0			22	72	2			4	13	9	
9		12	36	1			24	78	5			6	20	8	
10		14	42	2			26	85	4			8	27	7	
11	Mer	16	48	4		Sat	28	92	3		Jup	10	34	6	
12		18	54	6			30	99	0			12	41	5	
13		2	6	1			14	46	3			26	89	8	15
14		4	12	2			16	53	0			28	96	7	
15		6	18	4			18	59	8			30	103	8	
16	Sat	8	24	7		Jup	20	66	9		Mar	2	6	11	
17		10	30	10			22	73	4			4	13	10	
18		12	37	9			24	80	0			6	20	9	
19		26	80	0			8	26	8			20	69	4	
20		28	87	0			10	33	5			22	76	3	
21	Jup	30	93	6		Mar	12	41	2		Sun	24	83	2	
22		2	6	3			14	46	7			26	90	1	15
23		4	12	6			16	53	5			28	97	0	15
24		6	18	11			18	60	7			30	104	2	
25		20	63	9			2	6	9			14	48	8	
26	Mar	22	69	8		Sun	4	13	6		Ven	16	55	7	15
27		24	76	2			6	20	3			18	62	7	
28		26	82	5			8	27	1			20	69	8	
29		28	89	5			10	34	0			22	76	7	
30		30	96	8			12	40	0			24	83	7	

	Cancer					Leo					Virgo					
	Numbers	Years	Months	Days		Numbers	Years	Months	Days		Numbers	Years	Months	Days		Degrees
Sat	20	69	9		Jup	2	6	11		Mar	14	46	5			1
	22	76	8			4	13	10			16	53	8			2
	24	83	8			6	20	5			18	59	0			3
	26	90	9			8	27	7			20	66	7			4
	28	97	9			10	34	6			22	72	0			5
Jup	30	104	10		Mar	12	41	5		Sun	24	78	11			6
	14	49	0			26	89	7			8	26	3	15		7
	16	56	0			28	96	6			10	32	5			8
	18	63	0			30	103	9			12	39	3			9
	20	69	11			2	6	10			14	45	7			10
Mar	22	76	6		Sun	4	13	8		Ven	16	52	0			11
	24	83	9	5		6	20	6			18	58	4			12
	8	27	11			20	68	4			2	6	3	15		13
	10	34	10	15		22	75	1			4	12	5			14
	12	41	10			24	81	8			6	19	3	15		15
Sun	14	48	9		Ven	26	88	5		Mer	8	25	3			16
	16	55	9			28	95	0			10	31	7			17
	18	62	8			30	101	7			12	38	11	15		18
	2	6	11			14	47	4			26	82	4			19
	4	13	3			16	54	1			28	88	11			20
Ven	6	20	10		Mer	18	60	8		Sat	30	94	8			21
	8	27	10			20	67	4			2	6	3	15		22
	10	34	5			22	73	11			4	12	6	20		23
	12	41	8			24	80	5			6	18	9	7		24
	26	90	1	15		8	27	5			20	62	5			25
Mer	28	97	6		Sat	10	33	5		Jup	22	68	5			26
	30	104	0			12	40	1			24	74	4	15		27
	2	6	11			14	46	8			26	80	3			28
	4	13	10			16	52	11			28	86	2			29
	6	20	10			18	59	10			30	92	2			30

Table 2: Chart (κανόνιον δεύτερον καὶ πλινθίον).

Degree	Libra Numbers	Libra Minutes[1]	Libra Years	Libra Months	Scorpio Numbers	Scorpio Minimum	Scorpio Years	Scorpio Months	Sagittarius Numbers	Sagittarius Minutes	Sagittarius Years	Sagittarius Months
1	0	20	1	0	12	20	34	4	24	20	63	1
2	2	40	8	2	14	40	40	3	26	40	69	0
3	5	0	15	3	17	0	48	9	29	0	74	11
4	7	20	22	4	19	20	53	4	1	20	3	1
5	9	40	29	4	21	40	59	4	3	40	9	0
6	12	0	36	3	24	0	65	11	6	0	15	4
7	24	20	76	3	6	20	17	4	18	20	46	11
8	26	40	80	0	8	40	23	8	20	40	51	0
9	29	0	87	9	11	0	29	11	23	0	58	8
10	1	20	4	11	13	20	36	0	25	20	64	7
11	3	40	11	9	15	40	42	5	27	40	70	5
12	6	0	18	10	18	0	48	7	30	0	76	1
13	18	20	55	10	0	20	0	11	12	20	31	0
14	20	40	62	6	2	40	7	2	14	40	37	2
15	23	0	69	9	5	0	13	3	17	0	43	1
16	25	20	76	8	7	20	19	7	19	20	49	7
17	27	40	83	8	9	40	25	10	21	40	55	0
18	30	0	90	9	12	0	32	0	24	0	60	11
19	12	20	37	4	24	20	64	9	6	20	16	1
20	14	40	44	11	26	40	70	10	8	40	21	11
21	17	0	51	6	29	0	76	10	11	0	27	10
22	19	20	58	1	1	20	3	6	13	20	33	8
23	21	40	65	6	3	40	9	8	15	40	39	3
24	24	0	72	3	6	0	15	10	18	0	46	4
25	6	20	19	11	18	20	48	1	0	20	0	10
26	8	40	26	0	20	40	54	1	2	40	6	8
27	11	0	33	0	23	0	60	1	5	0	12	7
28	13	20	40	6	25	20	65	11	7	20	18	6
29	15	40	47	11	27	40	71	11	9	40	24	4
30	18	0	54	4	30	0	78	0	12	0	30	2

1 The Greek label for the minutes column is συγ, probably an abbreviation of συγκείμενοι ἀριθμοί: συγκείμαι would be understood in the active sense, as "lie together", that is, the minutes that belong together with the integer value.

Capricorn				Aquarius				Pisces				
Numbers	**Minutes**	**Years**	**Months**	**Numbers**	**Minutes**	**Years**	**Months**	**Numbers**	**Minutes**	**Years**	**Months**	**Degree**
6	20	16	7	18	20	46	7	0	20	0	11	1
8	40	21	9	20	40	52	7	2	40	6	10	2
11	0	27	6	23	0	58	6	5	0	13	9	3
13	20	33	6	25	20	64	6	7	20	19	9	4
15	40	40	6	27	40	70	6	9	40	26	1	5
18	0	45	0	30	0	76	5	12	0	32	6	6
0	20	0	11	12	20	31	2	24	20	66	7	7
2	40	6	3	14	40	37	2	26	40	72	7	8
5	0	13	11	17	0	42	11	29	0	79	0	9
7	20	18	7	19	20	49	7	1	20	3	11	10
9	40	24	7	21	40	55	7	3	40	9	11	11
12	0	30	1	24	0	65	10	6	0	16	6	12
24	20	61	3	6	20	16	9	18	20	48	1	13
26	40	66	2	8	40	22	6	20	40	57	3	14
29	0	72	3	11	0	28	6	23	0	64	2	15
1	20	3	6	13	20	35	6	25	20	70	3	16
3	40	9	9	15	40	40	9	27	40	77	4	17
6	0	15	0	18	0	47	6	30	0	84	4	18
18	20	46	10	0	20	0	10	12	20	35	1	19
20	40	52	1	2	40	6	1	14	40	41	6	20
23	0	58	1	5	0	11	2	17	0	48	2	21
25	20	62	2	7	20	19	3	19	20	55	1	22
27	40	70	6	9	40	25	9	21	40	61	11	23
30	0	72	8	12	0	31	3	24	0	68	11	24
12	20	31	3	24	20	64	1	6	20	18	2	25
14	40	37	9	26	40	70	9	8	40	25	0	26
17	0	43	6	29	0	76	8	11	0	31	11	27
19	20	49	8	1	20	3	0	13	20	38	4	28
21	40	54	9	3	40	9	9	15	40	45	8	29
24	0	60	11	6	0	16	0	18	0	52	8	30

(continued)

Table 2 (continued)

	Aries				Tau				Gem			
Degree	**Numbers**	**Minutes**	**Years**	**Months**	**Numbers**	**Minutes**	**Years**	**Months**	**Numbers**	**Minutes**	**Years**	**Months**
1	12	20	36	1	24	20	78	2	6	20	21	7
2	14	40	42	2	26	40	85	11	8	40	29	6
3	17	0	51	4	29	0	93	8	11	0	36	8
4	19	20	57	5	1	20	4	4	13	20	45	1
5	21	40	64	6	3	40	11	11	15	40	55	1
6	24	0	61	7	6	0	19	7	18	0	61	10
7	6	20	18	3	18	20	59	7	0	20	1	2
8	8	40	26	0	20	40	63	9	2	40	8	6
9	11	0	33	5	23	0	76	10	5	0	17	3
10	13	20	39	6	25	20	83	4	7	20	22	3
11	15	40	47	5	27	40	91	2	9	40	33	4
12	18	0	54	6	30	0	99	6	12	0	41	5
13	0	20	1	3	12	20	40	9	24	20	84	8
14	2	40	8	1	14	40	48	7	26	40	92	2
15	5	0	15	9	17	0	56	5	29	0	100	3
16	7	20	22	6	19	20	64	2	1	20	4	7
17	9	40	29	4	21	40	72	1	3	40	12	8
18	12	0	37	8	24	0	80	2	6	0	20	9
19	24	20	75	7	6	20	21	2	18	20	63	6
20	26	40	82	6	8	40	27	11	20	40	71	8
21	29	0	90	5	11	0	37	4	23	0	79	9
22	1	20	4	4	13	20	44	8	25	20	81	9
23	3	40	11	4	15	40	52	8	27	40	96	1
24	6	0	18	3	18	0	60	7	30	0	104	2
25	18	20	58	3	0	20	1	1	12	20	42	9
26	20	40	65	2	2	40	9	0	14	40	51	1
27	23	0	73	2	5	0	16	11	17	0	59	2
28	25	20	80	01	7	20	24	10	19	20	67	4
29	27	40	88	4	9	40	32	9	21	40	75	6
30	30	0	96	1	12	0	40	9	24	0	82	8

Cancer					Leo					Virgo					
Numbers	**Minutes**	**Years**	**Months**		**Numbers**	**Minutes**	**Years**	**Months**		**Numbers**	**Minutes**	**Years**	**Months**		**Degree**
18	20	63	11		0	20	1	2		12	20	41	11		1
20	40	72	1		2	40	9	2		14	40	48	7		2
23	0	80	3		5	0	17	2		17	0	56	1		3
25	20	88	4		7	20	25	3		19	20	63	10		4
27	40	96	8		9	40	33	4		21	40	71	4		5
30	0	104	11		12	0	41	4		24	0	78	11		6
12	20	43	1		24	20	83	10		6	20	20	9		7
14	40	51	4		26	40	91	10		8	40	28	4		8
17	0	59	1		29	0	99	7		11	0	35	11		9
19	20	67	7		1	20	4	7		13	20	43	2		10
21	40	75	9		3	40	12	7		15	40	50	11		11
24	0	83	2		6	0	20	6		18	0	58	4		12
6	20	22	1		18	20	62	8		0	20	1	1		13
8	40	30	3		20	40	70	6		2	40	8	7		14
11	0	38	2		23	0	77	4		5	0	16	1		15
13	20	46	6		25	20	85	2		7	20	23	6		16
15	40	54	7		27	40	93	11		9	40	30	7		17
18	0	62	11		30	0	101	8		12	0	38	7		18
0	20	1	7		12	20	41	9		24	20	77	3		19
2	40	9	1		14	40	49	6		26	40	84	10		20
5	0	16	4		17	0	57	4		29	0	91	7		21
7	20	25	4		19	20	65	1		1	20	4	10		22
9	40	33	2		21	40	72	6		3	40	11	9		23
12	0	41	1		24	0	80	6		6	0	18	9		24
24	20	84	4		6	20	21	3		18	20	57	2		25
26	40	92	3		8	40	28	11		20	40	64	3		26
29	0	100	4		11	0	36	11		23	0	71	4		27
1	20	4	7		13	20	44	5		25	20	78	4		28
3	40	12	8		15	40	52	2		27	40	84	4		29
6	0	20	9		18	0	59	9		30	0	96	3		30

On the terms in the first table, see the critical apparatus in Pingree's edition: they appear in a very corrupt form, but the system can be easily reconstructed, thanks to its regularity and the witness of F 20.

Unlike Pingree, I attempt to reproduce the exact layout of the tables in the manuscript, including the separations between the groups of columns representing a zodiacal sign, as well as the horizontal divisions every three rows. Both aspects are far from irrelevant, since (1) the spaces between the sign-columns in the first table allowed for the insertion of the astrological terms, and (2) the horizontal dividing lines reflect the structure of the numbers in both tables.[2]

To continue with questions of formatting, both tables are divided into quarters, one per page, in such a way that the two halves comprising the second and the first set of six zodiacal signs, respectively (more on this awkward ordering below), each occupy the verso and the subsequent recto and are thus visible at the same time. In both tables, the column of zodiacal degrees appears in each quarter, always in the outermost column.

F 16. The tables themselves: Valens VIII (end)

Type of fragment: Tables for calculating the length of life from the ascendant degree. The first table contains a system of terms added in the intercolumn spaces at some point between the introduction of the tables in Critodemus' work and Valens' activity.

2 For the sake of comparison, it is worth noting that nearly all the Oxyrhynchus astronomical tables have divisions every two lines, the same as the oldest manuscripts of Ptolemy's *Handy Tables*; the tables of Ptolemy's *Almagest*, on the other hand, have horizontal divisions every three lines. In most cases this makes no difference, but for the table of rising times in the second book of the *Almagest* horizontal rulings every three rows are clearly more appropriate, since the signs are divided into three sections of 10° each. On the Oxyrhynchus papyri, see the commented edition/translation, Jones 1999; for the manuscripts of Ptolemy's *Handy Tables*, cf. Vat. gr. 1291 (ninth century), and Leiden BPG 78 (ninth century) for Ptolemy's *Almagest*, cf. Par. gr. 2389 (ninth century).

General description of the tables

Both tables have the same format and work in the same way. The most obvious difference is that the first table contains a system of terms written in the intercolumn spaces which is different from the Egyptian system used by Critodemus in F 10. The tables are formed by 30 rows corresponding to the 30° of the signs, and by 12 columns corresponding to the 12 signs. More specifically, the latter consist of 12 blocks of columns, each containing a first column that displays a broken sequence of even numbers between 2 and 30, and a second subblock that displays the length of life of the person whose ascendant is in the corresponding degree (given by the sign-column and the degree-row). The internal divisions of the latter block separate the fractional parts of the year, always expressed in the sexagesimal system, as is typical in ancient astronomy. From Valens' explanation (F 17), it seems that he proposed the calculation of the fractional parts as well as the times of life for other latitudes to his students: the tables were then probably completed (only for the latitude of Alexandria) in the course of the manuscript transmission. F 17 explains the rationale of the 'numbers' column in both tables, and F 18 shows how the lengths of life are to be calculated, both those corresponding to the latitude of Alexandria, as in the present tables, and for other climata.

The values for the length of life are found by multiplying the number assigned to the corresponding degree by the length of daylight in time-degrees of the day when the Sun is at this longitude (that of the corresponding degree), and dividing the result by 60. The system of the signs is interpreted so that Aries 8° is the vernal equinox instead of Aries 0°. This was a traditional feature of Babylonian astronomy that was adopted by some Greek authors.[3] Thus, it is the day length corresponding to Aries 8° (and its opposite Libra 8°) that is 180 time-degrees, not Aries 0° (and Libra 0°); this is true at all latitudes. Therefore, if we (like Valens) use a zodiac centered at 0°, we must make the conversion by adding 8° before entering the table. So, for example, if our ascendant is Libra 0°, we go to Libra 8°; we then multiply the 28 in the first column of the first table by the 180 time-degrees, and divide by 60, which equals 84. These are the years of life, as shown in the corresponding column.[4] The procedure is the same for the second table, substituting the corresponding numbers in the first columns.

---

3 Cf. Thrasyllus, *CCAG* VIII 3, 99: αἱ τροπαὶ γίνονται οὐ περὶ πρώτην μοῖραν τοῦ ζῳδίου, ὥς τινες νομίζουσιν, ἀλλὰ περὶ η̅ ("The turns do not happen around the first degree of the sign, as many assume, but around the 8th"). Cf. *HAMA*, 596, for evidence of other astrologers' knowledge of this norm, including Geminos, Manilius, Valens, and Firmicus.

4 Another example: the maximum daylight at any latitude is at Cancer 0° in the standard system. If we have an ascendant at Cancer 0°, at Alexandria the daylight is 210 time-degrees. Then we go

The only essential difference between the two tables is found in the "numbers" column. The second table is simply a variation of the first, arrived at by slightly changing these numbers such that the distances inside the sets of six numbers remain equal, but changing the minimum number to 0;20 (i.e., 0.33) instead of 2: with the new numbers, lower lengths of life are obtained, such as 1 year of life for the first entry (at the latitude of Alexandria; cf. 6 years and 1 month in the first table). Obtaining low life expectancies could in fact have been the aim of constructing the second table. Another interesting aspect that will be discussed below is that these numbers in both tables form squares of 5 × 5 sets of numbers, each set comprising six numbers, in which exactly five elements are found in every row and column: this is called a Latin square.

Why the tables begin with Libra and how this relates to the terms

An obvious feature of both tables that remains to be addressed is the fact that they begin not with the sign of Aries, but with Libra, the sign in opposition. There is a plausible explanation for this: the first degrees of the first sign of the table are made to correspond to the lowest numbers: 2, 4, 6, 8, 10, 12. Because of these low numbers, the resulting life lengths here are among the shortest given by the table. As it turns out, the traditional "Egyptian" term formed by the first 6° of Aries is ruled by Jupiter, whereas the term of the first 6° of Libra belongs to Saturn. Thus, it would make little sense, when designing a system for the length of life, for the beginning of Aries to correspond to the lowest life expectancy. A way to solve this would be to modify the structure of the table, having it begin with the second half of the signs, that is, with Libra.

The alternative terms attached to the first table, beginning with the Sun and extending 5° each, are written in the rows corresponding in each case to the first degree of each set of five, in the empty intercolumn spaces between signs.[5] They were likely introduced at a later stage of the transmission, using these intercolumn spaces with no horizontal rulings. At the same time, we can probably rule out that the system was devised independently and then transferred to the table, since the

to Cancer 8° in the first table, which shows the number 16, and so 16 × 210/60 = 56 years. This will change at a different latitude, giving more years at higher latitudes.

5 The terms are marked with star symbols, but this was probably a late ancient development, since in the ancient astronomical and astrological papyri we only find symbols for the Sun and the Moon.

terms also form a Latin square obtained from rotation, thus taking inspiration from the table.[6]

Attaching the tables to the work of Critodemus

Valens does not clearly present the tables as belonging to the work of Critodemus in the main text. This is however indicated in the headers of III [6] and IX 9 (virtually the same chapter copied twice), which are explicit in attributing the table to Critodemus:

> περὶ ἐχθρῶν ἀστέρων καὶ κλιμακτερικῶν τόπων περὶ τὸ ᾱ ὄργανον Κριτοδήμου / περὶ τῶν ἐχθρῶν τόπων καὶ ἀστέρων. περὶ κλιμακτηρικῶν τόπων πρὸς τὸ ᾱ ὄργανον.
>
> On malefic stars and critical places with respect to the 1st table of Critodemus / On the malefic places and stars. On critical places with respect to the 1st table of Critodemus

We have confirmation that this is the table reproduced above, since in this chapter Valens uses the astrological terms (F 20) that are attached to the table labeled as κανόνιον ᾱ. Valens seems to acknowledge Critodemus as the origin of the table later in book IX (9.27):

> Τούτων οὕτως ἐχόντων νῦν καὶ περὶ τῆς τῶν ἀπογωνίων δυνάμεως ἀρκτέον. τὴν μὲν πῆξιν Κριτόδημος ἐποιήσατο, τὴν δὲ εἴσοδον πρότερον αὐτὸς ἀνευρών, διασαφήσας δὲ ἐν ἑτέροις βιβλίοις καὶ νῦν δὲ ἀκριβέστερον ἐξευρὼν ἐπιδιαγραφήσω.
>
> This being so, discussion of the power of the *apogonia* must begin. Critodemus made the construction, and I invented the method before,[7] having made it clear in other books; and now, after a more detailed analysis, I will delineate it.

The reference is quite obscure, but there is an important clue. The word used by Valens to define what Critodemus did is πῆξις ("construction"). While this can generally mean calculation (as in the title of VIII 3: Πῆξις ὡροσκοπούσης μοίρας), in the main text Valens only uses it to indicate the construction or calculation of tables.[8] In particular, the method of the three factors, which Valens also describes

---

6 It is perhaps also significant that the terms begin with the Sun at Libra 1°, since the Sun was perceived as the center of the moving stars, in a similar way that this point of Libra is exactly at the middle of the zodiac.

7 Probably meaning before in book VIII, since he talks of "other books".

8 Notably those of Critodemus in book VIII 1.1: Τὸ ᾱ ὄργανον τὴν πῆξιν ἔχει ἀπὸ ᾱ ἕως μοίρας ϛ′ τοιαύτην . . . 1.10: Ἵνα δὲ συντομωτέραν τὴν πῆξιν δηλώσωμεν . . . But also the table of rising times at I 6.22: τὴν πῆξιν τοῦ ἀναφορικοῦ.

as πῆξις,[9] takes up the middle section of book VIII. First (VIII 6), Valens explains the construction of the table associated with this method, which he calls "of *apogonia*" (VIII 7.10, IX 9.38), and follows it with a long section (VIII 7) providing numerous horoscope examples. The table is not given, but Valens explains it sufficiently well for the structure to be understandable. For each degree, it seems to have featured the accumulated time-degrees (from the beginning of Aries), like any table of rising times, and in three subsequent columns it gave three factors, the first containing the time-degrees corresponding to the sector extending 30° (this would be the first factor), the second containing the time-degrees of the sector extending 60° (second factor), and the third containing the time-degrees of the sector extending 90° (third factor).[10] Then, given the ascending degree, one makes a prediction of the length of life by choosing one or the sum of several of these factors (understood as years of life, or in some cases months or days). It is therefore a variation of the *tetartemorion* method, in which one chooses, at most, the time-degrees corresponding to the next 90°. Thus, from what Valens says here in book IX – that, concerning the power of the *apogonia*, Critodemus made the construction and Valens found the method – and from the structure of book VIII, one could infer that he interpreted his "three-factor" method as his own improvement on Critodemus' tables. Indeed, at the most basic level, the input and the aim of the two kinds of tables are the same, to provide a value for the length of life given the ascendant degree.

A relevant question is whether the second table is also the work of Critodemus. A first argument in favor of this possibility is simple: Valens makes no use of the second table beyond explaining its construction, whereas he gives many examples of calculations with real horoscopes using the first table (F 19). This would be odd if the second table was his own contribution. Secondly, at the end of his ninth and last book, Valens refers to the construction of two tables for the determination of the longitude of the Moon, which he similarly labels κανόνες τῶν $\overline{\beta}$ πλινθίων ("tables of the two charts", IX 19, cf. Appendix III). Their position at the end of the book, their number, and their bizarre nature suggest that Valens in part created them to imitate the two tables for the length of life from Critodemus' treatise.

9 VIII 6.1: ἡ δὲ πῆξις τῶν ἀναφορῶν καὶ τῶν τριῶν ὅρων.

10 The name *apogonia* may derive from the fact that in this method one chooses from three classical angles: 90° (three signs), 60° (two signs), or 30° (one sign).

The word πλινθίον

It is noteworthy that both tables are called both κανόνιον and πλινθίον. Specifically, the first is titled "first table: chart" (κανόνιον α̅· πλινθίον) and the second "second table and chart" (κανόνιον δεύτερον καὶ πλινθίον). Whereas κανόνιον, as well as the non-diminutive κανών, is the usual word for a numerical table in Valens, πλινθίον only occurs once elsewhere in his treatise, to announce the two mentioned tables at the end of book IX, where they are bizarrely designated with a similar double expression: "tables of the two charts" (κανόνες τῶν β̅ πλινθίων).

More generally, there are few words attested to designate numerical tables in ancient Greek. Here are the most obvious:

- Valens: κανών/κανόνιον, ὄργανον
- Ptolemy: κανών/κανόνιον, καταγραφή[11]
- Nicomachus: διάγραμμα

A crucial instance for our purposes is found in the work of a less evident candidate, the Alexandrian Biblical exegete Philo (first century AD), who mentions a figure showing the three basic means (arithmetic, geometric, and harmonic), calling it first διάγραμμα, then πλινθίον, and then finally both, while expressing his doubts about the proper designation: "the diagram or the chart or whatever it should be called" (τὸ διάγραμμα ἢ πλινθίον ἢ ὅ τι χρὴ καλεῖν, *Opif.* 107).[12]

The word πλινθίς and its diminutive πλινθίον were used to refer to rectangular-shaped objects, sometimes composed of smaller rectangles forming an orthogonal grid. There are examples from geography (rectangular sections of the Earth, Strabo II 1.35), from ornithomancy (sections of the heavens: Plut. *Rom.* 22.1), from sundial-making (sections of the heavens again: Vitr. IX 8.1), or from board games (a sort of draughts, Pollux *Onom.* 9.98). Thus, it is not surprising that it came to designate a table of numbers containing orthogonal rulings. The meager evidence that we have seems to suggest that, by the second half of the second century AD, the time of Ptolemy and Valens, πλινθίς had been replaced by the words κανών/

11 καταγαφή for tables is only used in the *Harmonics* I, in late ancient additions: cf. Tolsa 2018.
12 Philo Alex. *Opif.* 107: "It [the number 7] is not just perfect, but it is also, so to speak, most harmonious and in a way the source of a most beautiful diagram, which comprises the whole of harmonics – the fourth, the fifth, the octave – and all the ratios – the arithmetic mean, the geometric, and the harmonic. The chart consists of these numbers: six, eight, nine, twelve" (ἔστι δὲ οὐ τελεσφόρος μόνον, ἀλλὰ καὶ ὡς ἔπος εἰπεῖν ἁρμονικωτάτη καὶ τρόπον τινὰ πηγὴ τοῦ καλλίστου διαγράμματος, ὃ πάσας μὲν τὰς ἁρμονίας, τὴν διὰ τεττάρων, τὴν διὰ πέντε, τὴν διὰ πασῶν, πάσας δὲ τὰς ἀναλογίας, τὴν ἀριθμητικήν, τὴν γεωμετρικήν, ἔτι δὲ τὴν ἁρμονικὴν περιέχει. τὸ δὲ πλινθίον συνέστηκεν ἐκ τῶνδε τῶν ἀριθμῶν, ἓξ ὀκτὼ ἐννέα δώδεκα).

κανόνιον and ὄργανον. In fact, from his double designations, it is possible to deduce that Valens himself was not thoroughly familiar with the word. In turn, the attestations in Philo and Nicomachus could indicate that in the first century, the prevailing designation for a numerical table was διάγραμμα, while πλινθίον was also in use.

Other instances are found in various kinds of numerological tables in several manuscripts, including the so-called circle of Petosiris, which often feature the name πλινθίς/πλινθίον. Unfortunately, it is difficult to date this material, but the late Hellenistic period seems to be the most probable original context (see discussion of F 17 below for a comparison between one of these squares and Critodemus' tables).[13]

F 17. Structure of the "numbers" column: Valens VIII 1–2

Type of fragment: Valens' explanation of the construction of the first column of the tables, formed by broken sequences of six evenly distributed numbers.

<α̅.> Πῆξις τοῦ α̅ ὀργάνου
1 Τὸ α̅ ὄργανον τὴν πῆξιν ἔχει ἀπὸ α̅ ἕως μοίρας ς̅[a] τοιαύτην. **2** τῇ α̅ μοίρᾳ τοῦ Ζυγοῦ παράκειται ἀριθμὸς β̅, τῇ β̅ δ̅, τῇ γ̅ ς̅, τῇ δ̅ η̅, τῇ ε̅ ι̅, τῇ ς̅ ι̅β̅, τουτέστι παραύξησις μοιρῶν β̅. **3** εἶτα ἀπὸ τῆς ζ̅ μοίρας συνδέσμου λύσις, παραύξησις προσθέσεως μοιρῶν ι̅δ̅, καὶ γίνονται ἐπὶ τῆς μοίρας ζ̅ μοῖραι κ̅ς̅, ἐπὶ τῆς η̅ μοίρας κ̅η̅, ἐπὶ τῆς θ̅ μοίρας λ̅, ἐπὶ τῆς ι̅ β̅, ἐπὶ τῆς ι̅α̅ δ̅, ἐπὶ τῆς ι̅β̅ ς̅. **4** ἔπειτα πάλιν συνδέσμου προσθέσεως μοῖραι ι̅δ̅, καὶ γίνονται ἐπὶ τῆς ι̅γ̅ μοίρας κ̅, ἐπὶ τῆς ι̅δ̅ μοίρας κ̅β̅, ἐπὶ τῆς ι̅ε̅ μοίρας κ̅δ̅, ἐπὶ τῆς ι̅ς̅ μοίρας κ̅ς̅, ἐπὶ τῆς ι̅ζ̅ μοίρας κ̅η̅, ἐπὶ τῆς ι̅η̅ μοίρας λ̅. **5** εἶτα συνδέσμου προσθέσεως μοῖραι ι̅δ̅, καὶ γίνονται μοῖραι μ̅δ̅, ἐξ ὧν ἀφελὼν τὰς λ̅ λοιπαὶ ι̅δ̅· αὗται ἔσονται ἐπὶ τῆς ι̅θ̅ μοίρας, ἐπὶ τῆς κ̅ ι̅ς̅, ἐπὶ τῆς κ̅α̅ ι̅η̅, ἐπὶ τῆς κ̅β̅ κ̅, ἐπὶ τῆς κ̅γ̅ κ̅β̅, ἐπὶ τῆς κ̅δ̅ κ̅δ̅. **6** καὶ πάλιν συνδέσμου προσθέσεως μοῖραι ι̅δ̅ καὶ γίνονται λ̅η̅, ἐξ ὧν ἀφελὼν τὰς λ̅ λοιπαὶ η̅· αὗται ἔσονται ἐπὶ τῆς κ̅ε̅, ἐπὶ τῆς κ̅ς̅ ι̅, ἐπὶ τῆς κ̅ζ̅ ι̅β̅, ἐπὶ τῆς κ̅η̅ ι̅δ̅, ἐπὶ τῆς κ̅θ̅ ι̅ς̅, ἐπὶ τῆς λ̅ ι̅η̅. **7** ὥστε κατὰ μὲν μοίρας ς̅ συνδέσμου λύσις ἔσται προστιθεμένων τῶν ι̅δ̅ μοιρῶν καὶ ἐφεξῆς τῶν τῆς παραυξήσεως β̅ ἐπὶ πάντων ζῳδίων. **8** ἕξει οὖν ὁ μὲν Ζυγὸς ἐπὶ τῆς α̅ μοίρας παρακειμένας β̅, ἐπὶ δὲ τῆς λ̅ μοίρας ι̅η̅. ὁμοίως δὲ τοῖς αὐτοῖς ἀριθμοῖς τοῦ Ζυγοῦ ὅ τε Λέων καὶ οἱ Ἰχθύες ὠργανοθέτηνται.
9 εἶτα ἀκολούθως Σκορπίος ἕξει ἐπὶ μὲν τῆς α̅ μοίρας ι̅δ̅ καὶ τῆς παραυξήσεως τῶν β̅ προστιθεμένων γενήσονται ἕως τῆς ς̅ μοίρας κ̅δ̅ καὶ ὁμοίως τῶν συνδέσμων

13 Neugebauer and Saliba 1989, 189, suggest a Hellenistic/early Roman date based only on the ascription to Nechepsos/Petosiris and Pythagoras. Hippolytus of Rome (*Against all heresies* IV 14, second/third century AD) partly describes the system, but this does not imply that it was a contemporary practice. The example given (Hector) is the same as in the manuscripts.

λυομένων ἕξει τῇ μοίρᾳ λ̅ παρακειμένας λ̅. τοὺς δὲ αὐτοὺς ἀριθμοὺς ἕξει ὅ τε Κριὸς καὶ ἡ Παρθένος.
10 Ἵνα δὲ συντομωτέραν τὴν πῆξιν δηλώσωμεν πρὸς τό τινας καὶ ὅλον τὸ ὄργανον μνημονεύειν, οὕτως λογιζόμεθα ἐπὶ προσθέσεως. **11** ἐπὶ τῆς πρώτης μοίρας τοῦ Ζυγοῦ παρακειμένους ἔχει β̅. **12** ταύταις προσέθηκα ι̅β̅ τοῦ κύκλου καὶ γίνονται ι̅δ̅· ταύτας ὁ Σκορπίος ἐπὶ τῆς α̅ μοίρας ἕξει. **13** πάλιν τὰς ι̅δ̅ προσέθηκα ταῖς ι̅β̅ καὶ γίνονται κ̅ς̅· ταύτας ὁ Τοξότης ἕξει ἐπὶ τῆς α̅ μοίρας. **14** καὶ ἐφεξῆς ἀνὰ ι̅β̅ προστιθέντες εὑρήσομεν τοῦ ἑξῆς ζῳδίου ταύτας παρακειμένας τῇ α̅ μοίρᾳ· αἷς ἐπισυνθέντες τὰς ι̅β̅ καὶ λύοντες συνδέσμους τῇ τῶν ι̅δ̅ μοιρῶν συνθέσει ὅλον τὸ ὄργανον συμπήξομεν.
15 ὅσας οὖν ὁ Τοξότης ἔχει παρακειμένας, τοσαύτας καὶ Ταῦρος· ὅσας ὁ Ὑδροχόος, τοσαύτας καὶ ὁ Καρκίνος· ὅσας ὁ Αἰγόκερως, τοσαύτας καὶ οἱ Δίδυμοι. καὶ ἔσται τούτων ἰσοδυναμία μὲν καὶ ἀλληλουχία, διάφορος δὲ διὰ τὰς ἀνωμαλίας τῶν ἀναφορῶν. **16** ἔστι δὲ τὸ ὄργανον τοῦτο ἔχον καὶ τὰ ἔτη παρακείμενα τοῖς ἀριθμοῖς καὶ ταῖς μοίραις ὑποδείγματος χάριν, τὴν δὲ ἀκρίβειαν καὶ[b] ἕκαστον κλίμα καὶ τὰς τῶν τόπων ἀλλαγὰς οἵ γε νοῦν ἔχοντες εὐκατάληπτον ἕξουσι.

β′. Πῆξις τοῦ δευτέρου ὀργάνου φυσική
1 Τῆς περὶ τὸ ὄργανον οἰκονομίας προκειμένης ἀναγκαῖον ἔσχον καὶ τὴν καταγωγὴν τῆς πήξεως ὑποτάξαι. **2** ἡ μὲν οὖν τῶν συνδέσμων πρόσθεσις, τουτέστιν αἱ ι̅δ̅, σημαίνουσι τὰ φῶτα τῆς Σελήνης, αἱ δὲ <ι̅>β̅ τῆς παραυξήσεως δακτύλους Ἡλίου· δὶς δὲ τὰ ι̅δ̅ γίνονται κ̅η̅, κύκλος Σελήνης.
3 ἐπεὶ οὖν α̅ μοίρᾳ τοῦ Ζυγοῦ παράκεινται β̅, ἐκ τούτων ἀφεῖλον μοῖραν α̅ μ̅· λοιπὰ γίνονται ο κ̅, ὅ ἐστι μέγεθος τρίτον. **4** ταῦτα ἐπὶ τὸν ξ γίνεται ρ̅η̅· ταῦτα ἀναλύω ἕως τῶν ξ′ καὶ γίνεται μύρια καὶ ὀκτακόσια. **5** ταύτας ποιῶ φ̅μ̅ εἰς κύκλον ἕνα ἥμισυ καὶ γίνεται ο κ̅· ταῦτα ἔσται παρακείμενα ἐπὶ τῆς τοῦ Ζυγοῦ α̅ μοίρας. **6** ταῦτα γ′ γίνεται ξ̅, ὅπερ ἐστὶν ἐνιαυτὸς εἷς.
7 προσεπικατάξομεν δὲ καὶ τοὺς λοιποὺς ἐνιαυτοὺς οὕτως, προστιθέντες ἑκάστῳ ἔτει β̅ κ̅. **8** ἔσονται οὖν ἐπὶ τῆς β̅ μοίρας τοῦ Ζυγοῦ β̅ μ̅, ἐπὶ τῆς γ̅ ε̅ ο, ἐπὶ τῆς δ̅ ζ̅ κ̅, ἐπὶ τῆς ε̅ θ̅ μ̅, ἐπὶ τῆς ς̅ ι̅β̅ ο. **9** εἶτα ἐπὶ τοῦ συνδέσμου προσθέντες τὰ ι̅δ̅ φῶτα τῆς Σελήνης τοῖς ι̅β̅ ο ἕξομεν κ̅ς̅ ο· ἐκ τούτων ἀφείλαμεν α̅ μ̅, καὶ λοιπὸν ἔσονται ἐπὶ τῆς ζ̅ μοίρας κ̅δ̅ κ̅. **10** ἐπὶ τῆς η̅ μοίρας παλινδρομήσομεν δύο κ̅, καὶ ἔσονται κ̅ς̅ μ̅, ἐπὶ τῆς θ̅ κ̅θ̅ ο, ἐπὶ τῆς ι̅ α̅ κ̅, ἐπὶ τῆς ι̅α̅ γ̅ μ̅, ἐπὶ τῆς ι̅β̅ ς̅ ο. **11** ὥστε ἐπὶ τὸν σύνδεσμον καὶ τὸν πρῶτον ἀριθμὸν ἀφαιρεῖν δεῖ α̅ μ̅, ἐπὶ δὲ τῆς παραυξήσεως προστιθέναι β̅ κ̅. **12** τῷ δὲ αὐτῷ τρόπῳ καὶ ἐπὶ τῶν λοιπῶν ζῳδίων οἰκονομεῖν δεῖ. καθάπερ δὲ καὶ ἐπὶ τοῦ α̅ ὀργάνου εὑρίσκομεν τοῦ ἑξῆς ζῳδίου τὸν παρακείμενον ἀριθμὸν τῇ α̅ μοίρᾳ, οὕτως καὶ ἐπὶ τοῦ δευτέρου ὀργάνου εὑρήσομεν.

a ς̅ correxi, θ̅ codd.
b καί codd., καθ’ Pingree, sugg. Kroll

1. Construction of the first table

1 The first table from 1° to 6° is constructed as follows: **2** next to the 1st degree of Libra is number 2, next to the 2nd 4, next to the 3rd 6, next to the 4th 8, next to the 5th 10, next to the 6th 12, that is, an increment of 2. **3** Then from the 7th degree there is a break of the sequence, an increment of 14, and we have for the 7th degree 26, for the 8th degree 28, for the 9th degree 30, for the 10th degree 2, for the 11th degree 4, for the 12th degree 6. **4** Then again the 14 of the sequence break, and we have for the 13th degree 20, for the 14th 22, for the 15th 24, for the 16th 26, for the 17th 28, for the 18th 30. **5** Then again the 14 of the sequence break, and we have 44, from which, taking out the 30, 14 remain. This is the number for the 19th degree, for the 20th 16, for the 21st 18, for the 22nd 20, for the 23rd 22, for the 24th 24. **6** Then again the 14 of the sequence break and we have 38, from which, taking out the 30, 8 remain. This is the number for the 25th degree, for the 26th 10, for the 27th 12, for the 28th 14, for the 30th 18. **7** So every 6 degrees there will be a break of the sequence, 14 degrees are added, and then successively the 2 of the increment, in all signs. **8** Therefore, Libra will have the number 2 in the 1st degree of Libra and 18 in the 30th. Leo and Pisces have been set up with the same numbers as Libra.[14]

9 Then Scorpio will have 14 next to the 1r degree of Scorpio, and with the increment of 2 to be added up to the 6th degree, they become 24. And similarly the sequences are broken, and it will have 30 next to the 30th degree. Aries and Virgo will have the same numbers.

10 In order to show the construction more concisely for memorizing the particulars and the whole table, we calculate by the increments: **11** It has 2 is next to the first degree of Libra. **12** To this I have added the 12 of the circle of signs[15] for a total of 14. Scorpio has this next to the 1st degree. **13** Again, I have added 12 to this 14 for a total of 26: Sagittarius has this next to the 1st degree. **14** And adding 12 in succession we will find the number corresponding to the 1st degree of each sign. By adding 12 and by breaking the sequences with the addition of 14, we will construct the entire table. **15** The numbers of Sagittarius will be the same as those of Taurus; those of Aquarius will be the same as those of Cancer; those of Capricorn will be the same as those of Gemini. In one respect, there will be similar power and mutual support,[16] but the table will be different because of the inequality of the rising times. **16** This table also has the years next to the numbers and the degrees as an example. Intel-

14 This is because 12 is added to move from one sign to the next, information which is given shortly after.

15 This is mnemotechnics.

16 Between these pairs of signs having an identical column of “numbers”.

ligent students will easily get the precise calculations, as well as every clima and change in location.[17]

2. Natural[18] construction of the second table.

1 After the arrangement of the table I think it is necessary to append the background of the construction. **2** The increase between the sequences, 14, indicates the illuminations of the Moon, while the increase of 12 indicates the digits of the Sun;[19] two times 14 equals 28, the Moon's period.[20]

3 Since 2 is next to the first degree of Libra 1, I have subtracted 1;40 from it: the rest is 0;20, which is a magnitude of one-third. **4** This distributed into 60 yields 180.[21] I split this up into sixtieths for a result of 10,800.[22] **5** I divide this into 540°, that is, into one and a half circles, and the result is 0;20.[23] This number will be placed beside the 1st degree of Libra. **6** This 1/3 becomes 60, which is equivalent to one year.[24]

7 We will calculate the remaining years as follows, adding 2;20 to each: **8** there will be, for the 2nd degree of Libra, 2;40; for the 3rd 5;0, for the 4th 7;20, for the 5th 9;40, for the 6th 12;0. **9** Then at the break of the sequence we add the 14 illuminations of the Moon to the 12;0 which give 26;0; from this we subtract 1;40, and the rest will be, for the 7th degree, 24;20. **10** Then for the 8th degree we advance again 2 and 20 min, and this makes 26;40, for the 9th 29;0, for the 10th 1;20, for the 11th 3;40, for

---

17 The final assertion about the students' capacity to obtain exactness supports the literal interpretation of the preceding point, according to which the table only shows the years: the months and days could have been added later in the course of the transmission after Valens.

18 Used as "exact" by Valens: cf., e.g., πᾶσα γὰρ ἀγωγὴ συγκρινομένη πρὸς ἑτέραν καὶ τὸν ἔλεγχον λαβοῦσα φυσικὴν καὶ ἀκριβῆ καὶ τὴν θεωρίαν ἐνδείκνυται. (VI 2.8, "every method when combined with another and taking a test gives a physical and exact theory").

19 The illuminations of the Moon (φῶτα τῆς Σελήνης) and the digits of the Sun (δάκτυλοι τοῦ ἡλίου) are Valens' mnemotechnical strategies to remember the numbers 14 and 12, respectively. The illuminations were equivalent to days in the lunar period (14 is the half period, or the whole period from new Moon to full Moon): cf. Plut. *Is. Os.* 369B, Dorotheus V 5.16–22 (=Hephaestio III 2.6); the digits of the Sun, which are actually the same number as the digits of the Moon, were used to measure the magnitude of eclipses: 12 was the total diameter.

20 This paragraph is valid for both tables.

21 There are 180 one-thirds in 60 (the operation is 60/[1/3]).

22 There are 10,800 sixtieths in 180 (=180 × 60).

23 10,800 divided by 540 (=360 + 180) gives 20, which Valens interprets as 20 min. This obviously makes no mathematical sense. Valens is trying to deduce the reason why the first number should be 0;20, without realizing (or unwilling to reveal) that it is arbitrary. The only reasons for choosing this starting point (0;20) are that it is close enough to 0 to give short lifetimes (see Table 2 in F 16), and the resulting numbers and the constant increase are round enough.

24 This is the length of life given in the table: 0;20×180 (the approximate daylight time at Libra 1°) = 60, which, divided by 60, gives 1 year of life.

the 12th 6;0. **11** So in the next sequence we subtract 1;40 from the first number, and the increase is 2;20. **12** Use this procedure in the same way for the rest of the signs. In the same way as in the first table we find the number next to the first degree of the next sign, we will find it in the second table.[25]

Form of the tables in Valens

Valens in VIII 1.16 (see note ad loc) notes that the first table only indicates the years and that it is the job of the student to obtain the exact calculations and the life lengths for other climata. In any case, the fact that the life lengths are calculated using the norm Aries 8° (see above), to which Valens must convert his longitudes, confirms that they were already given in Critodemus' book. Only in the course of Valens' manuscript transmission were the tables completed with the columns for the fractional parts of the years, following Valens' exhortation.

Properties of the numbers

In the first table, we begin from 2 and add 2 for each degree for the first 6°. Then we add 14 to begin the next group of 6, and so on, always following the rule that $30 + n = n$. Now, to obtain the number beginning the next column, one adds 12 to the first number of the present column.

This is the way to obtain the 'numbers' column, but the question of whether there is any kind of rationale in these numbers can only be answered by zooming out, considering these groups of six numbers as elements.[26] Using letters in alphabetical order to denote the groups of six from the beginning of the table, we see that the first groups are A = (2 4 6 8 10 12) and B = (26 28 30 2 4 6). The first half of the preceding group (the first three elements) is repeated as the second half of the following group. The remaining groups of six in the same sign are arranged in the same way so that in order to build the next group of six we first fill the last three positions with the first set of three in the preceding group of six and then complete the three first elements: C = (20 22 24 26 28 30), D = (14 16 18 20 22 24), etc. This way, all possible groups of three – (2, 4, 6), (8, 10, 12), (14, 16, 18), (20 22 24), (26 28 30) – appear exactly twice in each sign. The fact that the sets of three are never broken is probably related to the dividing lines every 3°. These five groups of three numbers which repeat exactly

25 That is, adding 12 to the number corresponding to the first degree of the preceding sign.
26 King 1989 already showed this.

twice in each sign are only a feature of the first table: when we look at the second table, we see that there are ten different groups of three in each sign, each appearing just once (which are the same ten groups of three appearing in the other signs).

Considering the groups of six as elements, both tables form a structure known as a Latin square. A Latin square is formed of $n \times n$ symbols in which every row and every column contain the n different elements.[27] Looking at the tables, it is easy to see that at the beginning of each successive sign, we have the penultimate group of six of the preceding sign, and that the order is the same everywhere:

Lib	Sco	Sag	Cap	Aqu
A	D	B	E	C
B	E	C	A	D
C	A	D	B	E
D	B	E	C	A
E	C	A	D	B

After the fifth column (corresponding to Aquarius), the next sign (Pisces) has exactly the same structure as the first (Libra), and so on.

There is a multitude of possible Latin squares of the fifth order. The present case is a rotation – the simplest kind – since all columns use the same order, beginning from the fourth element of the previous column. Another possibility would have been to begin with the second, or the third, or the fifth element, and in all cases we would obtain a Latin square.

But there are multiple other possibilities, some of them involving simple recursive methods. One example is the *sextina*, a medieval poetical composition containing six-line stanzas, in which the last six words of the first stanza are repeated in a different and precise order as the last six words in the remaining stanzas. The arrangement is shown below, with the nth column representing the final words in the nth stanza:

A	F	C	E	D	B
B	A	F	C	E	D
C	E	D	B	A	F
D	B	A	F	C	E
E	D	B	A	F	C
F	C	E	D	B	A

27 After Euler's use of the term "carré latin" for a specific 7 × 7 square fulfilling this condition in his attempt to solve a problem on the existence of the so-called Graeco-Roman squares, consisting in a special combination of two Latin squares. The label comes from the fact that he initially used Latin letters for one of the squares and Greek for the other: Euler 1782, 91; see also Andersen 2013.

The procedure here consists in beginning the next stanza (=column) with the final word of the preceding stanza (for the second column: F), then using the first (A), then the fifth (E), then the second (B), and so on – jumping up and down toward the center.

There exist a myriad of combinations, including those that do not use recursivity, the number rising greatly with the size of the square. Rectangles reminiscent of Latin squares of the simpler kind, obtained by rotation like the one formed by the sets of six numbers of our tables, appear naturally as representations of Greek systems of astrological terms such as the one that Ptolemy calls Chaldean (*Tetrabiblos* I 21):[28] for each sign, the first term is the planetary ruler of its triangle (the triangles are Aries–Leo–Sagittarius, Taurus–Virgo–Capricorn, etc.); the second term is the planetary ruler of the next triangle; and so on. Thus, the system is the following (Jupiter is the planetary ruler of the first triangle, Venus of the second, Saturn and Mercury of the third, Mars of the fourth):

Aries	**Taurus**	**Gemini**	**Cancer**
Jupiter	Venus	Saturn	Mars
Venus	Saturn	Mercury	Jupiter
Saturn	Mercury	Mars	Venus
Mercury	Mars	Jupiter	Saturn
Mars	Jupiter	Venus	Mercury

The rotation works by beginning each column with the second element of the previous column. The only distortion is that there are two planetary rulers for the third triangle (Saturn and Mercury), so there is a leap in the fourth column.[29]

We have seen that the terms attached to Critodemus' table are also structured by rotation. Unlike the systems examined above, they are not defined through rulerships, but this is precisely why their representation can form a perfect Latin square, imitating the structure of the table for the length of life. The only rules for this system are to: (1) use the natural order of the planets, discarding the Moon but not the Sun; (2) begin with the Sun, perhaps conceived of as the center among the moving celestial bodies, in the same way as the beginning of Libra is the center of

28 Although it was probably a Hellenistic creation; by contrast, the one he called "Egyptian" was of Babylonian origin (Jones and Steele 2011).

29 A similar structure is formed by the terms announced (but not used elsewhere) by Valens at III 6. Here the order is obtained in the same way, only the Sun (coruler of the first triangle) and the Moon (coruler of the second triangle) are considered too: Aries has Sun, Jupiter, Venus, Moon, Saturn, Mercury, Mars; Taurus has Venus, Moon, etc.

the zodiac; and (3) build the rotation by taking the second element from the previous column.

Libra	Scorpio	Sagittarius	Capricorn	Aquarius	Pisces
Sun	Mars	Jupiter	Saturn	Mercury	Venus
Mars	Jupiter	Saturn	Mercury	Venus	Sun
Jupiter	Saturn	Mercury	Venus	Sun	Mars
Saturn	Mercury	Venus	Sun	Mars	Jupiter
Mercury	Venus	Sun	Mars	Jupiter	Saturn
Venus	Sun	Mars	Jupiter	Saturn	Mercury

Interestingly, the second half of the zodiac presents the inverted order (Sun, Venus, Mercury, etc.), while also beginning with the Sun. A probable reason is to break the symmetry: otherwise, diametrically opposed degrees in the zodiac would always belong to the same term.

A natural question that arises in this context is whether the appearance of such squares was just the unconscious result of applying simple rules for the terms, or whether authors were conscious of the properties of the squares. At least in the case of the terms in the first table of Critodemus, one can suggest that the author used the arrangement intentionally. The terms in Critodemus’ first table are not tied to any set of rules beyond the very square structure they form. Even if they represent the simplest kind of Latin square – obtained by the simplest rotation – we can say that its author, inspired by the patterns of some systems of astrological terms, was conscious of the Latin square structure of the table for the length of life and imitated it with a simple Latin square in his own system of terms.

Graeco-Roman squares outside astrology

We can relate the Greek astrological use of squares fulfilling special properties of balance with the tradition of magic squares – numerical squares with columns, rows and diagonals adding up to the same number – in the medieval Arabic and Indian cultures, ultimately deriving from ancient China.[30] More closely related, from a geographical perspective, to the squares in Critodemus’ work is the Greek isopsephic poetry of the early imperial age, in which each line gives the same number when adding up the numbers represented by the letters of the Greek

30 See Tolsa 2020. See Andersen 2013 for the use of Latin squares in these traditions, too.

alphabet.[31] Also, the first attestation of the SATOR square, an inscription of 5 × 5 letters featuring a whole sentence that can be read both horizontally and vertically and forward and backward (palindrome), is found among the grafitti of Pompeii (i.e., predating the eruption of Vesuvius in AD 79):[32]

S A T O R
A R E P O
T E N E T
O P E R A
R O T A S

Such ingenious word plays have obvious similarities with the Latin squares of our tables for calculating the length of life. However, both phenomena are specifically literary (or paraliterary), and from this point of view astrological tables and systems of terms are quite another thing. Isopsephic poems and Sator squares were clearly far more difficult to create than the Latin squares in our tables, whose construction is more straightforwardly mechanical. A better analogy with our tables is a square of numbers used for predicting victory in a duel or contest between two individuals – a sort of numerological version of the rock-paper-scissors game – found in a series of manuscripts containing various kinds of numerological procedures.[33] The square in question follows a letter allegedly addressed by Pythagoras to his son Telauges:

> Πυθαγόρας Τηλαύγῃ χαίρειν
>
> Πολλὰ παθὼν καὶ πολλὰ πειράσας ἐπέσταλκά σοι τόδε βιβλίον ἔχον ἐν ἑαυτῷ πλινθίδα πάνυ χαριεστάτην· ἐντυχὼν γὰρ εἰς αὐτὴν διὰ τῶν ὑποκειμένων γραμμάτων εἴσεται τά τε ἐνεστῶτα καὶ τὰ προγεγονότα καὶ <τὰ> αὖθις ἐσόμενα. ὑπέταξα οὖν πλινθίδα ἐννεάδος δοκιμαζομένην τρόπῳ τοιῷδε (. . .)

---

31 Consider the poet Leonidas of Alexandria, formerly an astrologer (as he himself discloses, *Anth. Pal.* IX 344) who composed an isopsephic poem for Nero's mother Agrippina (*Anth. Pal.* VI 329); cf. also the isopsephic lines Νέρων / ἰδίαν μητέρα ἀπέκτεινε, reported by Suetonius (*Nero* 39.2) to have been painted on the Roman walls. Isopsephic poetry was already a Hellenistic phenomenon, but in Roman imperial times it seems to have penetrated a wider cultural milieu; cf. Apollonius of Perge *apud* Pappus II 21.1, for which see Netz 2009, 52.

32 There is another such magic square among the Pompeii graffiti, albeit not forming a full sentence: ROMA / OLIM / MILO / AMOR: cf. Benefiel 2013, 69. A similar square survives in a Greek inscription from Smyrna, from the second century AD (ibid.), although it does not contain a palindrome.

33 For example, Par. gr. 2009, 2v; Par. gr. 2892, f. 2r.

> Pythagoras to Telauges,
>
> Greetings. Having experienced and attempted much, I have sent you this little book containing a most graceful square (*plinthion*). Entering in it the letters as shown below one will know the present, the past, and that which will come. I have then set up the square of the ennead which is to be tried in the following way (. . .)

The text goes on to explain that in a contest between two individuals, one begins by reducing the contestants' names to numbers between 1 and 9 using the Greek alphanumerical system. In this system – the usual way to write numbers in ancient Greek manuscripts – the first nine letters were used for the ones (α – θ), the next nine for the tens (ι – ϙ), and the last nine for the hundreds (ρ – ϡ), incorporating old, obsolete characters to complete the second and third series. For each of the letters of the contestants' names, one takes the corresponding number in the alphanumerical system, discarding the eventual 0's, and adds them up: in the example referred to here, ΕΚΤΩΡ is 5 + 2 + 3 + 8 + 1 = 19. This number is then reduced *modulo* 9 (i.e., we take the remainder of the division by 9):

$$19 = 2 \times 9 + 1$$

So, for Ἕκτωρ we enter the square using number 1, and we do the same with the name of the other contestant. Finally, we look up the intersecting cell, and the number that bears a mark wins (a N of νίκη in the original Greek). If the contestants have the same number, the one who began the fight wins if the numbers are odd, whereas his opponent wins if they are even.[34]

34 This system is also explained by Hippolytus of Rome, *Ref. Her.* IV 14.

	W	W			W	W			W	W			W	W		att.	
9	1	9	2	9	3	9	4	9	5	9	6	9	7	9	8	9	9
W			W	W			W	W			W	W		def.			W
8	1	8	2	8	3	8	4	8	5	8	6	8	7	8	8	8	9
	W	W			W	W			W	W		att.			W	W	
7	1	7	2	7	3	7	4	7	5	7	6	7	7	7	8	7	9
W			W	W			W	W		def.			W	W			W
6	1	6	2	6	3	6	4	6	5	6	6	6	7	6	8	6	9
	W	W			W	W		att.			W	W			W	W	
5	1	5	2	5	3	5	4	5	5	5	6	5	7	5	8	5	9
W			W	W		def.			W	W			W	W			W
4	1	4	2	4	3	4	4	4	5	4	6	4	7	4	8	4	9
	W	W		att.			W	W			W	W			W	W	
3	1	3	2	3	3	3	4	3	5	3	6	3	7	3	8	3	9
W		def.			W	W			W	W			W	W			W
2	1	2	2	2	3	2	4	2	5	2	6	2	7	2	8	2	9
att.			W	W			W	W			W	W			W	W	
1	1	1	2	1	3	1	4	1	5	1	6	1	7	1	8	1	9

W, winner; att., attacker wins; def., defender wins.

As we see, the square is designed by alternating the victories in both the rows and the columns so as to guarantee the most equal balance of the two possibilities.[35] It is, then, a sort of equally balanced randomizer. Consequently, all rows and columns add up to the same numbers, as in our tables. The rock–paper–scissors could be presented in a similar way:

	W	W			
r	p	r	s	r	r
W					W
s	p	s	s	s	r
			W	W	
p	p	p	s	p	r

r, rock; p, paper; s, scissors.

35 The only problem is that, since the size is odd (9), in the case of equal numbers, there are five cells indicating the victory of the first contestant and just four for the opponent.

However, it is not insignificant that the outcome of the game was presented in a square. It seems that a certain fascination emerged with these shapes and the balanced properties of the numbers or elements they contained, even if the squares also had an explanatory function.[36]

The second table: construction and rationale

It is easy to see that, using the second table, the shortest life spans are close to 1 year, while in the first table they were around 6 years. This modification, which more closely matches the patterns of mortality in premodern societies, was surely the motivation for the table's construction. Valens, however, does not use this table for any horoscope example, unlike the first one (cf. F 19), perhaps because the numbers are more complicated, making the length of life more difficult to calculate.

The construction of the second table, however, is straightforward: we modify in exactly the same way all the groups of six of the first table, leaving the last element untouched and subtracting 0;20 from the penultimate element, 2 × 0;20 from the fourth, 3 × 0;20 from the third, 4 × 0;20 from the second, and 5 × 0;20 from the first:

In this way, the basic structure – the Latin square of 5 × 5 formed by the five groups of six numbers in each sign for five successive signs – is preserved, while maintaining the characteristic constant increase within the groups of six, which is now

36 By contrast, Ptolemy in *Tetrabiblos* I 21 refuses to draw a table to display the "Chaldean terms" because he deems them simple enough to be comprehended without such a device. This is the ultimate reason why we have so few tables: Graeco-Roman writing gave greater privilege to writing over other presentational devices such as tables, drawings, or diagrams. Cf. Chapter 2 on Valens' own criticism of Critodemus' tables.

2;20. However, unlike the first table, in which for each column we had five different sets of three numbers appearing twice, we now have ten different groups of three numbers appearing once.

F 18. Calculation of the length of life: Valens VIII 5.1–13

Type of fragment: Valens' own explanation of the calculation of the life lengths – basically his own approximative method, perhaps so that the reader-astrologer did not need to carry the table (the 'numbers' column was easy to deduce) or produce tables for each latitude.

<ε̅.> Εἴσοδος τῶν προκειμένων β̅ ὀργάνων.
1 Τὸ πρότερον ὄργανον ἁρμόζει πρὸς χρόνους ζωῆς, τὴν σύστασιν ἔχον καὶ τὴν εἴσοδον ἐκ τῆς τοῦ ὡροσκόπου μοίρας· καθ' ἣν γὰρ ἂν εὑρεθῇ, ταύτης τὸ παρακείμενον κατὰ τὸ κλίμα μέγεθος ι̅β̅πλασιάσαντες καὶ τὸ λ' λαβόντες τοσαῦτα ἔτη φήσομεν τὰς παρακειμένας μοίρας μερίζειν καὶ τοσοῦτον διάστημα ἀπέχειν τὸν θάνατον τῆς ζωῆς. **2** ὁμοίως δὲ καὶ τὸ ξ' τοῦ ι̅β̅πλασιασθέντος λαβόντες λογιούμεθα ἑκάστην μοῖραν τοῦ ζῳδίου τοσαῦτα ἔτη μερίζειν· ἐὰν δέ πως ἐπὶ συνδέσμου καταντήσῃ ἡ ὡροσκοποῦσα μοῖρα, τὰ γεννηθέντα ὀλιγοχρόνια ἔσται.
3 ἔπειτα εἰσελθόντες εἰς τὸ ὄργανον κατὰ τὴν τοῦ ὡροσκοποῦντος ζῳδίου μοῖραν θεωροῦμεν τίς ἀριθμὸς παράκειται, καὶ τοῦτον συγκρίναντες πόστον μέρος ἐστὶ τῆς ἑξηκοντάδος, τοσοῦτον ἐκ τοῦ ι̅β̅πλασιασθέντος ἀριθμοῦ ἡγησόμεθα πλῆθος βιωσίμων ἐτῶν. **4** δεῖ μέντοι λογίζεσθαι τὸν παρακείμενον [καὶ] ταῖς μοίραις ἀριθμὸν τῶν ἐτῶν, πρῶτον ὥρας εἶτα ἡμέρας εἶτα μῆνας εἶτα ἐνιαυτούς.
5 ἄλλως τε καὶ ἐπεὶ β̅ ὁ ἀριθμὸς παράκειται ταῖς μοίραις ἐν τῷ ὀργάνῳ, δεῖ σκοπεῖν [τούτων] τὴν ἡμίσειαν τοῦ χρόνου τοῦ ἐπιβάλλοντος τῇ μοίρᾳ κατὰ τὰς ζῳδιακὰς καὶ κλιματικὰς διαφοράς. **6** οἷον ἐπὶ τῆς τοῦ Ζυγοῦ α̅ μοίρας παράκειται ἀριθμὸς β̅. 7 τὰ δὲ β̅ τῆς ἑξηκοντάδος ἐστὶ τριακοστόν· τὸ δὲ λ' τῶν ρ̅π̅ (τουτέστι τοῦ μεγέθους τοῦ Ζυγοῦ) ἔστιν ϛ̅, ὃ κεῖται[a] ἑκάστῃ <τῶν> μοιρῶν. **8** ἐὰν <δὲ> τοσούτων ἐτῶν λογισώμεθα, αἱ λ̅ μοῖραι ρ̅π̅ ἔτη μεριοῦσιν, ἅπερ ἀδύνατόν ἐστιν ἀνθρώπῳ βιῶσαι. **9** ἐὰν <δὲ> τὸ ξ' τῶν ρ̅π̅ λάβωμεν, εὑρήσομεν γ̅, ἃ μία μοῖρα μερίζει· τρὶς οὖν τὰ λ̅ γίνεται ϟ̅. **10** ἢ καὶ τῶν ρ̅π̅ ἐξ ἡμισείας λαβόντες (ἅπερ ἐστὶ τὰ αὐτά) γίνονται ϟ̅. **11** φήσομεν <οὖν> τὸν Ζυγὸν τὰ τέλεια ἔτη τοσαῦτα μερίζειν κατὰ τὴν ἐπιβάλλουσαν μοῖραν τοῦ μεγέθους.
12 ὁμοίως δὲ καὶ ἐπὶ τῶν λοιπῶν ζῳδίων τὸ παρακείμενον ἑκάστῃ μοίρᾳ μέγεθος ι̅β̅πλασιάσαντες καὶ τούτου τὸ ξ' ἢ τὸ ἥμισυ λαβόντες εὑρήσομεν τὰ ἐλάχιστα ἢ τὰ τέλεια ἔτη. **13** ἑκάστη δὲ μοῖρα ἑκάστου ζῳδίου διαφόρους χρόνους παραυξήσεως

ἔχει, ὅθεν αἱ στιγμαὶ ἢ ῥοπαὶ τῶν ὡροσκοπουςῶν καὶ παραλλαγαὶ τῶν μοιρῶν πολλὴν δύναμιν κέκτηνται.

a ὃ κεῖται Pingree: συγκεῖται Kroll; οὐ κεῖται MSS

5. Method for the 2 appended tables

1 The first table is fitting for the length of life, and it has its essence and use from the degree of the ascendant. Wherever this is found, we multiply by 12 the hourly magnitude next to it in the correct clima, then take 1/30 of this amount, and we say that the degrees found there allot that same number years and that the time of death is that number of years away.[37] **2** In the same way, taking 1/60 of the multiplication by 12 we will calculate each degree of the sign as allotting that many years.[38] If the degree of the ascendant is at a sequence break, the native will be short-lived.[39]

3 Then entering the table at the degree of the ascendant sign, we see which number is next to it, and we determine what part of 60 this number is; we take the same part of the result of the number multiplied by 12; and we will consider the result to be the years of life.[40] **4** It is necessary to count the number of years next to the degrees first as hours, then as days, then as months, then as years.[41]

5 In another way: when the number 2 appears next to the degrees in the table, it is necessary to examine one-half of the time corresponding to the degree, taking into account the difference of clima and sign. **6** For example: the number 2 is next to the 1st degree of Libra, **7** and 2 is 1/30 of 60. 1/30 of 180 (that is, the magnitude of Libra) is 6,[42] which applies to each degree. **8** If we calculate from that many years,

37 This is only true of the degrees that have a number corresponding to 2 in the table, such as Libra 1°. The general procedure consists in multiplying the number of the 'numbers' column by the hourly magnitude, by 12, and then dividing by 60.

38 Correctly speaking, not "each degree of the sign" but each unit in the 'numbers' column. See the example below: Valens is attempting to make a shortcut to the table. If the degrees with number 2 (like Libra 1°) give 6 years, then each number unit gives 3 years; thus, the degrees with corresponding number 30 give 30 × 3 = 90 years. The obvious problem is that the length of day changes through the 30° of the sign; in this shortcut, one takes as a very rough approximation the length of day corresponding to the beginning of the sign.

39 Perhaps implying that when the ascendant degree is at a sequence break, we should choose the minimum years.

40 Although Valens is quite obscure on this point, it is the correct procedure and is valid for every clima. By the 'numbers' which is to be multiplied by 12, he refers to the hourly magnitude so as to obtain an expression of daylight in time-degrees.

41 This seems to mean that for Valens, the quantities given as years in the table should be interpreted as hours, days, or months in the case of short-lived infants.

42 This is the standard, correct procedure.

the 30 degrees will allot 180 years, which is impossible to live for a person.[43] **9** So if we take 1/60 from 180, we will find 3 as the amount which 1 degree[44] will allot: three times 30 is 90. **10** Or one-half of 180 is 90 (which is the same thing). **11** So we will say that Libra allots this maximum of years, according to the applicable degree of the magnitude.[45]

12 Similarly, for the rest of the signs we multiply the magnitude next to each degree by 12, and take 1/60 or the half to find the minimum or the maximum years.[46] **13** Each degree of each sign has different times with increase; hence the moments within the hours and the changing of the degrees have great effects.[47]

How the years are calculated, and Valens' shortcut

It can appear confusing that in this chapter Valens mainly explains his own approximate method. Instead of calculating the correct values, Valens devises a way to dispense with most of the heavy work. The exact procedure is disclosed between the presentation of his method and his example for the sign of Libra (3):

> εἰσελθόντες εἰς τὸ ὄργανον κατὰ τὴν τοῦ ὡροσκοποῦντος ζῳδίου μοῖραν θεωροῦμεν, τίς ἀριθμὸς παράκειται, καὶ τοῦτον συγκρίναντες, πόστον μέρος ἐστὶ τῆς ἑξηκοντάδος, τοσοῦτον ἐκ τοῦ ι̅β̅πλασιασθέντος ἀριθμοῦ ἡγησόμεθα πλῆθος βιωσίμων ἐτῶν.
>
> Then entering the table at the degree of the ascendant sign, we see which number is next to it, and we determine what part of 60 this number is; we take the same part of the result of the number multiplied by 12; and we will consider the result to be the years of life.

So, we take the number (e.g., 2 in Libra 1°), then divide by 60, and then multiply by the length of daylight. For the latitude of Alexandria as in the table, the daylight

43 The problem here arises from the confusion between degrees and number-units, which has appeared before in the text. Valens first argues that if Libra 1° gives 6, then Libra 30° ought to give 30 × 6 = 180. But this is not true, since the relevant quantity is the one in the 'numbers' column. Libra 1° has 2, and therefore each of these units gives approximately 3 years; then Libra 9°, which corresponds with 'number' 30, gives nearly 3 × 30 = 90 years.

44 That is, one number-unit. Now Valens seems to proceed correctly, despite confusing the 180 obtained erroneously before with the 180 time-degrees corresponding to Libra 1°.

45 That is, within Libra, for which we establish the magnitude of 180 time-degrees of day length. Depending on the 'number', the years can reach 90.

46 Half of the day length to obtain the maximum life span is right (e.g., in Libra 180/2 = 90) but a 60th for the minimum is wrong; it should be a 30th (e.g., in Libra 180/30 = 6).

47 That is, since it all depends on the ascendant degree, and this changes rapidly through the hour (roughly 15° per hour), it is important to be precise in determining the time of the ascendant. See Tolsa 2017c on this problem and the ways in which ancient astrologers dealt with it.

time is 183.89 at Libra 1° with the equinox at Aries 8°: then, the result is 6.12 years, that is, 6 years 1 month.

Valens is more interested in his shortcut than the result itself, but this is hardly surprising; indeed, he rarely copies faithfully the procedures of other astrologers, but rather creatively reinvents, combines, or adapts methods, of which we have already seen several examples. Here, he looks for a way to deduce the years of life from the calculation of the shortest lifetime in the sign. This corresponds to the degree of the sign with number 2 in the 'numbers' column: so in Libra, using an approximate length of daylight of 180 (that of the equinox), one gets 2/60 × 180 = 6. Then, to obtain the maximum length of life for this sign, Valens uses the degree with a corresponding number of 30 and calculates 180/60 × 30 = 90.

Of course, this is just an approximation, because the length of daylight changes throughout the sign, but it will still give roughly correct results. For example, Libra 19° has an associated number of 14, and 180/60 × 14 = 42, which is relatively close to the 40 in the table. With this procedure, one daylight time is used for all degrees of the sign, so a simple table of rising times by sign is sufficient, apart from the table with the 'numbers' column described in the previous fragment.

F 19. An example from Valens: Valens VIII 8.14–26

Type of fragment: A horoscope example from Valens illustrating his use of the first table for calculating the length of life. The chapter includes three other horoscope examples which I omit. The present example (the second one in the text) is clear enough and the others are used in exactly the same way; they would add little to our knowledge. The horoscope is for clima 3 (so the lengths of life given in Table 1 do not apply here), and it involves a correction the ascendant degree through a method which Valens has described in VIII 3, and which is probably not taken from the work of Critodemus.

14 Ἄλλη. Οὐεσπασιανοῦ ἔτος ζ̅, Ἐπιφὶ κ̅ε̅ εἰς τὴν κ̅ς̅, ὥρα νυκτερινὴ γ̅, κλίμα γ̅. **15** Ἥλιος Καρκίνῳ κ̅ζ̅ μ̅γ̅, Σελήνη Ἰχθύσι ιβ ν̅β̅, πανσέληνος Ἐπιφὶ κ̅β̅ ὥρᾳ ἡμερινῇ γ̅ Αἰγόκερῳ κ̅<̅δ̅>̅. **16** ἀπὸ πανσεληνιακῆς ἡμέρας τε καὶ ὥρας ἐπὶ τὴν γενεθλιακὴν ἡμέραν τε καὶ ὥραν γίγνονται ἡμέραι γ̅ ὧραι ι̅β̅, αἵπερ εἰσὶ τοῦ ἀπὸ σεληναίου δρόμου ἐπὶ σύνοδον (τουτέστι τῶν ι̅ε̅) ε′ λ′. **17** τοῦτο ἀφεῖλον ἀπὸ τοῦ μεγέθους τοῦ παρακειμένου τῇ τοῦ Αἰγόκερω μοίρᾳ κ̅, ὅπερ ἐστὶ ι̅β̅ κ̅, καὶ λοιπαὶ γίγνονται θ̅ ι̅β̅. **18** τοῦτο μέρος ὥρας ἔσται. **19** ἐψήφισα οὖν ὥρας β̅ καὶ προσέθηκα τὸ μέρος <καὶ> ἐγκλίματος τ̅ζ̅, γίγνονται τ̅μ̅ ν̅ε̅. **20** ταύτας εὗρον ἐν τῷ ἐγκλίματι περὶ τὴν κ̅θ̅ τοῦ Ὑδροχόου. **21** καὶ προσέθηκα τὰς η̅ καὶ γίγνεται ὁ ὡροσκόπος Ἰχθύων μοῖραι ζ̅. **22** ταύτας ἔχων εἰσῆλθον εἰς τὸ ὄργανον κατὰ τὴν ζ̅ μοῖραν τῶν Ἰχθύων καὶ εὗρον παρακειμένας κϛ, αἵπερ εἰσὶν τῆς ἑξηκοντάδος γʹ ιʹ. **23** τὸ δὲ παρακείμενον μέγεθος τῇ τοῦ Ὑδροχόου μοίρᾳ κθ ἐτύγχανεν ιγ ἐγγύς. **24** ταῦτα ποιῶ ἐπὶ τὸν ιβ, γίγνονται ρνϛ. **25** εἶτα λαμβάνω τὸ γʹ καὶ τὸ ιʹ, γίγνονται ξη. **26** ἐτελεύτα κατ' ἀρχὰς τῷ ξθ ~ ἔτει.

14 Another: Vespasian year 7, Epiphi 25 to 26, 3rd hour of the night, clima 3. **15** Sun in Cancer 27° 43', Moon in Pisces 12° 52', full Moon Epiphi 22 in the 3rd hour of the day in Capricorn 24°. **16** From the day and hour of the full Moon to the day and hour of birth are 3 days 12 hours, which is 1/5 1/30 of the lunar period up to the new Moon (which is of 15 days).[48] **17** I subtracted this from the magnitude entered at the 20th degree of Capricorn,[49] which is 12;20, and the result is 9;12.[50] **18** This will be the fraction of the hour. **19** I calculated the hours, 2, and added the fraction and the accumulated rising time, 307, for a total of 340;55.[51] **20** I found this quantity in the accumulated rising times at Aquarius 29°.[52] **21** I add the 8°, and the Ascendant becomes Pisces 7°.[53] **22** Having this, I went to the table at Pisces 7°, and found 26 next to it, which is 1/3 1/10 of 60.[54] **23** The hourly magnitude at Aquarius 29° was approximately 13. **24** I multiply this by 12, and I get 156. **25** Next I take 1/3 and 1/10 of this and get 68. **26** He/she died halfway through the 69th year.

---

48 That is, 7/30 expressed as a sum of simple fractions. Valens first applies the method of correction of the ascendant, which consists in astrologically (not astronomically) determining the exact time within the third hour. This is done by finding the period elapsed since the previous syzygy in relation to the whole period until the next syzygy (7/30), and then subtracting this fraction from the end of the reported hour.

49 Valens should look up the magnitude at Capricorn 27°, not 20°. This possibly indicates that Valens' table was not arranged by degrees, but divided into sections of 10°, like the one in Ptolemy's *Almagest* (II 8).

50 It should be 9;27 (using the given value). He seems to use the hourly magnitude 12 instead of 12;20 (12−7/30×12 = 9;12). Indeed, the hourly magnitude at Capricorn 20° is 12, not 12;20; see the previous note and the table below. Capricorn 30° has instead a magnitude of ca. 12;20, so maybe he was looking there now.

51 With the numbers given (using 12;20 as the magnitude and 9;12 as the fraction), it should be 340;52.

52 This is the new ascendant degree, calculated with a table of rising times (for clima 3, see below), after subtracting a portion of the last hour determined through the astrological method described.

53 This addition of 8° reflects the fact that the tables for calculating the length of life use the norm Vernal Equinox = Aries 8° (not Aries 0°, which Valens uses regularly).

54 That is, 13/30, and 26 is the number in the 'numbers' column of Table 1, for Pisces 7°. Now Valens applies the calculation for the length of life: he looks up the hourly magnitude of Aquarius 29° (the ascendant without adding the 8°), multiplies it by 12, and by 26/60.

Context in Valens

The sequence of chapters in book VIII is somewhat bizarre. Valens begins by explaining the construction of the tables of Critodemus, then, in the middle of the book (VIII 6 and 7) develops and exemplifies at length another method for calculating the length of life through the table of rising times (the "three-sign method"), and toward the end he returns without previous notice to Critodemus' tables, first providing four horoscope examples (VIII 8), the second of which I provide here, which is somewhat clearer and more correct than the rest. Finally, Valens explains a method for finding the time of death using the terms attached to the first table (VIII 9, F 20), after which he copies the tables.[55]

Methods for the correction of the ascendant degree

There is a recurrent preoccupation in Valens with the accuracy of the ascendant degree. This value was in fact destined to be inaccurate, since the time was generally given as the integer hour of birth, as in "the third hour". Some reports indicated the beginning, middle, or end of the hour, but this was rare.[56] It was not, however, just a question of reports: knowing the time in Antiquity was difficult in itself, and it is estimated that sundials had, on average, a deviation of approximately 15 min.[57] Astrologers were evidently aware of this problem and, quite paradoxically, designed methods to deduce the correct degree from astrological considerations.[58] For example, Ptolemy (*Tetrabiblos* III 3) proposes to find the most probable ascending sign by the astronomical method, using the integer hour, and then to pick the degree of the planet having more relations of familiarity with the place of the preceding new or full Moon, regardless of the sign in which that planet was positioned.[59]

---

55 Just before the tables at the very end of the last chapter (VIII 9.24–39) we find a second passage on the construction of the first table, although it is incomplete and repetitive.

56 Cf. Tolsa 2017c.

57 On the unreliability of ancient time-measuring instruments, cf. Hannah 2009, 169 n. 79, with bibliography.

58 Astrology as a whole was attacked for the impossibility of precisely calculating the ascendant by Sextus Empiricus, *Against the astrologers* (*passim*). As for the paradoxical solution of the astrologers, Ptolemy is perfectly aware that the reliability of astrology is qualitatively inferior to that of astronomy (*Tetrabiblos* I 1), whereas Valens sometimes appears to argue that astrological methods can help correct astronomical tables: cf. Tolsa 2019b.

59 Many of these procedures appear in Paulus *Isag.* 33.

The method used by Valens here is not described in other sources, but it was probably not his own invention, since it appears to have been used in some papyrus horoscopes from both before and after his time.[60] There are not, however, sufficient grounds to attribute the method to Critodemus, and some hints in Valens' own work also suggest otherwise:

(1) This is not the only method for the correction of the ascendant offered in VIII 3 for the use of the table for the length of life: there is another one which is elsewhere cryptically associated with the astrologer Thrasyllus (IX 11).[61]

(2) Valens is keen on such procedures, and it would be odd if he did not use one here. In fact, he uses another ascendant-correcting method for his own doctrine for the length of life ("the three factors") in VIII 6–7.

(3) As shown above, in IX 9 Valens returns to the contents of book VIII, claiming that Critodemus created the construction (i.e., the tables) but that he himself has discovered a better method, apparently that of the "three factors". Here in book IX Valens gives two other methods for correcting the ascendant to be used in combination with the three factors, explicitly noting their use of the Sun and the Moon. At the end of the chapter he says, again quite cryptically, as in the case of Thrasyllus, "Critodemus in this method has also used the Sun, in a procedure which he has not published" (Κριτόδημος μὲν οὖν ἐπὶ ταύτῃ τῇ ἀγωγῇ κοινῶς τῷ Ἡλίῳ κέχρηται, μεθόδῳ δ' ἑτέρᾳ ἣν οὐκ ἐξέδοτο, 53). An apparently exhaustive summary of Thrasyllus' treatise is extant, in which no methods for the correction of the ascendant appear.[62] In fact, the method which Valens relates to Thrasyllus seems too technical and complex for what seems to be an introductory manual. The more probable conclusion is that Valens wished to transmit the idea that these methods were not his own invention, that is, that they had a certain tradition; however, since he was unable to find any authoritative treatise in which they were described, he strategically dropped the names of Thrasyllus and Critodemus.

The use of such numerical methods makes Valens' treatise unique, but it is difficult to believe that no other astrologers applied them. More likely is that astrologers

60 Cf. Tolsa 2017c.

61 It is offered as an alternative in the last paragraph of the chapter: VIII 3.24–27. Valens IX 11.2: "It is difficult and laborious to test other people's methods, especially those not from written books or powerful discourses" (δύσκολον μὲν οὖν καὶ ἐργῶδες ἀλλοτρίας δόξας ἐλέγχειν, καὶ ταῦτα μηδὲ διὰ γεγραμμένων βιβλίων μηδὲ διὰ λόγων ἐνεργητικῶν). And, after the exposition of the method, cryptically: "Thrasyllus made a scientific beginning and calculated the end" (ὁ Θράσυλλος καὶ τὴν ἀρχὴν φυσικὴν ποιησάμενος τὸ τέλος συνέπλεξεν, 10).

62 *CCAG* VIII 3, 99–101.

had few opportunities to discuss them in the manuals, in which they typically dealt with less technical material, partly in order not to scare away potential readers, but also following a convention of literary elegance – I again recall Valens' prejudices against Critodemus' tables, given in Chapter 2 – and perhaps also with the intention of keeping some of their tools to themselves. It is illustrative that manuals never explain such a basic, but not at all evident procedure as the use of the table of rising times to calculate the ascendant, which every astrologer necessarily mastered. Since they were seldom presented in manuals, these numerical methods could not be assigned to any astrological authorities.

Valens' correction of the ascendant in the given example

To follow the procedure described in the fragment, I will first explain how to determine the ascendant using the table of rising times. Taking the example given by Valens, imagine that the time of birth was exactly when the third hour of the night was completed. The Sun was in Cancer 27°: then the aim is to find the part of the zodiac that has risen from the horizon from sunset onward. We are told that 3 h of night have elapsed, which by definition means 3/12 of the whole nighttime. For this, we look at the point opposite the Sun, which is the one ascending in the East at sunset: Capricorn 27°. We go to the table of rising times and multiply the hourly magnitude at this point by 3: the hourly magnitude is the time-degrees (degrees on the equator) that rise in 1 (night) hour. Some tables of rising times provided this value in a column.[63] However, it can be easily found by dividing by 12 the accumulated time-degrees from the given point (for us, Capricorn 27°) through the following 6 signs (the time of the whole night). In the present case, the table is for clima 3. Valens used a table by 10° intervals (see Appendix II for a table of ascensions for the latitude of Alexandria by degrees):[64]

63 In the more complete ascensional tables, like the ones in Ptolemy's *Handy Tables*, the hourly magnitude was already given in a column for each degree. Cf. the simpler table in *Almagest* II 8, where this value is not present. This difference clearly reflects the practical astrological usage conceived for the *Handy Tables*.

64 For simplicity of calculation, I use decimal fractional parts, not sexagesimal like the ancients. The ascension times of each sign for the different climata can be found in *GH*, 4 (clima 3a); the time-degrees for each sign were divided into equal parts corresponding to the three 10° segments.

Zod. degree		Acc. t.-deg.	Hourly magn.
Aries	10	6.48	15.59
	20	12.97	16.17
	30	19.45	16.76
Taurus	10	27.34	17.11
	20	35.23	17.46
	30	43.11	17.81
Gemini	10	52.41	17.93
	20	61.70	18.05
	30	71	18.17
Cancer	10	81.70	18.05
	20	92.41	17.93
	30	103.11	17.81
Leo	10	115.22	17.46
	20	127.33	17.11
	30	139.44	16.76
Virgo	10	152.96	16.17
	20	166.48	15.59
	30	180	15
Libra	10	193.51	14.41
	20	207.03	13.83
	30	220.54	13.24
Scorpio	10	232.66	12.89
	20	244.77	12.54
	30	256.88	12.19
Sagittarius	10	267.58	12.07
	20	278.28	11.95
	30	288.99	11.83
Capricorn	10	298.28	11.95
	20	307.58	12.07
	30	316.87	12.19
Aquarius	10	324.76	12.54
	20	332.65	12.89
	30	340.54	13.24
Pisces	10	347.021	13.83
	20	353.50	14.42
	30	360	15

Now from Capricorn 27° to Cancer 27°, there are ca. 147 time-degrees,[65] which, divided by 12, give the hourly magnitude (12.25) which is multiplied by 3 to give 36.75. Then we go to the time-degrees of Capricorn 27°, approximately 315, and

65 To do this, one mentally extends the table beyond Pisces, adding 360° to the time-degrees of the following signs. So, Cancer 30 has 462.2 accumulated time-degrees.

add 36.75, to obtain 351.75. Finally, we look at what zodiacal degree corresponds roughly to this value, which is approximately Pisces 18°. This is the ascending point at the end of the third of the night.

But Valens applies a method that astrologically determines the exact point within this hour, from 2 completed hours to 3 completed hours. This method uses the time in days elapsed since the last syzygy (full or new Moon) in proportion with the whole lunar half-period (15 days): in this case, 3.5 days/15 = 7/30. This will now be the fraction of the hour (the hour contains 12.25 time-degrees, as we have calculated) to be subtracted from the 3 o'clock point:

$$351.75 - 7/30 \times 12.25 = 348.89$$

This corresponds approximately to Pisces 12°, which would be the corrected ascendant. In summary, the method works in such a way that:

(a) If birth took place in a syzygy (new or full Moon), the ascendant is correct (the given time is considered to indicate the completed hour).
(b) If birth took place half-way through the half-lunar period between one syzygy and the next, the correct point will correspond to a half hour (in our example, at two and a half hours of the night).
(c) For births progressively away from the last syzygy, the corrected time within the given hour will go back proportionally toward the beginning of that hour.[66]

Finding the length of life

We finally have the corrected ascendant degree. We have seen that, in the correct calculation, it would be Pisces 12°, but I will use Valens' mistaken ascendant for the sake of comparison with his own data: Aquarius 29°. Then, in order to use the 'numbers' column of Critodemus' first table (Valens only uses the first table in his examples), one adds 8° to convert to the tables' equinox norm: Aquarius 29° becomes Pisces 7° before entering the table. We look up the number given there, which is 26, but we cannot use the length of life given in the table, since it is calculated for Alexandria. For clima 3, one uses the hourly magnitude at Aquarius 29°, which, Valens says, is nearly 13 (we have 13.25 in our table), to multiply:

66 The example in Valens contains several mistakes that I describe in the footnotes. Also note that, instead of subtracting the time-degrees from the end point of the third hour, Valens adds 23/30 of the hourly magnitude to the end point of the second hour, which amounts to the same thing.

$$26 \times 13 \times 12/60 = 67.6 \text{ years of life}$$

Finally, Valens confirms the person’s death close to that year of life. All his examples in this book are of this kind: he selected from his archive the horoscopes which confirmed his methods. However, the large numbers of mistakes – in this chapter, all four examples contain several – raise suspicions that the data may have been tampered with, a practice occurring elsewhere in his manual.[67] A more legitimate strategy, if somewhat problematic for the consistency of the theory, is observed midway through his examples of the “three-factor” method in VIII 7, where Valens announces that for the rest of the horoscopes the method for the correction of the ascendant will be a slight variation of the method applied previously, which he actually puts to work with many more horoscopes.[68] Clearly he saw that these extra examples did not work with the first procedure, but a small change was enough to obtain good results.

F 20. Transits in opposition, with a double horoscope and Valens’ method: VIII 9.1–23 = III [6]

Type of fragment: Procedure for determining features of the death chart given the birth chart, using the terms attached to Critodemus’ Table 1 (F 15). This is probably Valens’ own method, taking as a point of departure a doctrine in Critodemus alluded to at the beginning of the passage. For the sake of clarity, I provide a representation of the double chart exemplifying the method.

<θ̄.> Περὶ τῶν ἐχθρῶν τόπων καὶ ἀστέρων. περὶ κλιμακτηρικῶν τόπων πρὸς τὸ ᾱ ὄργανον.
1 Σκοπεῖν δὲ δεῖ τοὺς ἐχθροὺς τόπους καὶ τοὺς ἀστέρας οὐ μόνον ἐπὶ τῶν ἄλλων, ἀλλὰ καὶ τοῦ ὡροσκόπου καὶ Ἡλίου καὶ Σελήνης· οὗτοι γὰρ καὶ ἐναντίοι ἐν ταῖς παρόδοις γενόμενοι τοὺς κλιμακτῆρας καὶ τοὺς θανάτους σημαίνουσιν. **2** οἷον ἐπὶ τοῦ Κρόνου τῶν μοιρῶν τὰς διαμετρούσας θεωρεῖν χρή τίνος εἰσὶ θεοῦ ὅρια, καθὼς ἐν τῷ ὀργάνῳ πρόκειται· κἀκεῖ Κρόνου ὄντος ἀποθανεῖται ἢ ἐν τοῖς τετραγώνοις ἢ ἰσαναφόροις, καθὼς ἂν καὶ ὁ χρόνος συντρέχῃ, ἐν τοῖς τετραγώνοις τοῦ ὡροσκόπου ἢ ἰσαναφόροις. **3** τὸ δ’ αὐτὸ καὶ ἐπὶ τῶν ἄλλων ἀστέρω. **4** ἐχθροὶ γάρ εἰσιν οἱ ἐκ

---

67 Cf. Tolsa 2019b.

68 VIII 7.108: ἔστι δὲ ἡ διάκρισις μετὰ πολλοῦ πόνου οὕτως ἡμῖν ἐζητημένη, ἣν καὶ ὑποτάξομεν δι’ ὑποδειγμάτων (“there is a revision searched by us with much labor, which we append below through examples”).

τῶν ἀντικειμένων μοιρῶν τῶν ὁρίων [οἱ] κύριοι· οὗτοι οὖν παραγενόμενοι ἐπὶ τοὺς τόπους τὰς ἀναιρέσεις σημαίνουσιν ἢ εἰς τὰ ἰσανάφορα τοῦ ὡροσκόπου.

5 Οἷον ἔστω Κρόνος Καρκίνου μοίρᾳ κα ὁρίοις Ἀφροδίτης· διαμετρεῖ Αἰγόκερως ὅρια Ἄρεως. **6** οὗτος ἦν Ταύρου μοίρᾳ κζ· ἐνθάδε Κρόνου ὄντος ἀποθανεῖται. Παρθένῳ ἀπέθανεν· τὸ γὰρ μοιρικὸν τετράγωνον τούτου.

7 Ζεὺς Σκορπίῳ μοίραις ιδ ὁρίοις Κρόνου. **8** ἡ δὲ ιδ τοῦ Ταύρου ἐστὶν ὅρια Κρόνου. οὗτος μὲν αὐτῷ ἐχθρὸς οὐ γίνεται. **9** ἔστιν οὖν τὸ ἰσανάφορον Σκορπίου Λέων, ἡ δὲ ιδ μοῖρα Λέοντος εἰσὶν ὅρια Ἡλίου. **10** ἐλθὼν οὖν ὁ Ζεὺς ἐπὶ τοὺς τόπους Ἡλίου ἢ ἐλθὼν εἰς τὸ ἰσανάφορον ἐκεῖ ἀνεῖλεν.

11 Ἄρης Ταύρου μοίρᾳ κζ ὁρίοις Ἡλίου. **12** αἱ δὲ αὐταὶ ἐν Σκορπίῳ ὅρια Ἡλίου· ἐχθρὸς δὲ αὐτοῦ οὐδεὶς γίγνεται. **13** ζητῶ οὖν τὰς Λέοντος μοίρας κζ ἢ ἐν τῷ ἰσαναφόρῳ, οἵ εἰσι Δίδυμοι κατὰ τὰς ὡριαίας διαστολάς. **14** εἰσὶ δὲ αἱ κζ Διδύμων ὅρια Ἀφροδίτης. **15** ἀποθανεῖται οὖν Ἄρεως ὄντος ἐν Σκορπίῳ ἢ Ἰχθύσι τοῖς ἰσανατόλοις ἢ τοῖς τούτων τετραγώνοις. **16** ἐὰν δὲ λογίσηταί τις τὰς κζ τοῦ Λέοντος, εὑρήσει Κρόνου ὅρια· Κρόνος δὲ ἦν ἐν Καρκίνῳ. **17** ἀποθανεῖται οὖν Ἄρεως ὄντος ἐν Καρκίνῳ ἢ Τοξότῃ ἢ τοῖς τούτων τετραγώνοις.

18 Ἀφροδίτη Σκορπίῳ μοίραις κζ ὁρίοις Ἡλίου. **19** αἱ διαμετροῦσαι Ταύρου κζ εἰσὶν ὅρια Ἡλίου· αὐτὸς δὲ ἑαυτῷ ἐχθρὸς οὐ γίνεται. **20** ζητῶ οὖν ἐν τῷ ἰσανατόλῳ τοῦ Σκορπίου τὰς κζ, εἰσὶ δὲ Ἑρμοῦ ὅρια. **21** ἀποθανεῖται οὖν Ἀφροδίτης οὔσης ἐν Παρθένῳ, ὅπου Ἑρμῆς, ἢ ἐν τοῖς τετραγώνοις. **22** τὸ δ' αὐτὸ καὶ ἐπὶ τοῦ Ἑρμοῦ ποιητέον.

23 Σκοπεῖν δὲ δεῖ καὶ τὰς κατακλίσεις πρὸς τὸν ἐναντίον τόπον[a] <καὶ τοὺς >[b] ἐπὶ τοῖς ἐχθροῖς τόποις ὄντας[c] καὶ τοὺς μηνιαίους καὶ ἡμερησίους καὶ ὡριαίους κλιμακτῆρας ποιοῦντας πρὸς τὴν τῆς Σελήνης τριακοντάδα, ἐξ ἧς ὁ ἐναντίος ἀστὴρ εὑρίσκεται.

a τόπον secl. Kroll
b καὶ τούς III [6]
c ὄντα S, ὄντας III [6]

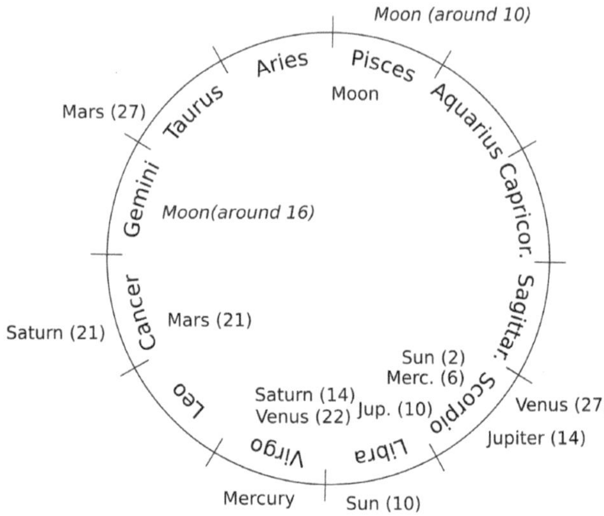

Birth chart outside/death chart inside.[69]

9. On the malefic places and stars. On critical places with respect to the first table. **1** One needs to examine the malefic places and stars, not only with respect to the other, but also with respect to the ascendant, the Sun, and the Moon: when these become opposite in the transits, they signify critical points and deaths.[70] **2** For example, with Saturn one must examine the opposite degree to discover to which god's terms it belongs to, as given in the table: when Saturn is in that degree,[71] in square aspect, or in signs of equal rising time he/she will die and, depending on whether the time is fitting, in square aspect with the ascendant or in signs of equal rising times. **3** The same for the other stars, since the rulers of the terms of the degrees in opposition are malefic. **4** They indicate destruction when they come to their places or to the places of the same rising times as the ascendant. **5** For example:

69 In the charts, the degrees indicated or implied by Valens are shown, not the recomputed ones, which are slightly different (see below). The data not given by Valens but deduced from the estimated date are shown in italics.

70 What follows is not exactly what has been announced here. We should perhaps assume that Critodemus had only these introductory concepts, and that the rest is from Valens.

71 That is, when Saturn is in the degree occupied by the planetary ruler of the terms (according to Table 1) opposite to Saturn in the birth chart.

Saturn in the 21st degree of Cancer, terms of Venus. In opposition is Capricorn, the terms of Mars; **6** this was in the 27th degree of Taurus. When Saturn is there, he/she will die. He/she died in Virgo, since it was square by degrees.[72]
7 Jupiter in the 14th degree of Scorpio, the terms of Saturn, **8** and Taurus 14° is also the terms of Saturn, which is not malefic for itself. **9** Leo has the same rising times as Scorpio,[73] and the 14th degree of Leo is in the terms of the Sun. **10** Therefore Jupiter, coming to the places of the Sun or coming to a sign of equal rising times, destroyed him/her there.
11 Mars in the 27th degree of Taurus, the terms of the Sun. **12** The same degree in Scorpio is in the terms of the Sun, and no star is malefic for itself. **13** So I investigate the 27th degree of Leo or in the sign of equal rising times, which is Gemini according to the rising times.[74] **14** The 27th of Gemini is the terms of Venus. **15** He/she will die when Mars is in Scorpio[75] or Pisces (the sign of equal rising times),[76] or in signs square with these. **16** If anyone checks Leo 27°, he/she will find the terms of Saturn. Saturn was in Cancer; **17** so he/she will die when Mars is in Cancer or in Sagittarius, or in the signs square with them.
18 Venus in the 27th degree of Scorpio, the terms of the Sun. **19** The opposite degree is Taurus 27°, terms of the Sun. Now the star is not malefic for itself. **20** So I search in the sign of equal rising times with respect to Scorpio at 27°, which is the terms of Mercury.[77] **21** He/she will die when Venus is in Virgo – where Mercury was found – or in the signs square with it. **22** The same method needs to be applied for Mercury.
23 It is necessary to inspect the times of lying-downs with respect to the opposite place, and the stars at hostile places and the ones producing monthly, daily and hourly critical points opposite from the thirty degrees of the Moon, from which the opposite star is found.[78]

---

72 Virgo is not in square with Taurus: thus, "by degrees" (μοιρικόν) probably means taking into account the degrees (rather than "exactly"), since Taurus 27° is very near Gemini, which is in square with Virgo.

73 The signs of equal rising times are Aries-Pisces, Taurus-Aquarius, Gemini-Capricorn, Cancer-Sagittarius, Leo-Scorpio, and Virgo-Libra (cf. Appendix II).

74 This makes no sense, because the sign of equal rising time with respect to Leo is again Scorpio (and with respect to Taurus, Aquarius).

75 Because Venus was in Scorpio 27 at birth (see below).

76 It should be Leo instead of Pisces.

77 Pisces and Cancer are the only signs with the 27th degree in the terms of Mercury according to Table 1. Therefore, Valens probably tampered with the procedure to match the theory to the data in the real death chart.

78 This concluding sentence resumes the introduction to the doctrine, which is most probably the original theory from Critodemus: namely, the necessity of inspecting the transits in the oppositions, here in particular defined with respect to the Moon. The structure and aim are similar to those of

Context in Valens

This passage has the peculiarity of being copied twice in Valens' compilation, here and in III [6]. Given the haphazard nature of the treatise, including frequent repetitions, it is possible that Valens himself copied the passage twice.[79] In fact, it would make sense that the passage was first written in book III. In this context, the method comes just after a discussion of possible destructive points in the method of *aphesis*, which is precisely attributed to Critodemus in the manuscript tradition (III 5.18–20 = F 12). In addition to their common concern with determining the time of death and the fact that both doctrines are in one way or another put in relation with Critodemus, quite specific technical details coincide, like the use of the signs of equal rising times. On the other hand, the present passage concerns Critodemus' tables for calculating the length of life, so book VIII would be an appropriate place: naturally, it is possible that Valens also thought so, and that he therefore decided to copy the passage there too.

Procedure in the horoscope example contrasted with the presentation

Let us first go through the example, and then tackle the problem of Valens' initial explanation. For each planet, in the natural order of the spheres (Saturn, Jupiter, Mars, Venus, and Mercury), Valens inspects the opposite position (e.g., Saturn Cancer 21° → Capricorn 21°) to find the terms of the latter point in Table 1. The planet ruling these terms (Mars) is then located in the birth chart (Taurus 27°) and the following prediction about the death chart is made: Saturn will be there when the native dies. Repeating this procedure for the other stars, we obtain their positions at death, which 'coincidentally' results in the real death chart.

In one circumstance, the procedure is altered: if the terms of the opposite position coincide with the terms of the actual position, then we take the sign of equal rising time, with respect to either of the two positions:

---

the concluding sentence in F 9, where Valens seems suddenly to have forgotten about his complicated procedure and returned to the initial trigger. Like there, he relates the doctrine of the transits to the time rulerships of the distributions, urging the astrologer to consider both aspects (transits and critical points) together in order to judge the severity of the time in question: in this case, he wisely suggests conducting the exploration at the time when the person falls sick (κατάκλισις).

79 *Pace* Pingree, who in his edition proposes to delete the first instance in book III. For example, we have seen how the beginning of Critodemus' *Horasis* is copied twice, too (F 3), annotated with very similar criticisms. Also, as has been mentioned, the construction of the 'numbers' in the tables for the length of life is explained at the beginning and the end of book VIII.

(a) Jupiter is in Scorpio 14, which has the same terms as the opposing Taurus 14: then, Valens looks at the sign of equal rising times as Scorpio: Leo. Leo 14 falls in the terms of the Sun: then death comes when Jupiter is in the place of the Sun.

(b) Mars is in Taurus 27, which has the same terms as the opposing Scorpio 27: now, Valens goes not to the sign of equal rising time as Taurus (Aquarius) but to the sign of equal rising times as Scorpio: Leo.

This inconsistency, if it may be called so, is probably the result of trying different possibilities in order to find a match with the death chart. In the case of the inspection of Venus (see note above), we find a clearer case of tampering to obtain a given position.

The date of the birth chart can be calculated from the information given by Valens:
Saturn in Cancer 21
Jupiter in Scorpio 14
Mars in Taurus 27
Venus in Scorpio 27
Mercury in Virgo (undetermined degree)

A good match for this is October 7, AD 2.[80] The positions of the luminaries on this date were:
Sun in Libra 10
Moon around Pisces 10

Again, enough data is disclosed to calculate the date of death. There is one and only one compatible stellar configuration within a lifetime from the time of birth, corresponding to October 27, AD 36, with the following positions (Valens' indications are given in parenthesis):[81]
Saturn in Virgo 12 (square with the place where Mars was, Scorpio 14)
Jupiter in Libra 10 (where the Sun was)
Mars in Cancer 10 (roughly where Saturn was, Cancer 21)
Venus in Virgo 28 (roughly where Mercury was, Virgo 22)
Mercury in Scorpio 6
Sun in Scorpio 2
Moon around Gemini 16

80 Peter 2001, 149. Positions on this date: Saturn in Cancer 14, Jupiter in Scorpio 12, Mars in Taurus 24, Venus in Scorpio 15, Mercury in Virgo 22.
81 Also calculated by Peter 2001, 149.

There seems to be a common pattern with F 9 above: horoscopes cast at birth and at death, probably reused from Critodemus – although not original, for reasons of date (in that case, for Nero, b. 37/d. 68) – for a more complicated variation devised by Valens himself, in both cases in combination with a table also found in Critodemus' work (in that case, the multiplication table). In both cases, the general idea is to involve the transits in consideration of critical times.

As in F 9, Valens' initial explanation in F 20 (1) differs from the procedure exemplified thereafter, and this again probably reflects an original doctrine of Critodemus:

> Σκοπεῖν δὲ δεῖ τοὺς ἐχθροὺς τόπους καὶ τοὺς ἀστέρας οὐ μόνον ἐπὶ τῶν ἄλλων, ἀλλὰ καὶ τοῦ ὡροσκόπου καὶ Ἡλίου καὶ Σελήνης· οὗτοι γὰρ καὶ ἐναντίοι ἐν ταῖς παρόδοις γενόμενοι τοὺς κλιμακτῆρας καὶ τοὺς θανάτους σημαίνουσιν.
>
> It is necessary to examine the hostile places and stars, not only with respect to the other stars, but also with respect to the ascendant, the Sun, and the Moon. When these come into opposition in their transits, they indicate critical points and deaths.

There is no talk of the terms here, which are so important for the procedure. One could argue that in this initial explanation the details are simply omitted, but the exclusive mention of the Sun, the Moon, and the ascendant is also in flagrant contradiction with the method that follows: in Valens' example, these elements are not even mentioned.

Looking at the charts, my guess is that this horoscope was originally inserted in Critodemus' work for the position of Saturn in the death chart (Virgo 14) in opposition with the Moon at birth (around Pisces 10): Saturn is one of the malefic stars; thus, its diametral positioning against the Moon can indicate death, according to these introductory statements. What Valens then seems to have attempted is to find a procedure which makes clear that some kinds of opposition among the other stars also indicate death. Thus, he found that if one considers the opposite point and then the position of the ruler of the terms of this point, this roughly agrees with the given death chart, even if the doctrine must involve several possibilities and, in some instances, even illegitimate adjustments.

Appendix I: timeline of ancient astrology

Date (approx.)	Historic events	Texts, theories, and authors
3000 BC	Writing culture in Mesopotamia and (shortly thereafter) Egypt	
2000 BC		First redaction of *Enūma Anu Enlil*: Mesopotamian astral omen lists Astronomical tomb ceilings in Egypt
1000 BC		First redaction of MUL.APIN: compilation of Mesopotamian astronomical knowledge
550 BC	Persian Babylon	
525 BC	Persian Egypt	
500 BC		Mesopotamian-style omen texts in Egypt (Vienna papyrus D 6286), Babylonian astronomical periods in Greece Invention of the zodiac in Babylon Babylonian mathematical-astronomical schemes Babylonian protohoroscopes (up to the first century BC)
312 BC	Seleucid Babylon	
305 BC	Ptolemaic Egypt	
175 BC	Antiochus IV founds Greek colony in Babylon	
150 BC		Astronomer Hipparchus (Rhodes, uses Babylonian eclipse data) Rising times Babylonian scheme for Alexandria by Hypsicles Horoscopic astrology (Nechepsos and Petosiris) in Egypt
100 BC		Posidonius (Rhodes, aware of astrology) **Critodemus** (of Rhodes?)
60 BC		Lucretius *De rerum natura* (no reference to horoscopic astrology)
50 BC		Varro (knows Nechepsos/Petosiris and Critodemus) and Cicero (writes in 44 BC that his friend Tarutius made a horoscope for Rome)

https://doi.org/10.1515/9783111329147-010

(continued)

Date (approx.)	Historic events	Texts, theories, and authors
30 BC	Roman Egypt	Vitruvius (Rome), transmits Varro
AD 14	Tiberius	Thrasyllus (Rhodes and Rome, astrologer friend of Tiberius), Manilius (Rome, astrological poet)
AD 70		Dorotheus of Sidon (astrological poet), Pliny *Natural History* (Rome)
AD 150		Claudius Ptolemy (Alexandria, astronomer and astrologer) and Vettius Valens (Alexandria, astrologer)
AD 400		Firmicus Maternus, Hephaestio of Thebes, Rhetorius (?) (astrologers)

Appendix II: basic astrological doctrines

Given data and astral chart

Astrologers' main working data consist of what is called the chart of the native, which is by default a birth chart featuring the positions of the celestial bodies (the planets and the luminaries) as well as that of certain significant points. The chart is almost never pictured in ancient manuals, or in the stand-alone charts of papyrus horoscopes,[1] but we can assume that astrologers often visualized it on an astrological board showing the zodiacal signs:

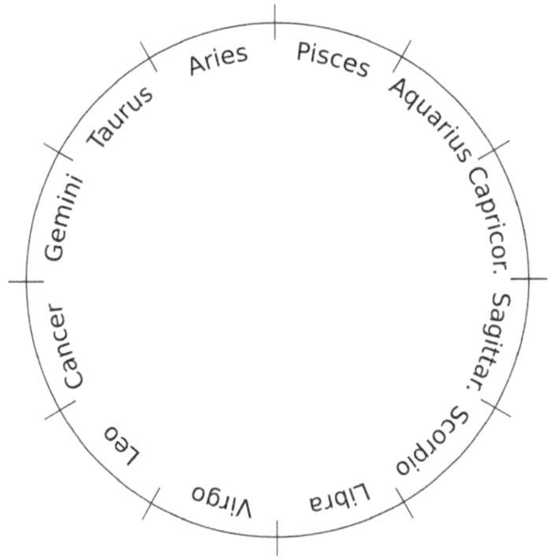

The circle of the zodiac represents the ecliptic, that is, the path of the Sun against the fixed stars, a great geocentric circle divided into 360° and distributed in 12 equal sections of 30° each, corresponding roughly to the constellations in the background. The Moon and the planets move approximately on this circle, too, because from a heliocentric perspective their orbital planes are roughly parallel with that of the Earth. The natural order of the signs runs counterclockwise, because that is the direction of the apparent motion of the Sun through the fixed stars along the year. The signs are also graduated counterclockwise, so that, for example, the first degree of Aries corresponds to the first 1/30th segment of Aries after the limit with Pisces.

The Moon also moves counterclockwise, but much faster than the Sun, because in approximately 29½ days on average she returns to the same relative position with the Sun (this is called the synodic period). While the Sun moves about 1° per

1 Most of the examples from Oxyrhynchus are published in Jones 1999.

https://doi.org/10.1515/9783111329147-011

day, the Moon advances on average about 13° per day. The planets also move counterclockwise most of the time, but they periodically stop and move clockwise, until they stop again and go back to normal. These are the so-called retrogradations (because they are interpreted as backward movements), a kind of parallax phenomenon caused by the motion of the Earth, our observing point. The duration and periodicity of the retrogradations depend on the planet: for example, Mars experiences retrogradations of between 60 and 80 days approximately every 2 years.

Crucial to astrologers are the so-called cardines or angles, that is, the four cuts with the zodiac of the Eastern and Western horizons, and with the perpendicular plane North–South:

- Ascendant: The point of the zodiac which is ascending on the Eastern horizon. It is ascending because the daily motion of the cosmos (the rotation of the Earth from a heliocentric perspective) makes the whole chart rotate clockwise throughout the day, keeping the horizon fixed.
- Descendant: The point of the zodiac opposite the ascendant, on the Western horizon.
- Midheaven: The point of the zodiac visible directly in a southerly direction (from the Northern hemisphere). It does not necessarily coincide with the point which is 90° before the ascendant, because of the inclination of the ecliptic (there is normally a longer section of ecliptic on one side or the other).
- Lower midheaven: The nonvisible point opposite the midheaven.

Note the differences between the chart, which is a schematic representation of the celestial configuration, and the sky picture. In the example below, the midheaven (the point where the ecliptic and the vertical line meet) is much closer to the descendant than from the ascendant.[2]

Ancient astrologers calculated these positions with the help of astronomical tables, often just considering the signs without specifying the degree: it is reasonable to interpret that these would have been the simpler and less expensive horoscopes, since the precise degrees are needed for many procedures.[3]

2 Generated through the software Skychart/Cartes du Ciel.

3 See the introduction to the different types of astronomical tables and the actual papyri edited and translated in Jones 1999.

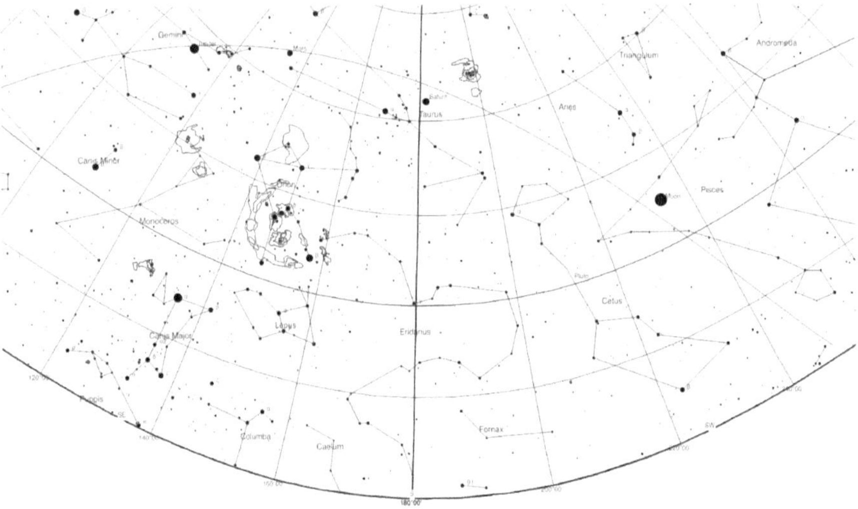

Astrological characteristics of the celestial bodies

To interpret the chart, astrologers assume a more or less established and shared corpus of doctrines defining the character of the celestial bodies and of the parts of the zodiac, which are frequently combined in order to evaluate the effect of the given position of a planet on a particular part of the zodiac. The detailed explanation of these doctrines usually forms the first part of ancient manuals, such as the first book of Ptolemy's *Tetrabiblos* or of Valens' *Anthologies*.[4]

Of course, the many areas into which the zodiac was divided implied frequent contradictions that astrologers tended to ignore. But let us begin with the celestial bodies. Their qualities derive mostly from the character of their associated divine figure. Here I outline just the most basic doctrines:[5]

- Beneficial (and humidifying) planets: Venus and Jupiter // harmful (and drying) planets: Mars and Saturn // ambiguous: Mercury
- Diurnal: Sun, Jupiter, Saturn // nocturnal: Moon, Venus, Mars // ambiguous: Mercury

4 For an extensive description of the ancient astrological doctrines, see Barton 1994.

5 Cf. Valens I 1 for associations of the celestial bodies with professions, and with parts of the body, often signaling diseases affecting these body parts.

- Human qualities: Sun = father, intelligence, authority; Moon = mother, conception, marriage;[6] Saturn = obstacles, grief; Jupiter = engendering, rank, justice; Mars = war, violence, banishment; Venus = desire, friendship, beauty, prosperity; Mercury = education, arts, religion.[7]

Relations between celestial bodies

The beneficial effect of a planet is usually interpreted to be enhanced by the presence of another beneficial planet in the same sign (conjunction), or at a certain angle, called **aspect**. The triangle, the relation between positions at a distance of four signs, is the positive aspect par excellence (e.g., the aspect between Capricorn and Taurus, or Aquarius and Gemini):

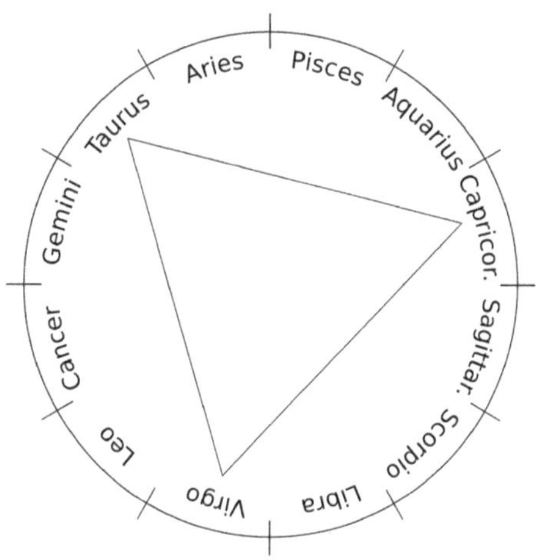

Triangle aspects between Taurus, Virgo, and Capricorn.

The hexagonal aspect is also beneficial, but to a lesser degree. The opposition and the square aspect, at a distance of three signs, are generally understood as

6 The androcentrism of ancient Greek societies is deeply embedded in the astrological system, hence marriage should be understood here as marriage to a woman, and similarly, the erotic interpretation of Venus is seen from the male point of view, so Venus is generally associated with women.

7 These qualities are used in particular to associate planets and human characters, as well as with professions: see, for example, the long lists in Valens *Anth.* I 1.

detrimental: they diminish the effect of beneficial planets and accentuate the effect of harmful planets.

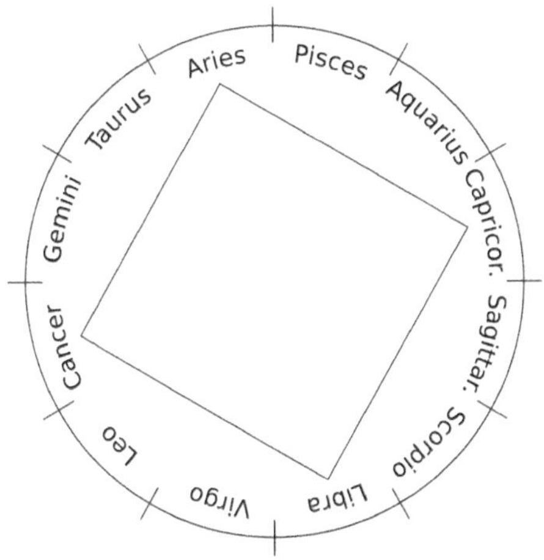

Square aspects of Aries and Libra (Capricorn and Cancer).

The remaining relations (contiguous signs and those at a distance of five signs) are said to be unaspected so that there is no influence between such signs.

Another kind of relation between the celestial bodies, which is naturally transferred to the zodiacal signs as in the case of the aspects, is between the **signs of equal rising times**, also called listening or beholding signs (Ptolemy *Tetr.* I 15).[8] In F 12 and F 20, above, these relations are used to establish destroying points, perhaps a natural application considering that the procedure for determining the length of life is often based on the rising times.

8 Ptolemy refers to signs that are equally distanced from the same tropical sign, thus Aries (and Libra) would be left without pairing, while Taurus would be joined to Pisces (both at a distance of one sign from Aries), and so on. The version I give is the one used, for example, by Valens in Critodemus F 20.

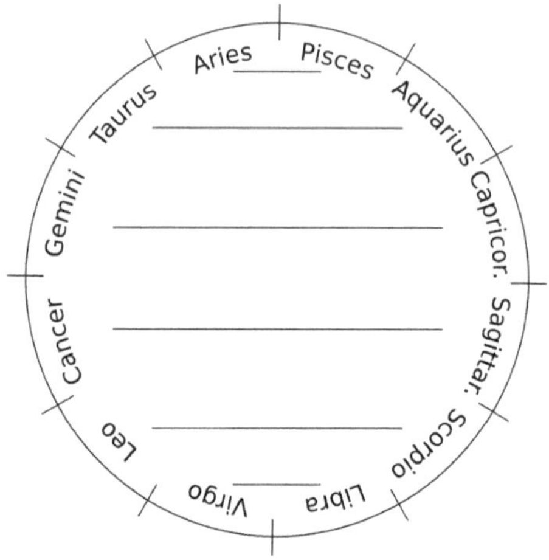

Signs that ascend in the same time (i.e., with equal rising times).

Relations between celestial bodies and parts of the zodiac

The parts of the zodiac in the several divisions are understood to be dominated (or ruled) by the planets and the luminaries (the Sun and the Moon), in such a way that the positive effects of the celestial bodies are enhanced when they are positioned in their own parts. The basic relation of domination is that of the signs (the so-called **domiciles**):

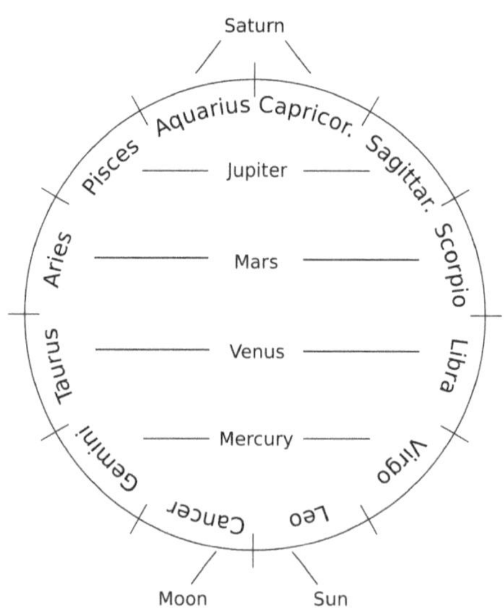

Another relation of domination is related to the four **triangles**. The signs are grouped in the four different equilateral triangles which can be formed in the zodiac, and rulers are defined for each of them by day and night:

Triangle	Ruler by day	Ruler by night
Aries–Leo–Sagittarius	Sun	Jupiter
Taurus–Virgo–Capricorn	Venus	Moon
Gemini–Libra–Aquarius	Saturn	Mercury
Cancer–Scorpio–Pisces	Mars and Venus	Mars and Moon

Another popular subdivision of the signs to which rulers were assigned is that of the so-called **terms**. Each one of the signs was divided into five to seven sections, depending on whether the system included the luminaries (one or both) or not: the subsections in each sign are then assigned to the different celestial bodies so that each one is ruler of one and only one subsection in a given sign. The subsections in each sign (the terms) need not be of the same length, and in different signs their lengths need not be the same, nor the order of the rulers. The most popular system was the one Ptolemy described as "Egyptian", which is the system used by Critodemus, the only author besides Valens (who was probably influenced by him) to give precise characteristics to each of these 60 terms (see Chapter 5).[9] However, the basic use was simple familiarity: as with the triangles and the signs, a planet in his own terms had its positive effects enhanced.

Another popular system of rulerships was that of the **exaltations**. Each celestial body was said to be exalted in a certain degree of the zodiac (although often the whole sign was taken), which again implied that when the star in question was situated in his own exaltation its positive effects were enhanced:

Star	Sign/degree
Sun	Aries 19°
Moon	Taurus 3°
Mercury	Virgo 15°
Venus	Pisces 27°
Mars	Capricorn 28°
Jupiter	Cancer 15°
Saturn	Libra 21°

9 Another system of terms, featuring the Sun but not the Moon, is given in the first table for calculating the length of life in Critodemus (cf. F 16, Chapter 8).

Astrological characterization of parts of the zodiac

We have already said that in addition to the relations of domination, the parts of the zodiac were also given certain characteristics. These can be divided between fixed parts (e.g., the terms or the zodiacal signs) and the movable system of the places, including the cardines.

The effects of the **signs** are divided according to several systems:[10]

- According to their relationship with the motion of the Sun: equinoctial (Aries and Libra), solstitial (Cancer and Capricorn), solid (the four signs following the equinoctial and solstitial), and bicorporeal (the other four)
- Masculine and feminine (alternate series beginning with Aries=masculine)
- According to the perceived characteristics of the signs: animal/human signs, water/earth signs[11]
- According to the nature of their ruling planet[12]

The system of **places** (or **houses**[13]), called **dodekatropos**, is movable in the sense that it is fixed not with respect to the signs but to the local horizon. This system evolved in complex ways in the medieval Islamic and European tradition, which took into account the fact, referred to above, that the midheaven does not necessarily fall in the middle part of the ecliptic above the horizon. Ancient astrologers generally defined it in a simple fashion: the signs of the chart were numbered from 1 to 12, beginning with the ascendant (in the sense of the signs), to form the places or houses. Of these, the first and the seventh – containing the ascendant and the descendant points, respectively – as well as the signs midway between these above and below the horizon (the fourth and the tenth) were called the cardines or angles. These four positions were understood to be the most effective houses, functioning generally as enhancers of positive planetary effects and also contributing particu-

10 There are effects attributed to parts of the signs corresponding to subgroups of the constellations, which poses the problem of the slow rotation of the zodiacal signs (if one understands them as beginning at the equinox, as Ptolemy did, for example) against constellations caused by the precession of the equinoxes. In antiquity, unlike in the present era, the displacement was not particularly remarkable, so the effects of the fixed stars could be identified with effects of the corresponding part of the signs: for example, in Valens *Anth.* I 2, these effects are included among the characteristics of the signs.

11 Some of these categories are not widely agreed upon. For example, for Valens (*Anth.* II 1.9), Cancer–Scorpio–Pisces is a wet triangle, but Ptolemy (*Tetr.* II 12) singles out only Aquarius, Virgo, and Capricorn as water signs.

12 For example, the signs ruled by Mercury produce men active in the field of arts and religion, just like Mercury, according to Valens *Anth.* I 2.

13 Not to be confused with the domiciles of the planets (see above).

lar nuances. The rest of the houses that bear an aspect (triangular or hexagonal) with the ascendant (3rd, 5th, 9th, and 11th) are also interpreted as positive.

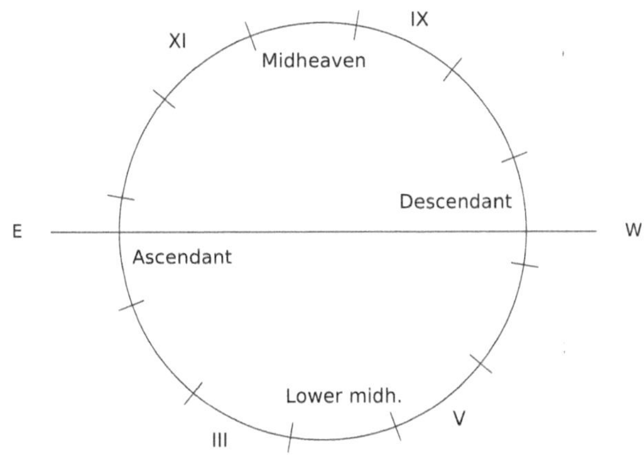

The houses also carried particular significations that provided the field of action for the effect of the planets therein: for example, the descendant was indicative of marriage, so the nature of the planets in that place had implications for the quality of marriage. These fields of influence show some kind of agreement with the relative power attributed to the houses so that bad houses are typically concerned with topics that have negative connotations: see the significations of the unaspected houses VI, VIII, and XII in the chart below:[14]

House	Field of action and special names
I	Life (**Ascendant**)
II	Commerce (Gate of Hell)
III	Brothers
IV	Family property (**Lower Midheaven**)
V	Children (Good Fortune)
VI	Diseases (Bad Fortune)
VII	Marriage **(Descendant)**
VIII	Death
IX	Religion, travel
X	Occupation, rank **(Midheaven)**
XI	Friends (Good Daimon)
XII	Enemies (Bad Daimon)

14 For more detailed significations, see Firmicus Maternus II 19 or Vettius Valens *Anth.* IV 12/IX 3.

Table of rising times: construction and use

As a necessary tool for the calculation of the ascendant, an approximate rising speed was determined for each of the zodiacal signs, depending on the clima (=latitude) where the birth took place. Now we have to picture the situation in which the horizon is fixed and the zodiacal circle turns clockwise with the daily motion of the cosmos so that over the course of a day the 12 signs have turned one complete revolution. The time of ascension was calculated in time-degrees, 360 time-degrees being the time of one complete revolution of the cosmos. One equinoctial hour, irrespective of the time of the year, was defined as 15 time-degrees so that the whole day contains 24 h. So, if all signs rose in the same time, they all would rise in 30 time-degrees, but, since the ecliptic is inclined with respect to the equator, some rise in more time-degrees, some in less. Thus, for the clima of Alexandria, 20 time-degrees were found for Aries and Pisces, the signs rising more rapidly (because that is the less vertical section of the ecliptic): 24 to Taurus and Aquarius, 28 to Gemini and Capricorn, and so on. In Neugebauer and Van Hoesen's presentation, with a_1 representing Aries, a_2 Taurus, and so on:[15]

$a_1 = a_{12} = 20$ time-degrees
$a_2 = a_{11} = 24$ time-degrees
$a_3 = a_{10} = 28$ time-degrees
$a_4 = a_9 = 32$ time-degrees
$a_5 = a_8 = 36$ time-degrees
$a_6 = a_7 = 40$ time-degrees

The scheme above is already a table of rising times for the signs, but for the precise calculation of the ascending degree, astrologers needed a table by degrees. This was simply obtained by dividing the rising time of the whole sign by 30.[16] In this approximating scheme, each degree in each sign rises in the same rising time (e.g., 0.67 for Aries) so that the rising times column more conveniently listed the accumulated rising times (e.g., for Aries 2, the rising times of the first 2° of Aries were added: 0.67 + 0.67).

---

15 This was called system A, because another system, called B, used a slightly different approximation. Cf. *GH*, 3.

16 With time, methods of spherical trigonometry were applied to the construction of tables of rising times such as the one in Ptolemy's *Syntaxis* (II 8). I am concerned here with the table obtained through simple numerical methods, imported from Babylon, which is the one attributed to Hypsicles (ca. 150 BC), which only began to be replaced by trigonometrical tables from the third century AD on. On this insight, see the dating of the astronomical tables related to Ptolemy's *Handy Tables* in the Oxyrhynchus papyri, edited in Jones 1999.

Points of the zodiac	Accumulated rising times
Aries 0	0
Aries 1	0.67 (=20/30)
Aries 2	1.33
...	
Aries 29	19.33
Taurus 0	20
Taurus 1	20.8 (=20 + 24/30)
Taurus 2	21.6
...	
Gemini 0	48
Gemini 1	48.93 (=48 + 28/30)
...	
Libra 0	180
...	
Scorpio 0	220
Scorpio 1	221.33
...	
Pisces 0	340
Pisces 1	340.67
...	

Then, to determine the time that it takes to rise from one point of the zodiac to another, astrologers only needed to calculate the difference between the corresponding accumulated rising times. For example, the sector from Taurus 2 to Gemini 1 takes 27.3 time-degrees to rise, that is, 1.82 h = 1 h 49 min (one adds 360 in cases extending beyond the table, e.g., the sector Pisces 1 to Aries 29 $\Longrightarrow$ 379.33 − 340.67 = 38.67 time-degrees).

The primary use of the ascensional table was to calculate the rising point, given the hour of the client's birth. This was, as a general rule, given in seasonal hours, which were defined as the divisions of the whole day from sunrise to sunset in 12 h (and from sunset to the next sunrise for night hours). For day births, the rising times corresponding to the hours elapsed from sunrise were easily calculated by considering that the 12 seasonal hours of daytime always equal the travel of 180 zodiacal degrees from sunrise to sunset. So, in the first place, one finds how many time-degrees correspond to one seasonal hour at that point of the year. That is, for the Sun in Taurus 1, 12 seasonal hours of daylight time = 221.33 (time-degrees in Scorpio 1) − 20.8 (time-degrees in Taurus 1) = 200.53 time-degrees $\Longrightarrow$ 1 seasonal hour = 16.71 time-degrees. Therefore, if the client was born at the end of the second seasonal hour, that is, 2 seasonal hours after sunrise, 16.71 time-degrees × 2 have elapsed since sunrise. This quantity is added to the accumulated rising times at the Sun's position: 20.8 + 2 × 16.71 = 54.22 time-degrees. We would then look for the degree of the zodiac corresponding to these accumulated time-degrees: this would be very close to Gemini 11. This is the ascending point. This calculation was much easier in tables directly providing the length in time-degrees of the seasonal hour for the longitude of the Sun in question. See below a table of rising times by degree featuring a column with the length of the seasonal hours for the latitude of Alexandria.

	Aries		Taurus		Gemini		Cancer		Leo		Virgo	
°	Accumulated time-degrees	Seasonal hour	Accumulated time-degrees	Seasonal hour	Accumulated time-degrees	Seasonal hour	Accumulated time-degrees	Seasonal hour	Accumulated time-degrees	Seasonal hour	Accumulated time-degrees	Seasonal hour
1	0.67	15.06	20.80	16.70	44.93	17.68	73.07	17.99	105.20	17.63	141.33	16.61
2	1.33	15.11	21.60	16.73	45.87	17.69	74.13	17.98	106.40	17.60	142.67	16.56
3	2.00	15.17	22.40	16.77	46.80	17.70	75.20	17.97	107.60	17.57	144.00	16.50
4	2.67	15.22	23.20	16.80	47.73	17.71	76.27	17.96	108.80	17.53	145.33	16.44
5	3.33	15.28	24.00	16.83	48.67	17.72	77.33	17.94	110.00	17.50	146.67	16.39
6	4.00	15.33	24.80	16.87	49.60	17.73	78.40	17.93	111.20	17.47	148.00	16.33
7	4.67	15.39	25.60	16.90	50.53	17.74	79.47	17.92	112.40	17.43	149.33	16.28
8	5.33	15.44	26.40	16.93	51.47	17.76	80.53	17.91	113.60	17.40	150.67	16.22
9	6.00	15.50	27.20	16.97	52.40	17.77	81.60	17.90	114.80	17.37	152.00	16.17
10	6.67	15.56	28.00	17.00	53.33	17.78	82.67	17.89	116.00	17.33	153.33	16.11
11	7.33	15.61	28.80	17.03	54.27	17.79	83.73	17.88	117.20	17.30	154.67	16.06
12	8.00	15.67	29.60	17.07	55.20	17.80	84.80	17.87	118.40	17.27	156.00	16.00
13	8.67	15.72	30.40	17.10	56.13	17.81	85.87	17.86	119.60	17.23	157.33	15.94
14	9.33	15.78	31.20	17.13	57.07	17.82	86.93	17.84	120.80	17.20	158.67	15.89
15	10.00	15.83	32.00	17.17	58.00	17.83	88.00	17.83	122.00	17.17	160.00	15.83
16	10.67	15.89	32.80	17.20	58.93	17.84	89.07	17.82	123.20	17.13	161.33	15.78
17	11.33	15.94	33.60	17.23	59.87	17.86	90.13	17.81	124.40	17.10	162.67	15.72
18	12.00	16.00	34.40	17.27	60.80	17.87	91.20	17.80	125.60	17.07	164.00	15.67
19	12.67	16.06	35.20	17.30	61.73	17.88	92.27	17.79	126.80	17.03	165.33	15.61
20	13.33	16.11	36.00	17.33	62.67	17.89	93.33	17.78	128.00	17.00	166.67	15.56
21	14.00	16.17	36.80	17.37	63.60	17.90	94.40	17.77	129.20	16.97	168.00	15.50
22	14.67	16.22	37.60	17.40	64.53	17.91	95.47	17.76	130.40	16.93	169.33	15.44
23	15.33	16.28	38.40	17.43	65.47	17.92	96.53	17.74	131.60	16.90	170.67	15.39
24	16.00	16.33	39.20	17.47	66.40	17.93	97.60	17.73	132.80	16.87	172.00	15.33
25	16.67	16.39	40.00	17.50	67.33	17.94	98.67	17.72	134.00	16.83	173.33	15.28
26	17.33	16.44	40.80	17.53	68.27	17.96	99.73	17.71	135.20	16.80	174.67	15.22
27	18.00	16.50	41.60	17.57	69.20	17.97	100.80	17.70	136.40	16.77	176.00	15.17
28	18.67	16.56	42.40	17.60	70.13	17.98	101.87	17.69	137.60	16.73	177.33	15.11
29	19.33	16.61	43.20	17.63	71.07	17.99	102.93	17.68	138.80	16.70	178.67	15.06
30	20.00	16.67	44.00	17.67	72.00	18.00	104.00	17.67	140.00	16.67	180.00	15.00

	Libra		Scorpio		Sagittarius		Capricorn		Aquarius		Pisces	
°	Accumulated time-degrees	Seasonal hour	Accumulated time-degrees	Seasonal hour	Accumulated time-degrees	Seasonal hour	Accumulated. time-degrees	Seasonal hour	Accumulated time-degrees	Seasonal hour	Accumulated time-degrees	Seasonal hour
1	181.33	14.94	221.20	13.30	257.07	12.32	288.93	12.01	316.80	12.37	340.67	13.39
2	182.67	14.89	222.40	13.27	258.13	12.31	289.87	12.02	317.60	12.40	341.33	13.44
3	184.00	14.83	223.60	13.23	259.20	12.30	290.80	12.03	318.40	12.43	342.00	13.50
4	185.33	14.78	224.80	13.20	260.27	12.29	291.73	12.04	319.20	12.47	342.67	13.56
5	186.67	14.72	226.00	13.17	261.33	12.28	292.67	12.06	320.00	12.50	343.33	13.61
6	188.00	14.67	227.20	13.13	262.40	12.27	293.60	12.07	320.80	12.53	344.00	13.67
7	189.33	14.61	228.40	13.10	263.47	12.26	294.53	12.08	321.60	12.57	344.67	13.72
8	190.67	14.56	229.60	13.07	264.53	12.24	295.47	12.09	322.40	12.60	345.33	13.78
9	192.00	14.50	230.80	13.03	265.60	12.23	296.40	12.10	323.20	12.63	346.00	13.83
10	193.33	14.44	232.00	13.00	266.67	12.22	297.33	12.11	324.00	12.67	346.67	13.89
11	194.67	14.39	233.20	12.97	267.73	12.21	298.27	12.12	324.80	12.70	347.33	13.94
12	196.00	14.33	234.40	12.93	268.80	12.20	299.20	12.13	325.60	12.73	348.00	14.00
13	197.33	14.28	235.60	12.90	269.87	12.19	300.13	12.14	326.40	12.77	348.67	14.06
14	198.67	14.22	236.80	12.87	270.93	12.18	301.07	12.16	327.20	12.80	349.33	14.11
15	200.00	14.17	238.00	12.83	272.00	12.17	302.00	12.17	328.00	12.83	350.00	14.17
16	201.33	14.11	239.20	12.80	273.07	12.16	302.93	12.18	328.80	12.87	350.67	14.22
17	202.67	14.06	240.40	12.77	274.13	12.14	303.87	12.19	329.60	12.90	351.33	14.28
18	204.00	14.00	241.60	12.73	275.20	12.13	304.80	12.20	330.40	12.93	352.00	14.33
19	205.33	13.94	242.80	12.70	276.27	12.12	305.73	12.21	331.20	12.97	352.67	14.39
20	206.67	13.89	244.00	12.67	277.33	12.11	306.67	12.22	332.00	13.00	353.33	14.44
21	208.00	13.83	245.20	12.63	278.40	12.10	307.60	12.23	332.80	13.03	354.00	14.50
22	209.33	13.78	246.40	12.60	279.47	12.09	308.53	12.24	333.60	13.07	354.67	14.56
23	210.67	13.72	247.60	12.57	280.53	12.08	309.47	12.26	334.40	13.10	355.33	14.61
24	212.00	13.67	248.80	12.53	281.60	12.07	310.40	12.27	335.20	13.13	356.00	14.67
25	213.33	13.61	250.00	12.50	282.67	12.06	311.33	12.28	336.00	13.17	356.67	14.72
26	214.67	13.56	251.20	12.47	283.73	12.04	312.27	12.29	336.80	13.20	357.33	14.78
27	216.00	13.50	252.40	12.43	284.80	12.03	313.20	12.30	337.60	13.23	358.00	14.83
28	217.33	13.44	253.60	12.40	285.87	12.02	314.13	12.31	338.40	13.27	358.67	14.89
29	218.67	13.39	254.80	12.37	286.93	12.01	315.07	12.32	339.20	13.30	359.33	14.94
30	220.00	13.33	256.00	12.33	288.00	12.00	316.00	12.33	340.00	13.33	360.00	15.00

Appendix III: tables in Valens IX and his emulation of Critodemus

Valens' eighth book primarily deals with the two large tables for calculating the length of life from Critodemus (F 15). At the end of his ninth and last book, Valens appended two tables that are not extant in the manuscript tradition, but whose appearance and use he describes in sufficient detail in the last chapter to allow a reconstruction (IX 19). These tables are relevant to this study insofar as they represent a curious imitation by Valens of the appearance of Critodemus' work.

To begin with, Valens presents the tables in the same redundant double terminology as the tables of book VIII (κανόνες τῶν β̄ πλινθίων, "tables of the two charts") and explicitly refers to their resemblance with those tables, noting that they are similar to one another in extension, again.[1] This would not be the only feature of Valens' treatise that can be interpreted as a conscious emulation of the structure of Critodemus' manual: as we have seen, both he and Firmicus introduce their book VII with the astrological oath, which in both authors is related to the doctrine of the distributions (see Chapter 3 for the oath, and Chapter 4 for the distributions).

It is possible that both Firmicus and Valens, who wrote eight and nine books, respectively, intended to write the same number of books as Critodemus. Critodemus probably wrote his *Horasis* in seven books, the last of which was introduced by the oath, and eventually an eighth book comprising the tables for calculating the length of life would have enlarged the treatise. Writing a work in the same number of books as an older authority was often understood as a sort of homage: for example, Claudius Ptolemy wrote his *Almagest* in 13 books, probably with the 13 books of Euclid in mind. Valens, in fact, composed his ninth book as a series of heavily disorganized (more so than elsewhere in his work, at least) doctrines, looking back at his former eight books as an exposition and explanation of all the doctrines relevant to Nechepsos (IX 1.1):

> Ὅσα μὲν ὁ θειότατος βασιλεὺς εἴρηκε Νεχεψὼ ὁ τὴν ἀρχὴν ποιησάμενος τῆς ιγ̄ βίβλου, ἐν τοῖς προσυντεταγμένοις ὑφ' ἡμῶν καὶ ἄλλων πόνοις κατημαξευμένα ἐστίν·
>
> That which the divine King Nechepsos said, beginning in his 13th book,[2] has been gone through in the previous studies by myself and others.

1 IX 19.2–3: "again they have been harmonized with one another in the same extension of range" (πάλιν ἐν ἴσῳ διαστήματι τοῦ πλάτους συνηρμόσθησαν).
2 Probably implying that the first book of the poem of Nechepsos and Petosiris was marked as the thirteenth book, perhaps conveying a sort of mystic outlook: cf. Heilen 2011, 24.

https://doi.org/10.1515/9783111329147-012

As we have seen, he begins the book by reflecting for the second time on the quality of Critodemus' work, quoting again its initial lines and imitating Critodemus' words for his own recollection of his arrival in Egypt (cf. Chapter 2). One could well hypothesize that these features suggest that Valens conceived his ninth book in part as an imitation of the works of Nechepsos and Critodemus, reflecting on and imitating their beginnings at the start of the book, and including his own tables at the end.

From the analysis of the tables themselves, one can in fact suspect that the desire to imitate was the main reason behind the creation of the tables. Both were large tables with roughly the same number of rows (49 and 55, respectively) designed to be used together to correct the position of the Moon at the hour of birth, presumably from positions given by day-to-day tables. The Moon's mean change in longitude per day is about 13°, varying approximately between a minimum of 11° and a maximum of 15°. The first table bizarrely treats these quantities as hours, and tabulates the intermediate points in differences of 5 min:

11;0   11;5   11;10   . . .   15;0   [49 rows in total]

The first table, then, is simply a long column containing these values, whereas the second tabulates daylight lengths for the days of the year. The second table seems more standard, and was probably, unlike the first one, not created *ad hoc* for this doctrine but reused from normal astrological practice. Here, it seems awkward that the table is computed for the 7th clima (not that of Alexandria); however, from what Valens says, it appears essential that this clima be used, because the difference between the maximum daylight time (19;30 time-degrees for this clima) and daylight at equinox (15 time-degrees) must present a similar number of intervals to the first table when divided in five-minute intervals. In addition to the column containing the daylight lengths, the table had other columns giving the amount of time-degrees corresponding to elapsed hours, which effectively functioned as a multiplication table that could have been useful for determining the ascendant point in combination with an anaphoric table providing the hourly magnitude throughout the year.[3]

3 With the layout explained by Valens, it seems that the rows only extend to half of the possible hourly magnitudes, from 19;30 down to 15;0. We must probably suppose, then, that for the remaining rows one goes to the hourly magnitude of the opposite point in the zodiac (e.g., using a column like the final column in the table below).

	2 h	3 h	4 h	5 h	6 h	7 h	8 h	9 h	10 h	11 h	12 h	
10;30	21;0	31;30	42;0	52;30	63;0	73;30	84;0	94;30	105;0	115;30	126;0	19;30
10;35	21;10	31;45	42;20	52;55	63;30	74;5	84;40	95;15	105;50	116;25	127;0	19;25
10;40	21;20	32;0	42;40	53;20	64;0	74;40	85;20	96;0	106;40	117;20	128;0	19;20
10;45	21;30	32;15	43;0	53;45	64;30	75;15	86;0	96;45	107;30	118;15	129;0	19;15
10;50	21;40	32;30	43;20	54;10	65;0	75;50	86;40	97;30	108;20	119;10	130;0	19;10
10;55	21;50	32;45	43;40	54;35	65;30	76;25	87;20	98;15	109;10	120;5	131;0	19;5
11;0	22;0	33;0	44;0	55;0	66;0	77;0	88;0	99;0	110;0	121;0	132;0	19;0
11;5	22;10	33;15	44;20	55;25	66;30	77;35	88;40	99;45	110;50	121;55	133;0	18;55
11;10	22;20	33;30	44;40	55;50	67;0	78;10	89;20	10;30	111;40	122;50	134;0	18;50
11;15	22;30	33;45	45;0	56;15	67;30	78;45	90;0	101;15	112;30	123;45	135;0	18;45
11;20	22;40	34;0	45;20	56;40	68;0	79;20	90;40	102;0	113;20	124;40	136;0	18;40
11;25	22;50	34;15	45;40	57;5	68;30	79;55	91;20	102;45	114;10	125;35	137;0	18;35
11;30	23;0	34;30	46;0	57;30	69;0	80;30	92;0	103;30	115;0	126;30	138;0	18;30
11;35	23;10	34;45	46;20	57;55	69;30	81;5	92;40	104;15	115;50	127;25	139;0	18;25
11;40	23;20	35;0	46;40	58;20	70;0	81;40	93;20	105;0	116;40	128;20	140;0	18;20
11;45	23;30	35;15	47;0	58;45	70;30	82;15	94;0	105;45	117;30	129;15	141;0	18;15
11;50	23;40	35;30	47;20	59;10	71;0	82;50	94;40	106;30	118;20	130;10	142;0	18;10
11;55	23;50	35;45	47;40	59;35	71;30	83;25	95;20	107;15	119;10	131;5	143;0	18;5
12;0	24;0	36;0	48;0	60;0	72;0	84;0	96;0	108;0	120;0	132;0	144;0	18;0
12;5	24;10	36;15	48;20	60;25	72;30	84;35	96;40	108;45	120;50	132;55	145;0	17;55
12;10	24;20	36;30	48;40	60;50	73;0	85;10	97;20	109;30	121;40	133;50	146;0	17;50
12;15	24;30	36;45	49;0	61;15	73;30	85;45	98;0	110;15	122;30	134;45	147;0	17;45
12;20	24;40	37;0	49;20	61;40	74;0	86;20	98;40	111;0	123;20	135;40	148;0	17;40
12;25	24;50	37;15	49;40	62;5	74;30	86;55	99;20	111;45	124;10	136;35	149;0	17;35
12;30	25;0	37;30	50;0	62;30	75;0	87;30	100;0	112;30	125;0	137;30	150;0	17;30
12;35	25;10	37;45	50;20	62;55	75;30	88;5	100;40	113;15	125;50	138;25	151;0	17;25
12;40	25;20	38;0	50;40	63;20	76;0	88;40	101;20	114;0	126;40	139;20	152;0	17;20
12;45	25;30	38;15	51;0	63;45	76;30	89;15	102;0	114;45	127;30	140;15	153;0	17;15
12;50	25;40	38;30	51;20	64;10	77;0	89;50	102;40	115;30	128;20	141;10	154;0	17;10
12;55	25;50	38;45	51;40	64;35	77;30	90;25	103;20	116;15	129;10	142;5	155;0	17;5
13;0	26;0	39;0	52;0	65;0	78;0	91;0	104;0	117;0	130;0	143;0	156;0	17;0
13;5	26;10	39;15	52;20	65;25	78;30	91;35	104;40	117;45	130;50	143;55	157;0	16;55
13;10	26;20	39;30	52;40	65;50	79;0	92;10	105;20	118;30	131;40	144;50	158;0	16;50
13;15	26;30	39;45	53;0	66;15	79;30	92;45	106;0	119;15	132;30	145;45	159;0	16;45
13;20	26;40	40;0	53;20	66;40	80;0	93;20	106;40	120;0	133;20	146;40	160;0	16;40
13;25	26;50	40;15	53;40	67;5	80;30	93;55	107;20	120;45	134;10	147;35	161;0	16;35
13;30	27;0	40;30	54;0	67;30	81;0	94;30	108;0	121;30	135;0	148;30	162;0	16;30
13;35	27;10	40;45	54;20	67;55	81;30	95;5	108;40	122;15	135;50	149;25	163;0	16;25

(continued)

	2 h	3 h	4 h	5 h	6 h	7 h	8 h	9 h	10 h	11 h	12 h	
13;40	27;20	41;0	54;40	68;20	82;0	95;40	109;20	123;0	136;40	150;20	164;0	16;20
13;45	27;30	41;15	55;0	68;45	82;30	96;15	110;0	123;45	137;30	151;15	165;0	16;15
13;50	27;40	41;30	55;20	69;10	83;0	96;50	110;40	124;30	138;20	152;10	166;0	16;10
13;55	27;50	41;45	55;40	69;35	83;30	97;25	111;20	125;15	139;10	153;5	167;0	16;5
14;0	28;0	42;0	56;0	70;0	84;0	98;0	112;0	126;0	140;0	154;0	168;0	16;0
14;5	28;10	42;15	56;20	70;25	84;30	98;35	112;40	126;45	140;50	154;55	169;0	15;55
14;10	28;20	42;30	56;40	70;50	85;0	99;10	113;20	127;30	141;40	155;50	170;0	15;50
14;15	28;30	42;45	57;0	71;15	85;30	99;45	114;0	128;15	142;30	156;45	171;0	15;45
14;20	28;40	43;0	57;20	71;40	86;0	10;20	114;40	129;0	143;20	157;40	172;0	15;40
14;25	28;50	43;15	57;40	72;5	86;30	10;55	115;20	129;45	144;10	158;35	173;0	15;35
14;30	29;0	43;30	58;0	72;30	87;0	101;30	116;0	130;30	145;0	159;30	174;0	15;30
14;35	29;10	43;45	58;20	72;55	87;30	102;5	116;40	131;15	145;50	160;25	175;0	15;25
14;40	29;20	44;0	58;40	73;20	88;0	102;40	117;20	132;0	146;40	161;20	176;0	15;20
14;45	29;30	44;15	59;0	73;45	88;30	103;15	118;0	132;45	147;30	162;15	177;0	15;15
14;50	29;40	44;30	59;20	74;10	89;0	103;50	118;40	133;30	148;20	163;10	178;0	15;10
14;55	29;50	44;45	59;40	74;35	89;30	104;25	119;20	134;15	149;10	164;5	179;0	15;5
15;0	30;0	45;0	60;0	75;0	90;0	105;0	120;0	135;0	150;0	165;0	180;0	

The procedure for the combined use of both tables, which is intended to correct the position of the Moon given by daily tables, is given here. The length of daylight for the day in question, as well as the motion of the Moon, are known. One begins with the second table and places one leg of a compass on the number corresponding to the daylight time of the day in question, then extending the other leg horizontally to the tabulated quantity of time-degrees corresponding to the hour of birth. Next, this measurement is transferred with the compass to the first table, placing one leg on the number corresponding to the motion of the Moon that day. Finally, we count how many entries are contained between the compass legs (vertically, either upward or downward), and then add this number to or subtract it from the degrees of the Moon to correct its position, perhaps understood as the longitude at sunset. In the end, of course, the outcome depends on the proportion between the width of the squares in the second table and the height of the squares in the first table: that is, on the purely graphical appearance of the tables.

References

Alvar 2008: Jaime Alvar, *Romanizing Oriental Gods*, Leiden/Boston.

Anagnostou-Canas 2017: Barbara Anagnostou-Canas, “Contrats et serments dans l’Égypte hellénistique et romain”, in M.-H. Marganne et al. (eds.), *En marge du Serment hippocratique. Contrats et serments dans le monde gréco-romain*, Liège, 51–65.

Andersen 2013: Lars Døvling Andersen, “Latin Squares”, in R. Wilson and J. J. Watkins, *Combinatorics: Ancient and Modern*, Oxford.

Andrade 2014: Nathanael Andrade, “Assyrians, Syrians and the Greek Language in the late Hellenistic and Roman Imperial Periods”, *Journal of Near Eastern Studies* 73, 299–317.

Asper 2009: Markus Asper, “The two cultures of mathematics in ancient Greece”, in E. Robson and J. Stedall (eds.), *The Oxford Handbook of the History of Mathematics*, Oxford, 107–132.

Bach 2013: Johannes Bach, “Berossos, Antiochos, und die Babyloniaka”, *Ancient West and East* 12, 157–180.

Bagnall and Frier 1994: Roger S. Bagnall and Bruce W. Frier, *The Demography of Roman Egypt*, Cambridge.

Barton 1994: Tamsyn Barton, *Ancient Astrology*, London.

Beaulieu 2018a: Paul-Alain Beaulieu, *A History of Babylon 2200 BC–AD 75*, Hoboken NJ.

Beaulieu 2018b: Paul-Alain Beaulieu, “Uruk Before and After Xerxes: The Onomastic and Institutional Rise of the God Anu”, in C. Waerzeggers and M. Seire (eds.), *Xerxes and Babylonia: the Cuneiform Evidence*, Leuven/Paris/Bristol, 189–206.

Benefiel 2013: Rebecca R. Benefiel, “Magic Squares, Alphabet Jumbles, Riddles, and More: the Culture of Word-Games among the Graffiti of Pompeii”, in M. Erler, Dorothee Gall et al. (eds.), *The Muse at Play. Riddles and Wordplay in Greek and Latin Poetry*, Berlin/Boston, 65–82.

Berdozzo 2011: Fabio Berdozzo, *Götter, Mythen, Philosophen. Lukian und die paganen Göttervorstellungen seiner Zeit*, Berlin/Boston.

Bernabé 1996: Alberto Bernabé, “La fórmula órfica ‘Cerrad las puertas, profanos’. Del profano religioso al profano en la materia”, *'Ilu* 1, 13–37.

Bernabé 2008: Alberto Bernabé, “Orfeo y Eleusis”, *Synthesis* 15, 13–36.

Bernabé 2009a: Alberto Bernabé, “Teogonías órficas”, in A. Bernabé and F. Casadesús (eds.), *Orfeo y la tradición órfica. Un reencuentro*, Madrid, 291–324.

Bernabé 2009b: Alberto Bernabé, “Las láminas de Olbia”, in A. Bernabé and F. Casadesús (eds.), *Orfeo y la tradición órfica. Un reencuentro*, Madrid, 537–546.

Bernabé and Casadesús 2009: A. Bernabé and F. Casadesús (eds.), *Orfeo y la tradición órfica. Un reencuentro*, Madrid.

Berti 2006: Irene Berti, “Now Let Earth be my Witness and the Broad Heaven Above, and the Down Flowing Water of the Styx…”, in E. Stavrianopoulou (ed.), *Ritual and Communication in the Graeco-Roman World*, Liège, 181–209.

Betegh 2004: Gábor Betegh, *The Derveni Papyrus: Cosmology, Theology and Interpretation*, Cambridge.

Bezza and Fumigalli (undated): Giuseppe Bezza and Marco Fumigalli, “L’ordinamento tolemaico dei confini”, http://www.cieloeterra.it/articoli.confini/confini.html, last accessed 08/06/2023.

Blanco Cesteros 2020: Miriam Blanco Cesteros, “The Paradox of a Magical Hymn: Reviewing the Poetic Compositions of the Greek Magical Papyri”, in A. Mastrocinque, J. E. Sanzo, and M. Scapini (eds.), *Ancient Magic*, Stuttgart, 257–286.

Boiy 2006: T. Boiy, *Late Achaemenid and Hellenistic Babylon*, Leuven.

https://doi.org/10.1515/9783111329147-013

Bos and Tzvi Langermann 2015: G. Bos and Y. Tzvi Langermann, *The Alexandrian Summaries of Galen's Critical Days*, Leiden.

Bouché-Leclercq 1899: Auguste Bouché-Leclercq, *L'astrologie grecque*, Paris.

Bowen and Todd 2004: *Cleomedes' Lectures on Astronomy*, Berkeley/Los Angeles/London.

Brisson 2015: Luc Brisson, "Prayer in Neoplatonism and the *Chaldean Oracles*", in J. Dillon and A. Timotin (eds.), *Platonic Theories of Prayer*, Leiden.

Bull 2018: Christian H. Bull, *The Tradition of Hermes Trismegistus: The Egyptian Priestly Figure as a Teacher of Hellenized Wisdom*, Leiden.

Burstein 1978: Stanley M. Burstein, *The Babyloniaca of Berossus*, Malibu CA.

Caballero Sánchez 2013: Raúl Caballero Sánchez, "Historia del texto del *Comentario anónimo al Tetrabiblos de Tolomeo*", *MHNH* 13, 77–198.

Cottrell and Ross 2019: Emily Cottrell and Micah Ross, "Persian astrology: Dorotheus and Zoroaster, according to the medieval Arabic sources (8th–11th century)", in P. B. Lurje (ed.), *Proceedings of the Eigth European Conference of Iranian Studies*, Saint Petersburg, 87–105.

Crist et al. 2016: Walter Crist, Anne-Elizabet Dunn-Vaturi, and Alex de Voogt, *Ancient Egyptians at Play: Board Games Across Boarders*, London.

De Breucker 2013: Geert de Breucker, "Berossos: his life and work", in J. Haubold, G. B. Lanfranchi, R. Rollinger and J. Steele (eds.), *The World of Berossos*, Wiesbaden, 15–30.

Dimitrov 2020: Bojidar Dimitrov, "'Fort. recte': witnesses to the text of Ptolemy's *Tetrabiblos* in its Near Eastern context", in D. N. Hasse et al. (eds.), *Ptolemy's Science of the Stars in the Middle Ages*, Turnhout.

Dyck 1981 Andrew R. Dyck, "Notes on the Epimerismoi Attributed to Herodian", *Hermes* 109, 225–235.

Escolano-Poveda 2022: Marina Escolano-Poveda, "Astrologica athribitana: Four demotic-hieratic horoscopes from Athribis (O. Athribis 17–36–5/1741 and ANAsh.Mus.D.O.633 reedited)", *Journal for the History of Astronomy* 53, 49–87.

Euler 1782: Leonhard Euler, "Recherches sur une nouvelle espèce de quarrés magiques", *Verhandelingen uitgegeven door het zeeuwsch Genootschap der Wetenschappen te Vlissingen* 9, 85–239.

Evans 1998: James Evans, *The History and Practice of Ancient Astronomy*, Oxford/New York.

Evans 2004: James Evans, "The Astrologer's Apparatus: A Picture of Professional Practice in Graeco-Roman Egypt", *Journal for the History of Astronomy* 35, 1–44.

Evans and Berggren 2007: James Evans and Lennart Berggren, *Geminos's Introduction to the Phenomena*, Princeton/Oxford.

Fantham 2002: Elaine Fantham, "Ovid's *Fasti*: Politics, History, and Religion", in B. W. Boyd, *Brill's Companion to Ovid*, Leiden, 197–233.

Ferella 2018: Chiara Ferella, "'A Path for Understanding': Journey Metaphors in (Three) Early Greek Philosophers", in Ch. Ferella and C. Breytenbach (eds.), *Paths of Knowledge. Interconnection(s) between Knowledge and Journey in the Greco-Roman World*, Berlin, 47–73.

Flügel 1872 Gustav Flügel, *Kitāb al-Fihrist*, Leipzig.

Fortenbaugh 2014: William W. Fortenbaugh, *Theophrastus of Eresus: Commentary Volume* 9.2. *Sources on Discoveries and Beginnings, Proverbs et al.* (Texts 727–741), Leiden.

Frommhold 2004: Katrin Frommhold, *Bedeutung und Berechnung der Empfängnis in der Horoskopie der Antike*, Münster.

Fujii 2013: Takashi Fujii, *Imperial Cult and Imperial Representation in Roman Cyprus*, Stuttgart.

Gabrielsen 2004: Vincent Gabrielsen, "Economic activity, maritime trade and piracy in the Hellenistic Aegean", *Revue des Études Anciennes* 103, 219–240.

Gansten 2012: Martin Gansten, "Balbillus and the method of *aphesis*", *GRBS* 52, 587–602.

García and Hernández Bermejo (undated): Expiración García and J. Esteban Hernández Bermejo, “Glossary” https://www.doaks.org/resources/middle-east-garden-traditions/resources/al-andalus-glossary/glossary, last accessed 08/06/2023.
Geller 2014: Markham J. Geller, “Berossos on Kos from the view of common sense geography”, in K. Geuss and M. Thiering, *Features of Common Sense Geography*, Zürich/Berlin, 219–228.
Glassner 2004: Jean-Jacques Glassner, *Mesopotamian Chronicles*, Atlanta.
Goyon 1993: Jean-Claude Goyon, “L’origine égyptienne des tablettes décanales de Grand (Vosges)”, in J.-H. Abry (ed.), *Les tablettes astrologiques du Grand (Vosges) et astrologie en Gaule romaine*, Lyon, 63–76.
Graf 2009: Fritz Graf, “Serious Singing: the Orphic Hymns as Religioius Texts”, *Kernos* 22, 169–182.
Grafton and Swerdlow 1985: A. T. Grafton and N. M. Swerdlow, “Technical Chronology and Astrological History in Varro, Censorinus, and others”, *CQ* 35.2, 454–465.
Hannah 2009: Robert Hannah, *Time in Antiquity*, New York.
Harvin and Totelin 2016: Gavin Arvin and Laurence Totelin, *Ancient Botany*, London.
Heilen 2010: Stephan Heilen, “Anubio reconsidered”, *Aestimatio* 7, 127–192.
Heilen 2011: Stephan Heilen, “Some metrical fragments of Nechepsos and Petosiris”, in I. Boehm and W. Hübner (eds.), *La poésie astrologique dans l’antiquité*, Lyon, 23–92.
Heilen 2015: Stephan Heilen, *Hadriani Genitura. Die astrologischen Fragmente des Antigonos von Nikaia*, Band 1 Edition und Übersetzung / Band 2 Kommentar, Berlin/Boston.
Herrero de Jáuregui 2010: Miguel Herrero de Jáuregui, *Orphism and Christianity in Late Antiquity*, Berlin.
Houlding 2010: Deborah Houlding, “The Transmission of Ptolemy’s Terms: An Historical Overview, Comparison and Interpretation”, http://www.skyscript.co.uk/pdf/Houlding_ptolemy_terms.pdf, 2010 (amended for online edition from *Culture and Cosmos* 11 [2007], 261–307), last accessed 08/06/2023.
Hübner 1998: Wolfgang Hübner, *Claudi Ptolemaei Opera quae Extant Omnia. Vol. III 1: ΑΠΟΤΕΛΕΣΜΑΤΙΚΑ*, Leipzig.
Hübner 2005: Wolfgang Hübner, “Sulla’s Horoscope?”, in G. Oestmann et al. (eds.), *Horoscopes and Public Spheres: Essays on the History of Astrology*, Berlin/New York, 13–36.
Hübner 2010: Wolfgang Hübner, *Manilius, “Astronomica” Buch V. Einführung, Text, Übersetzung und Kommentar*, Berlin/New York.
Hughes 1951: George R. Hughes, “A Demotic astrological text”, *Journal of Near Eastern Studies* 10, 256–264.
Jaccottet 2003: Anne-Françoise Jaccottet, *Choisir Dionysos: les associations dionysiaques, ou, La face cachée du dionysisme*, Vol. I Text / Vol. II Documents, Zürich.
Johnson 2004: William A. Johnson, *Bookrolls and Scribes in Oxyrhynchus*, Toronto.
Johnson 2012: William A. Johnson, *Readers and Reading Culture in the High Roman Empire: A Study of Elite Communities. Classical Culture and Society*, Oxford/New York.
Jones 1999: Alexander Jones, *Astronomical Papyri from Oxyrhynchus*, Philadelphia.
Jones and Steele 2011: Alexander Jones and John Steele, “A New Discovery of a Component of Greek Astrology in Babylonian Tablets: The ‘Terms’”, *ISAW Papers* 1, http://dlib.nyu.edu/awdl/isaw/isaw-papers/1/, last accessed 08/06/2023.
Jones and Greenbaum 2017: Alexander Jones and Dorian Greenbaum, “P.Berl. 9825: An elaborate horoscope for 319 CE and its significance for Greek astronomical and astrological practice”, *ISAW Papers* 12, http://dlib.nyu.edu/awdl/isaw/isaw-papers/12/, last accessed 08/06/2023.
Jones and Perale *forthcoming*: Alexander Jones and Marco Perale, “PSI inv. 3780+P.Med. inv. 124”, in *Papiri della Società Italiana* XX.

Jouanna 2012: Jacques Jouanna, “Rhetoric and Medicine in the Hippocratic Corpus: A Contribution to the History of Rhetoric in the Fifth Century”, in J. Jouanna (ed.) and N. Allies (translator), *Greek Medicine from Hippocrates to Galen: Selected Papers*, Leiden/Boston, 39–54.
Jursa 2018: Michael Jursa, “Xerxes: the Case of Sippar and the Ebabbar Temple”, in C. Waerzeggers and M. Seire (eds.), *Xerxes and Babylonia: the Cuneiform Evidence*, Leuven/Paris/Bristol, 63–72.
Karanika 2010: Andromache Karanika, “Inside Orpheus’ Songs: Orpheus as an Argonaut in Apollonius Rhodius’ *Argonautica*”, *GRBS* 50, 391–410.
Keyser 1994: Paul T. Keyser, “On Cometary Theory and Typology from Nechepso-Petosiris through Apuleius to Servius”, *Mnemosyne* 47, 625–651.
King 1989: David A. King, “Some Arabic Copies of Vettius Valens’ Table for Finding the Length of Life”, in G. Endress and M. Schmeink (eds.), *Symposium Graeco-Arabicum II: Bochum 3–5 März, 1987*, Amsterdam, 25–28.
Koch-Westenholz 1995: Ulla Koch-Westenholz, *Mesopotamian Astrology: An Introduction to Babylonian and Assyrian Celestial Divination*, Copenhagen.
Kosmin 2015: Paul Kosmin, “Seeing Double in Seleucid Babylonia: Rereading the Borsippa Cylinder of Antiochus I”, in A. Moreno and R. Thomas (eds.), *Patterns of the Past: Epitēdeumata in the Greek Tradition*, Oxford, 173–198.
Kronenberg 2009: Leah Kronenberg, *Allegories of Farming from Greece and Rome: Philosophical Satire in Xenophon, Varro, and Virgil*, Cambridge.
Kronenberg 2017: Leah Kronenberg, “Varro the Roman Cynic: The Destruction of Religious Authority in the *Antiquitates rerum divinarum*”, in J. König and G. Woolf (eds.), *Authority and Expertise in Ancient Scientific Culture*, Cambridge, 206–328.
László 2020: Levente László, “Rhetorius, Zeno’s Astrologer, and a Sixth-Century Astrological Compendium”, *DOP* 74, 329–350.
Lehoux 2002: Daryn Lehoux, “The Historicity Question in Mesopotamian Divination”, in J.M. Steele and A. Imhausen (eds.), *Under One Sky: Astronomy and Mathematics in the Ancient Near East* (*Alter Orient und Altes Testament* 297), Münster, 209–222.
Lightfoot 2021: J. L. Lightfoot, *Pseudo-Manetho. “Apotelesmatica”, books two, three, and six*, Oxford/New York.
Majercik 1989: Ruth Majercik, *The Chaldean Oracles. Text, Translation, and Commentary*, Leiden.
Martín Hernández 2006: Raquel Martín Hernández, “Orfeo científico”, *Estudios Clásicos* 129, 111–119.
Martín Hernández 2009: Raquel Martín Hernández, “Literatura mágica y pseudocientífica atribuida a Orfeo”, in A. Bernabé and F. Casadesús, *Orfeo y la tradición órfica. Un reencuentro*, Madrid, 437–458.
Martín Hernández 2015: Raquel Martín Hernández, *La ciencia de Orfeo*, Madrid.
Martínez Nieto 2009: Roxana B. Martínez Nieto, “Otros poetas próximos a Orfeo”, in A. Bernabé and F. Casadesús, *Orfeo y la tradición órfica. Un reencuentro*, Madrid, 549–576.
McNamara 2013: Charles McNamara, “Stoic Caricature in Lucian’s *de astrologia*: Verisimilitude as Comedy”, *Peitho* 1.4, 236–253.
Merkelbach and Totti 1990: Reinhold Merkelbach and Maria Totti, *Abrasax. Ausgewählte Papyri religiösen und magischen Inhalts*, 2 Vols., Wiesbaden.
Molina Moreno 2009: Francisco Molina Moreno, “La música de Orfeo”, in A. Bernabé and F. Casadesús, *Orfeo y la tradición órfica. Un reencuentro*, Madrid, 33–58.
Morand 2001: Anne-France Morand, *Études sur les* Hymnes orphiques, Leiden.
Murphy 2016: David J. Murphy, “‘By the Goose, by the Ram’. Socrates’ Other Unusual Oaths”, *SCO* 62, 15–51.
Nawotka 2017: Krzysztof Nawotka, *The Alexander Romance by Ps. Callisthenes: A Historical Commentary*, Leiden.

Nesselrath 2016: Heinz-Günter Nesselrath (ed.), *Gegen falsche Götter und falsche Bildung: Tatian. Rede an die Griechen*, Tübingen.

Netz 1999: Reviel Netz, *The Shaping of Deduction in Greek Mathematics: A Study on Cognitive History*, Cambridge.

Netz 2009: Reviel Netz, *Ludic Proof. Greek Mathematics and the Alexandrian Aesthetic*, Cambridge.

Neugebauer 1951: Otto Neugebauer, "The study of wretched subjects", *Isis* 42, 111.

Neugebauer 1954: Otto Neugebauer, "The Chronology of Vettius Valens' Anthologiae", *Harvard Theological Review* 47, 65–67.

Neugebauer 1975: Otto Neugebauer, *A History of Ancient Mathematical Astronomy*, Berlin/Heildelberg/New York.

Neugebauer and Saliba 1989: Otto Neugebauer and George Saliba, "On Greek Numerology", *Centaurus* 31, 189–206.

Neugebauer et al. 1981: Otto Neugebauer, Richard A. Parker and Karl-Th. Zauzich, "A Demotic Lunar Eclipse Text of the First century, B.C"., *Proceedings of the American Philosophical Society* 125, 312–327.

Obbink 2011: Dirk Obbink, "Orphism, Cosmogony, and Genealogy (Mus. fr. 14)", in M. Herrero de Jáuregui et al. (eds.), *Tracing Orpheus: Studies of Orphic Fragments*, Berlin/Boston.

Parker 1959: Richard A. Parker, *A Vienna Demotic Papyrus on Eclipse- and Lunar-Omina*, Providence.

Parker 1983: Robert Parker, *Miasma: Pollution and Purification in Early Greek Religion*, Oxford.

Peter 2001: Heinz Peter, *Kritodem. Testimonien- und Fragmentsammlung*, Lizentiatsarbeit Universität Zürich, Zürich.

Pinault 1992: Judy Robin Pinault, *Hippocratic Lives and Legends*, Leiden.

Pingree 1976: David Pingree, *Dorothei Sidonii carmen astrologicum*, Leipzig.

Pingree 1977: David Pingree, "Antiochus and Rhetorius", *Classical Philology* 72, 203–223.

Pingree 1978: David Pingree, *The* Yavanajataka *of Sphujidhvaja*, Cambridge MA.

Pingree 1986: David Pingree, *Vetti Valentis Anthologiarum libri novem*, Leipzig.

Quack 2002: Joachim Friedrich Quack, "Die Spur des Magiers Petese", *Chronique d'Égypte* 77, 76–92.

Rochberg 1998: Francesca Rochberg, *Babylonian Horoscopes*, Philadelphia.

Ross 2019: Micah T. Ross, "Demotic Horoscopes", in A. C. Bowen and F. Rochberg, *Hellenistic Astronomy. The Science in Its Contexts*, Leiden.

Rutherford 1997: Ian Rutherford, "Kalasiris and Setne Khamwas: A Greek Novel and Some Egyptian Models", *ZPE* 117, 203–209.

Ryholt 2002: Kim Ryholt, "Nectanebo's Dream or the Prophecy of Petesis", in A. Blasius and B. U. Shipper (eds.), *Apokalyptik und Ägypten: Eine kritische Analyse der relevanten Texte aus dem griechisch-römischen Ägypten*, Leiden, 221–241.

Ryholt 2011: Kim Ryholt, "New Light on the Legendary King Nechepsos", *Journal of Egyptian Archaeology* 97, 61–72.

Sherwin-Wite 1987: Susan Sherwin-White, "Seleucid Babylonia: A Case Study for the Installation and Development of Greek Rule", in A. Kuhrt and S. Sherwin-White (eds.), *Hellenism in the East*, Berkeley, 1–31.

Smith 2011: Andrew Smith, *Porphyrii Philosophi Fragmenta*, Leipzig.

Steele 2008: John M. Steele, *A Brief Introduction to Astrology in the Middle East*, London.

Steele 2013: John M. Steele, "The 'Astronomical' Fragments of Berossos in Context", in J. Haubold, G. B. Lanfranchi, R. Rollinger, and J. Steele (eds.), *The World of Berossos*, Wiesbaden, 107–122.

Steele 2017: John M. Steele, *Rising Time Schemes in Babylonian Astronomy*, Dordrecht.

Stinton 1977: T. C. W. Stinton, "Interlinear Hiatus in Trimeters", *Classical Quarterly* 27, 67–72.

Straus 2017: Jean A. Straus, "Les contrats d'apprentissage et d'enseignement relatives à des esclaves dans la documentation papyrologique grecque d'Égypte", in M.-H. Marganne et al. (eds.), *En marge du Serment hippocratique. Contrats et serments dans le monde gréco-romain*, Liège, 119–134.

Symons 2007: Sarah Symons, "A *star's year*: the annual cycle in the ancient Egyptian sky", in J. M. Steele, *Calendars and Years: Astronomy and Time in the Ancient World*, Oxford, 1–33.

Symons 2014: Sarah Symons, "Contexts and Elements of Decanal Star Lists in Ancient Egypt", in D. Bawanypeck and A. Imhausen (eds.), *Traditions of Written Knowledge in Ancient Egypt and Mesopotamia*, Münster, 91–122.

Tolsa 2017a: Cristian Tolsa, "Ptolemy's Savior God, 'Saving the Phenomena', and Plato's *Timaeus*", *Museum Helveticum* 74, 144–157.

Tolsa 2017b: Cristian Tolsa, "'New' Zodiacal Divisions in POxy LXI 4277", *ZPE* 203, 192–198.

Tolsa 2017c: Cristian Tolsa, "Time of Birth and Ascendant in the Papyrus Horoscopes", *ZPE* 204, 209–220.

Tolsa 2018: Cristian Tolsa, "The Table of Ptolemy's Terms", *Philologus* 168.2, 247–264.

Tolsa 2019a: Cristian Tolsa, "On the Origins of the Hippocratic *Oath*", *GRBS* 59, 621–645.

Tolsa 2019b: Cristian Tolsa, "Vettius Valens' Longitudes (1.18), Balbillus, and the Illusion of Astrology's Self-sufficiency", *GRBS* 59, 397–414.

Tolsa 2019c: Cristian Tolsa, "Astrometeorology by Syrus", *SCIAMVS* 20, 1–30.

Tolsa 2020: Cristian Tolsa, "The Earliest Arabic Magic Squares", *Suhayl* 18, 7–24.

Tolsa *forthcoming*.a: Cristian Tolsa, "Characterizing Lucian's *De astrologia*", in F. Mestre, P. Gómez and C. Tolsa (eds.), *Lucianea et pseudolucianea*, Barcelona.

Tolsa *forthcoming*.b: Cristian Tolsa, "Greek echoes of Late Hellenistic astrometeorology", in T. Tielemann and Giouli Korompili (eds.), *Meteorology Beyond Borders: Ancient and Modern Receptions*, Brill.

Tolsa *forthcoming*.c: Cristian Tolsa, "The astrological poem of Dorotheus and some of its transformations in the Arabic version of 'Umar ibn al-Farrukhān", *Aitia* 13.

van Bladel 2009: Kevin van Bladel, *The Arabic Hermes. From Pagan Sage to Prophet of Science*, Oxford.

van Bladel 2012: Kevin van Bladel, "The Arabic history of science of Abū Sahl ibn Nawbaḫt (fl. ca. 770–809) and its Middle Persian sources", in F. Opwis and D. Reisman (eds.), *Islamic Philosophy, Science, Culture, and Religion: Studies in Honor of Dimitri Gutas*, Leiden, 41–62.

van der Speck 2009: Robert van der Speck, "Multi-ethnicity and Ethnic Segregation in Hellenistic Babylon", in T. Derks and N. Roymans (eds.), *Ethnic Constructs in Antiquity: The Role of Power and Tradition*, Amsterdam, 101–15.

Verbrugghe and Wickersham 1996: Gerald P. Verbrugghe and John M. Wickersham, *Berossos and Manetho. Introduced and Translated*, Ann Arbor.

Vogt 2005: Sabine Vogt, "... er schrieb in Versen, und er tat recht daran", in T. Fögen (ed.), *Antike Fachtexte/Ancient Technical Texts*, Berlin/New York, 51–78.

Volk 2011: Katharina Volk, "Manilian Self-Contradiction", in S. J. Green and K. Volk (eds.), *Forgotten Stars. Rediscovering Manilius'* Astronomica, Oxford, 104–119.

Waerzeggers 2018: Carline Waerzeggers, "The Network of Resistance: Archives and Political Action in Babylonia Before 484 BC", in C. Waerzeggers and M. Seire (eds.), *Xerxes and Babylonia: the Cuneiform Evidence*, Leuven/Paris/Bristol, 89–134.

West 1982: M. L. West, *Greek Metre*, Oxford.

West 1983: M. L. West, *The Orphic Poems*, Oxford.

Willi 2013: Andreas Willi, "Epicharmus, Simonides, and the 'invention' of the Greek alphabet", *Museum Helveticum* 70, 129–140.

Zellmann-Rohrer 2023: Michael Zellmann-Rohrer, "The Chronokratores in Greek astrology, in light of a new papyrus text: Oxford, Bodl. MS gr. class. B 24 (P) 1–2", *Harvard Studies in Classical Philology* 112, 465–502.

Index of select Greek and Latin words from the texts

Greek

https://doi.org/10.1515/9783111329147-014

Latin

Index of authors

https://doi.org/10.1515/9783111329147-015

General index

https://doi.org/10.1515/9783111329147-016